new

917.7 Am

America's byways of the Midwest

STACKS

WITHDRAWN

AMERICA'S BYWAYS™ OF THE
MIDWEST

come
CLOSER *to the heart and soul of*
your AMERICA

America's Byways Series | MOBIL TRAVEL GUIDE

We gratefully acknowledge our inspection team for their efficient and perceptive evaluations of the establishments listed in this book and the establishments for their cooperation in showing their facilities and providing information about them. Thanks also go to the National Scenic Byways Program and the coordinators of the individual Byways for all their help and support in the coordination of this project.

VICE PRESIDENT, PUBLICATIONS: **Kevin Bristow**
MANAGING EDITOR: **Pam Mourouzis**
MANAGER OF PUBLISHING PRODUCTION SERVICES: **Ellen Tobler**
CONCEPT AND COVER DESIGN: **ABS Graphics, Inc. Design Group**
EDITOR: **Tere Drenth**
PRINTING ACKNOWLEDGEMENT: **North American Corporation of Illinois**

Copyright © 2004 EMTG, LLC. All rights reserved. Except for copies made by individuals for personal use, this publication may not be reproduced in whole or in part by any means whatsoever without prior written permission from Mobil Travel Guide, 1460 Renaissance Drive, Suite 401, Park Ridge, IL 60068; 847/795-6700; info@mobiltravelguide.com.

Mobil, Exxon, and Mobil Travel Guide are trademarks of Exxon Mobil Corporation or one of its subsidiaries. All rights reserved. Reproduction by any means, including, but not limited to, photography, electrostatic copying devices, or electronic data processing is prohibited. Use of information contained herein for solicitation of advertising or listing in any other publication is expressly prohibited without prior written permission from Exxon Mobil Corporation. Violations of reserved rights are subject to prosecution.

The information contained herein is derived from a variety of third-party sources. Although every effort has been made to verify the information obtained from such sources, the publisher assumes no responsibility for inconsistencies or inaccuracies in the data or liability for any damage of any type arising from errors or omissions.

Neither the editors nor the publisher assumes responsibility for the services provided by any business listed in this guide or for any loss, damage, or disruption in your travel for any reason.

ISBN: 0-9727-0228-8

Manufactured in the United States of America.

10 9 8 7 6 5 4 3 2 1

America's Byways Series | MOBIL TRAVEL GUIDE
Table of Contents

America's Byways of the Midwest

MAPS

Illinois
The Great River Road A5
The Historic National Road A6
Lincoln Highway A7
Meeting of the Great Rivers Scenic Route A8
Ohio River Scenic Byway A9

Iowa
Loess Hills Scenic Byway A10

Michigan
Woodward Avenue (M-1) A11

Minnesota
Edge of the Wilderness A12
Historic Bluff Country Scenic Byway A13
Minnesota River Valley Scenic Byway A14
North Shore Scenic Drive A15
The Grand Rounds Scenic Byway A16

Missouri
Crowley's Ridge Parkway A17
Little Dixie Highway of the Great River Road A18

North Dakota
Sheyenne River Valley Scenic Byway A19

Ohio
Amish Country Byway A20
CanalWay Ohio Scenic Byway A21

South Dakota
The Native American Scenic Byway A22
Peter Norbeck Scenic Byway A23

FEATURED BYWAY A24

A WORD TO OUR READERS A25

OVERVIEW OF THE NATIONAL SCENIC BYWAYS PROGRAM A27

INTRODUCTION A29

continued on next page

America's Byways Series | MOBIL TRAVEL GUIDE
Table of Contents

AMERICA'S BYWAYS OF THE MIDWEST

Illinois
The Great River Road 1
The Historic National Road 13
Lincoln Highway 19
Meeting of the Great Rivers Scenic Route 27
Ohio River Scenic Byway 35

Indiana
The Historic National Road 41
Ohio River Scenic Byway 53

Iowa
The Great River Road 61
Loess Hills Scenic Byway 71

Michigan
Woodward Avenue (M-1) 79

Minnesota
Edge of the Wilderness 91
The Grand Rounds Scenic Byway 99
The Great River Road 109
Historic Bluff Country Scenic Byway 127
Minnesota River Valley Scenic Byway 133
North Shore Scenic Drive 141

Missouri
Crowley's Ridge Parkway 151
Little Dixie Highway of the Great River Road 157

North Dakota
Sheyenne River Valley Scenic Byway 163

Ohio
Amish Country Byway 169
CanalWay Ohio Scenic Byway 177
The Historic National Road 193
Ohio River Scenic Byway 201

South Dakota
The Native American Scenic Byway 215
Peter Norbeck Scenic Byway 221

Wisconsin
The Great River Road 231

The Great River Road

also covers IA * MN * WI

MIDWEST A5

# The Historic National Road

also covers IN * MD * OH * PA * WV

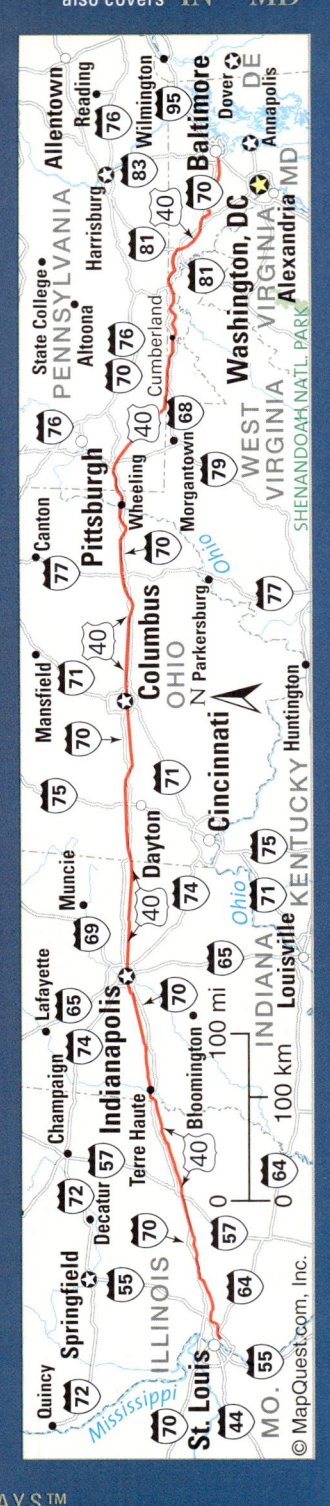

Lincoln Highway IL

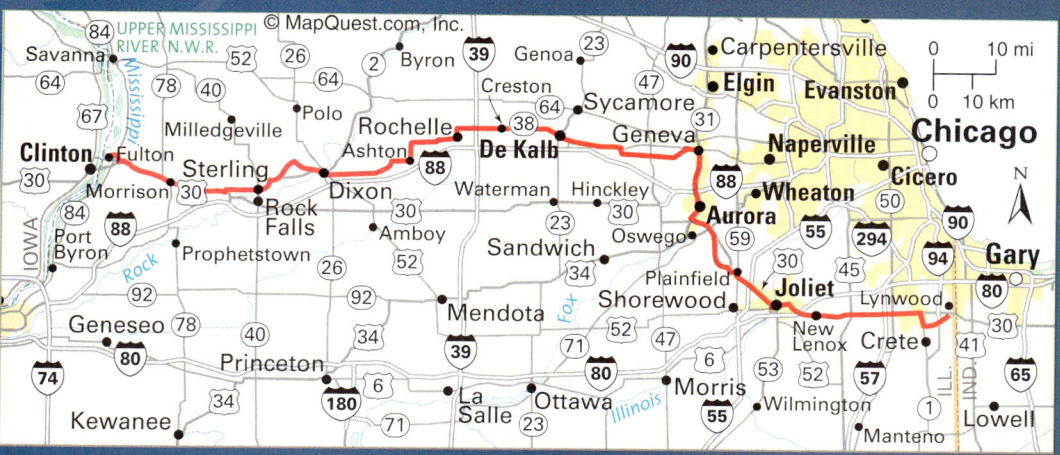

IL — Meeting of the Great Rivers Scenic Route

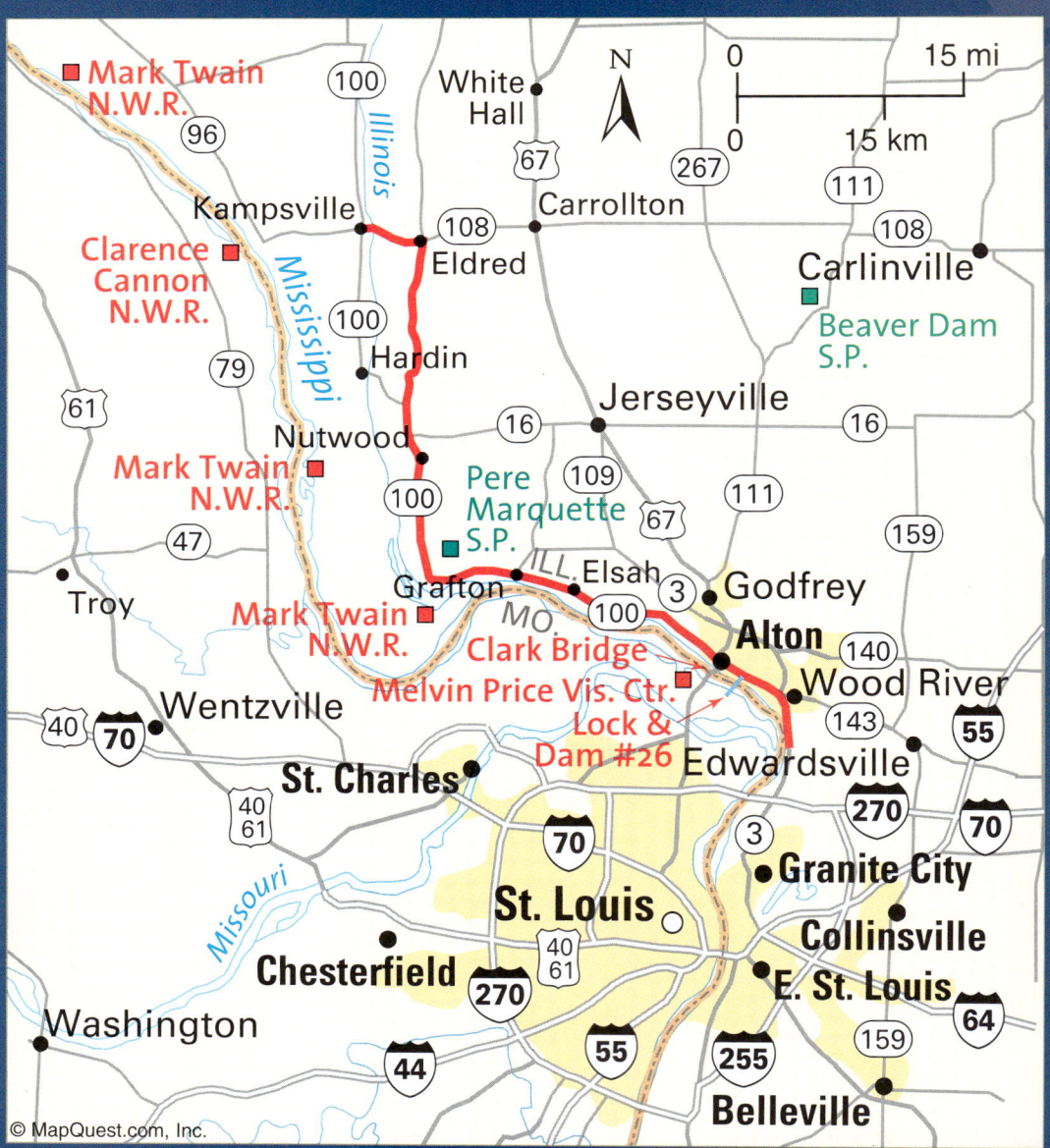

Ohio River Scenic Byway IL

also covers IN • OH

IA Loess Hills Scenic Byway

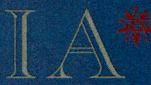

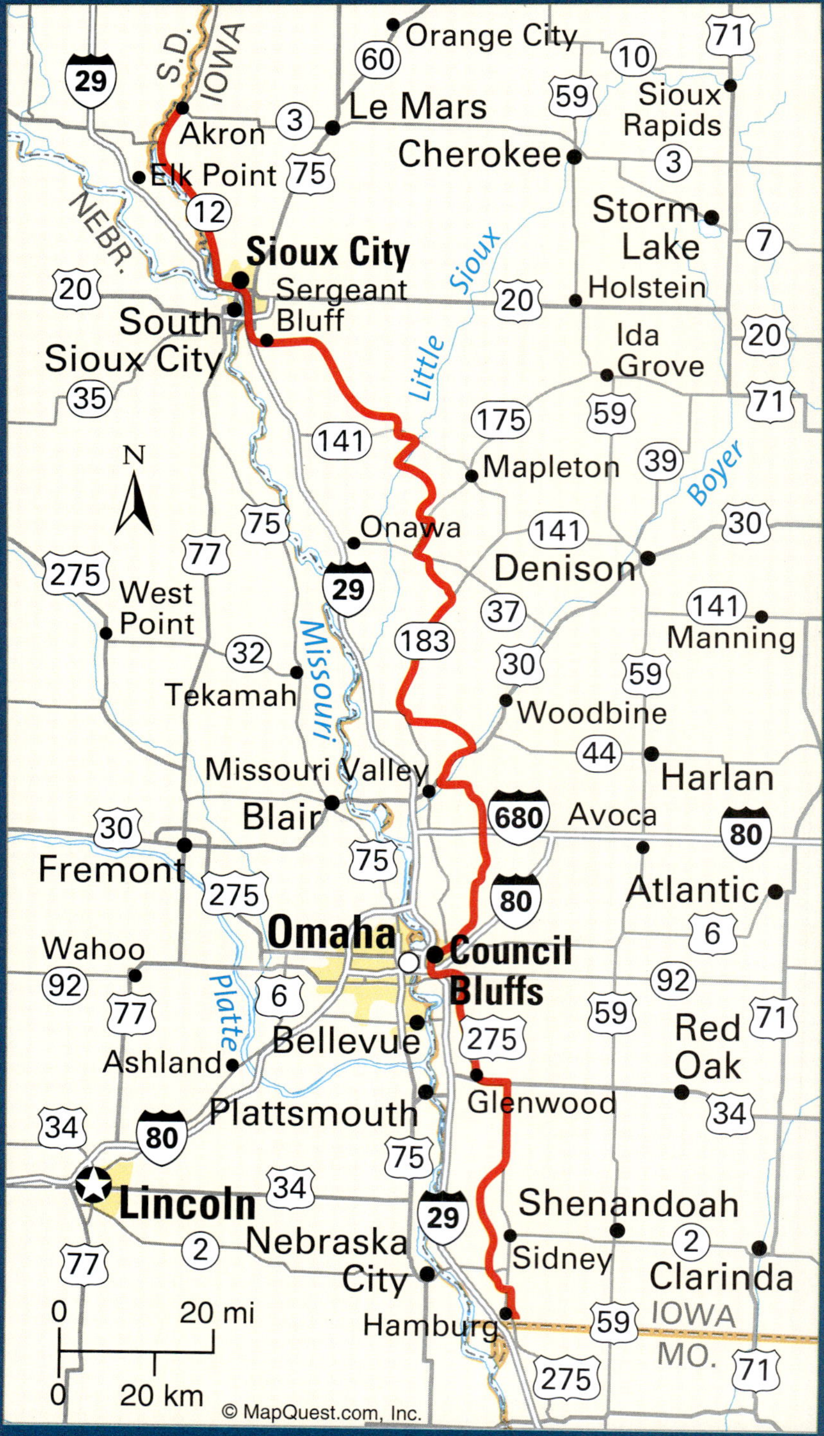

Woodward Avenue (M-1)

MN Edge of the Wilderness

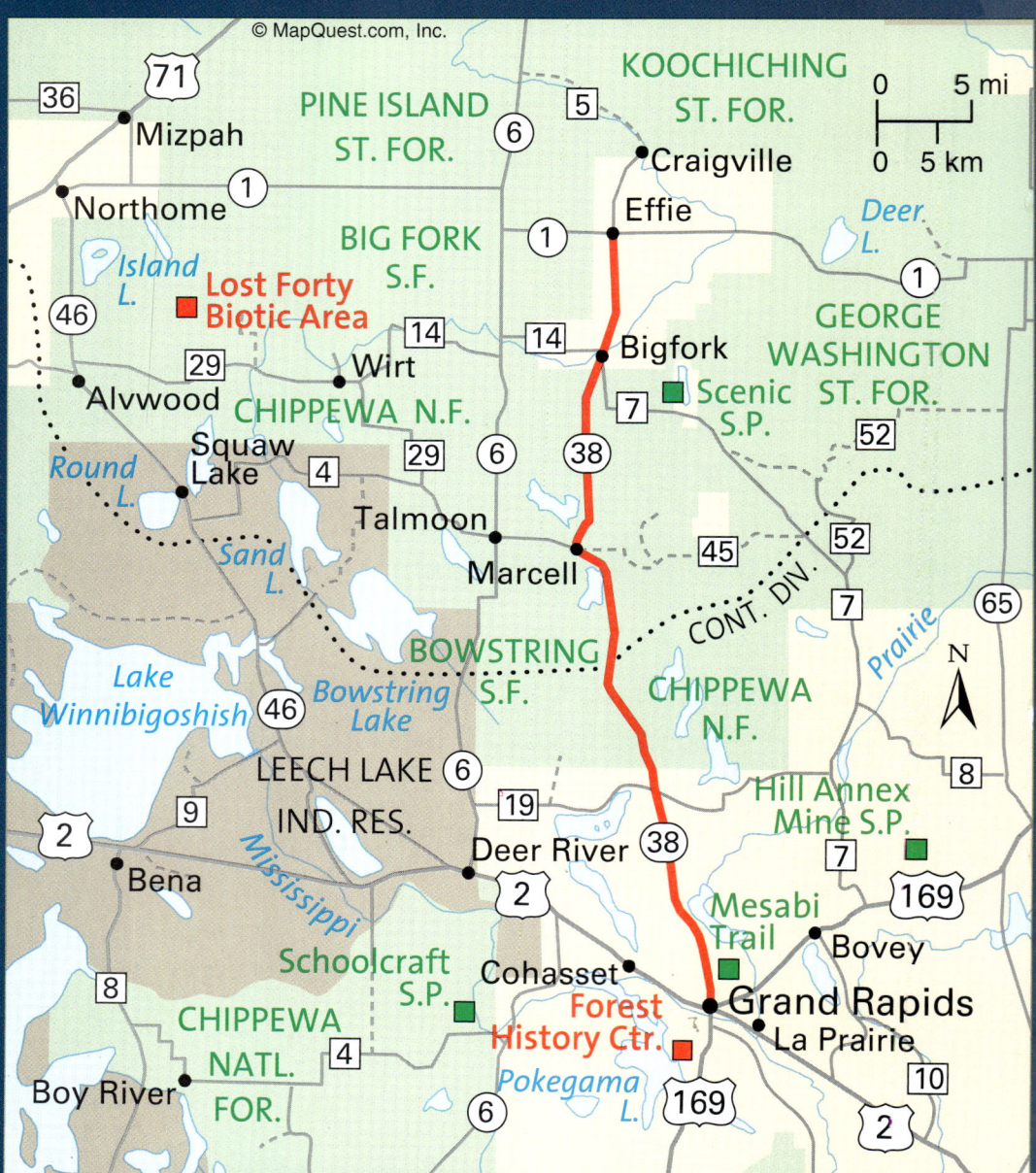

Historic Bluff Country Scenic Byway

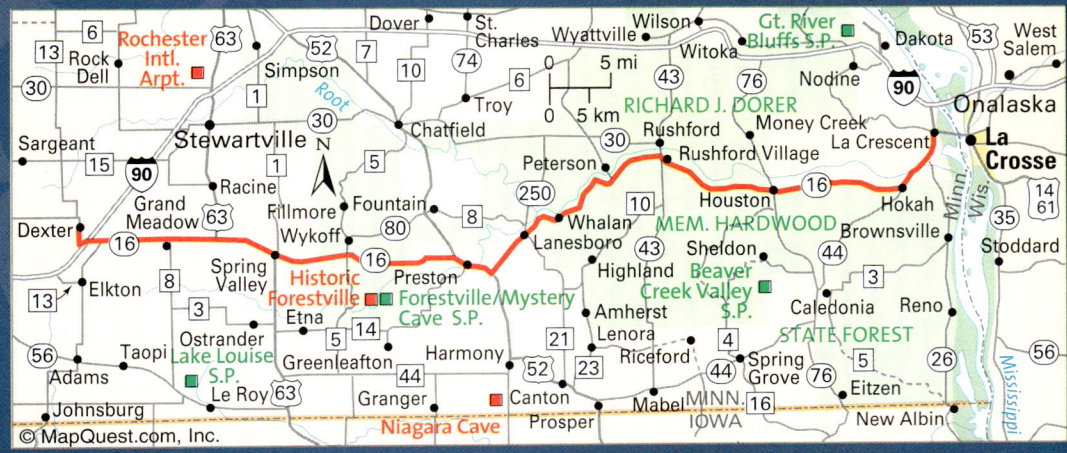

Minnesota River Valley Scenic Byway

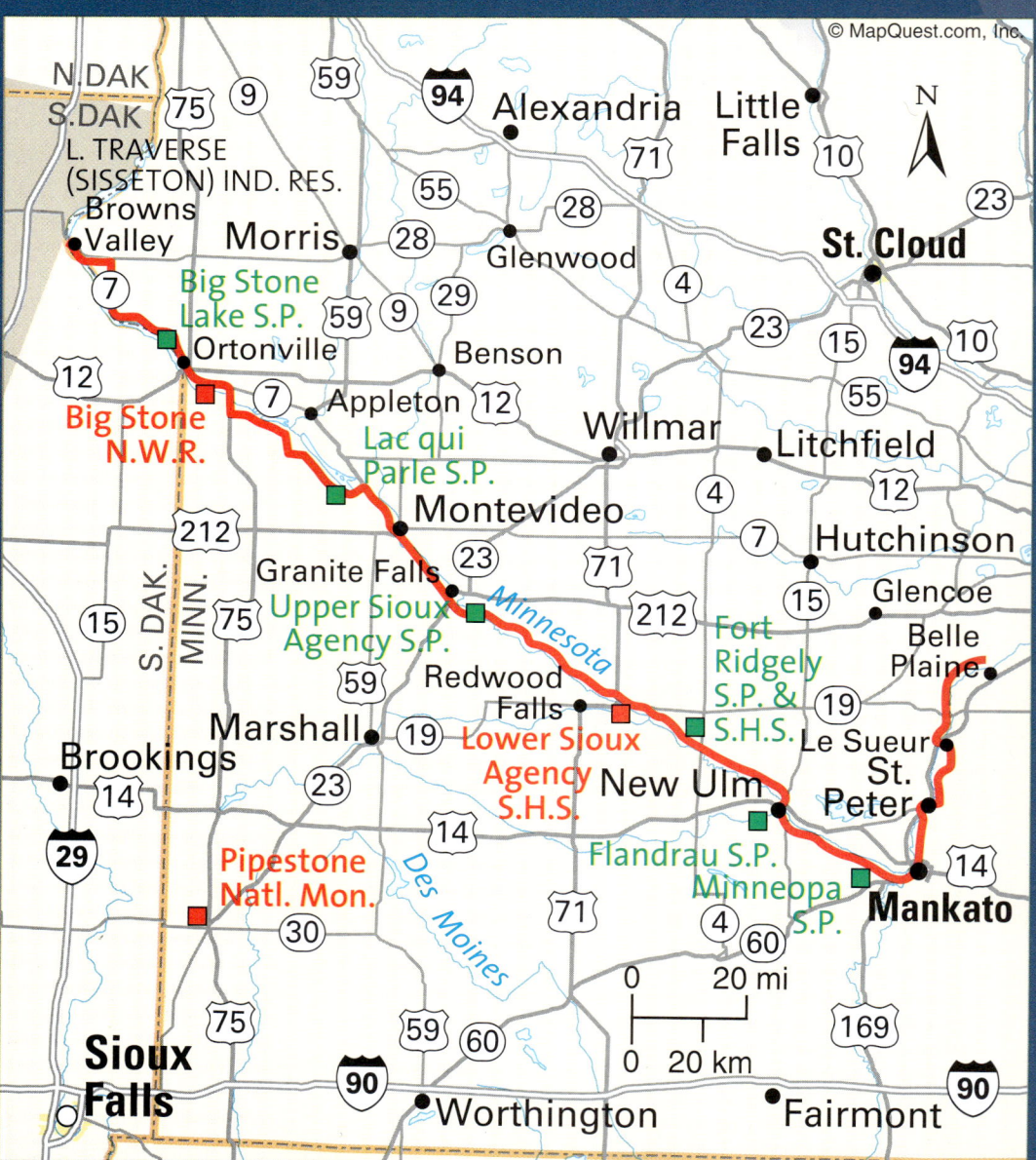

North Shore Scenic Drive

MIDWEST A15

MN The Grand Rounds Scenic Byway

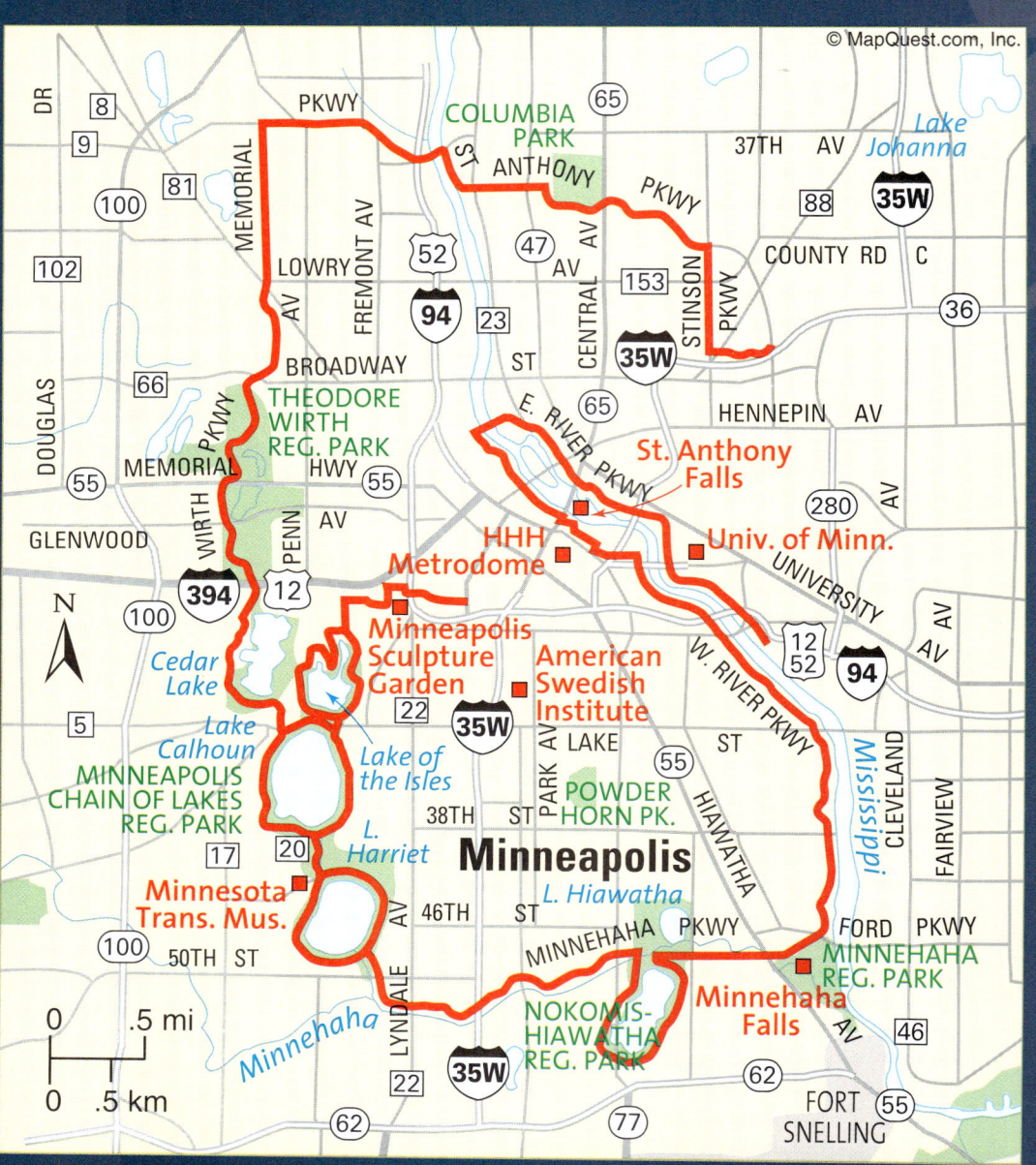

A16 AMERICA'S BYWAYS™

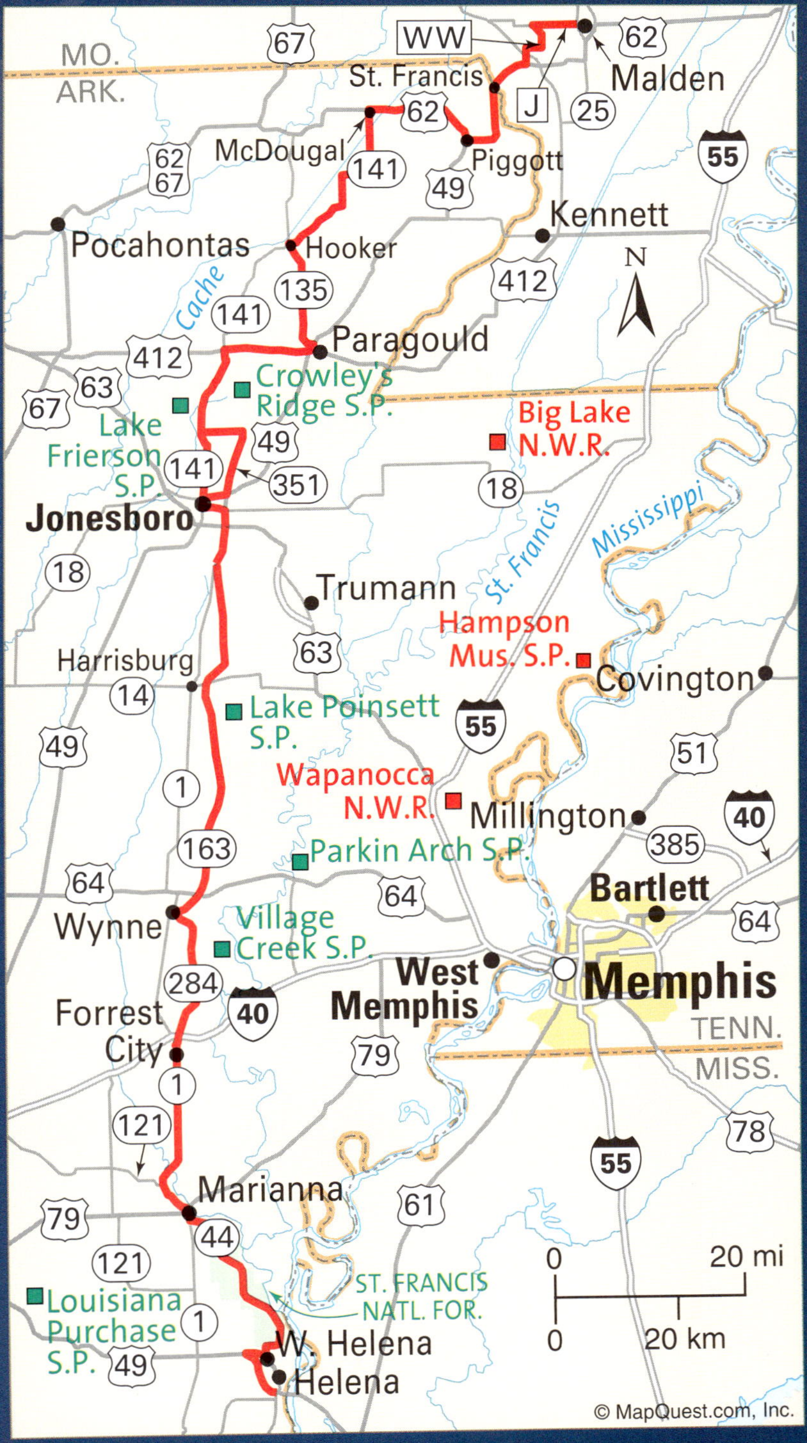

Crowley's Ridge Parkway

MO — Little Dixie Highway of the Great River Road

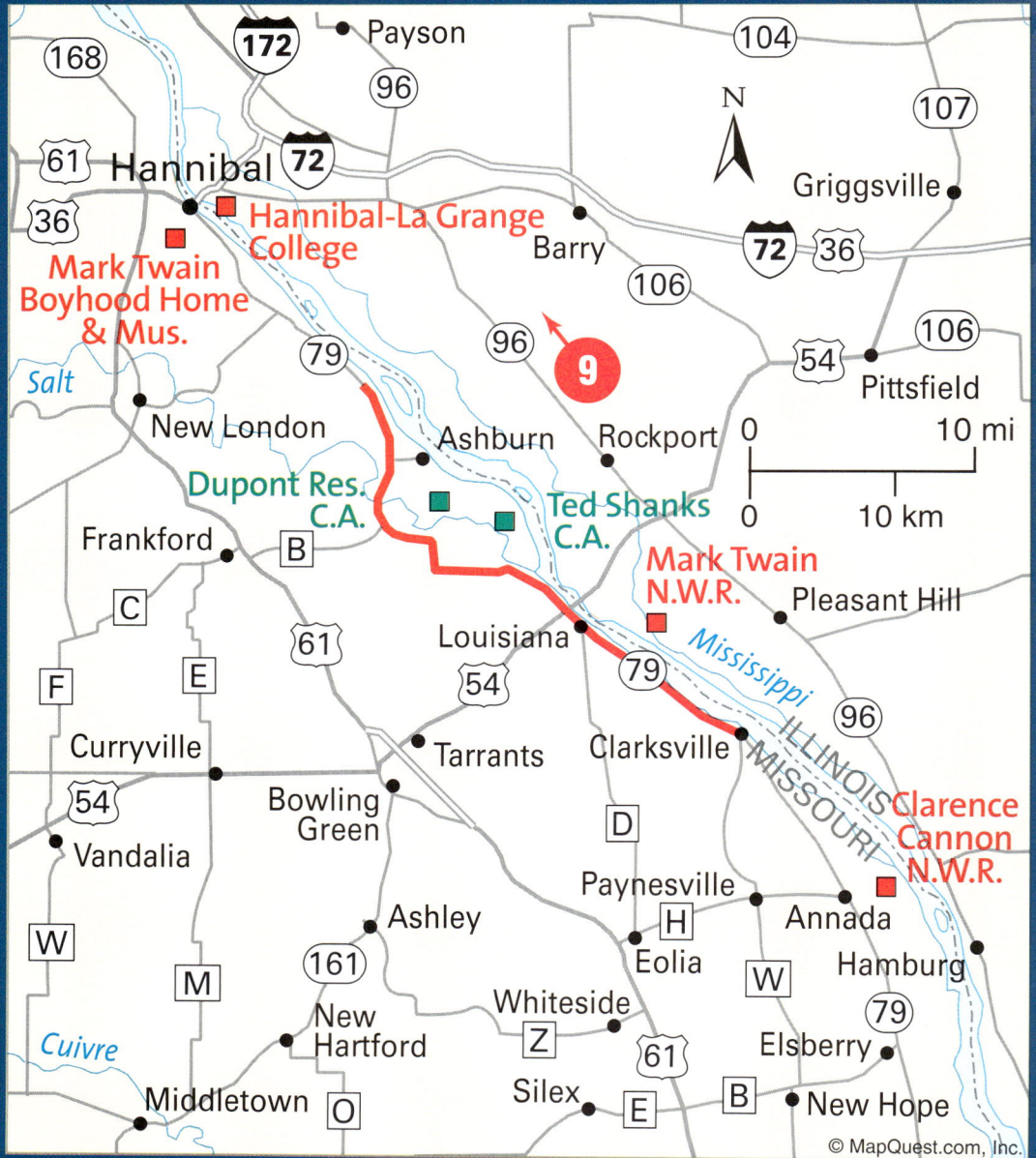

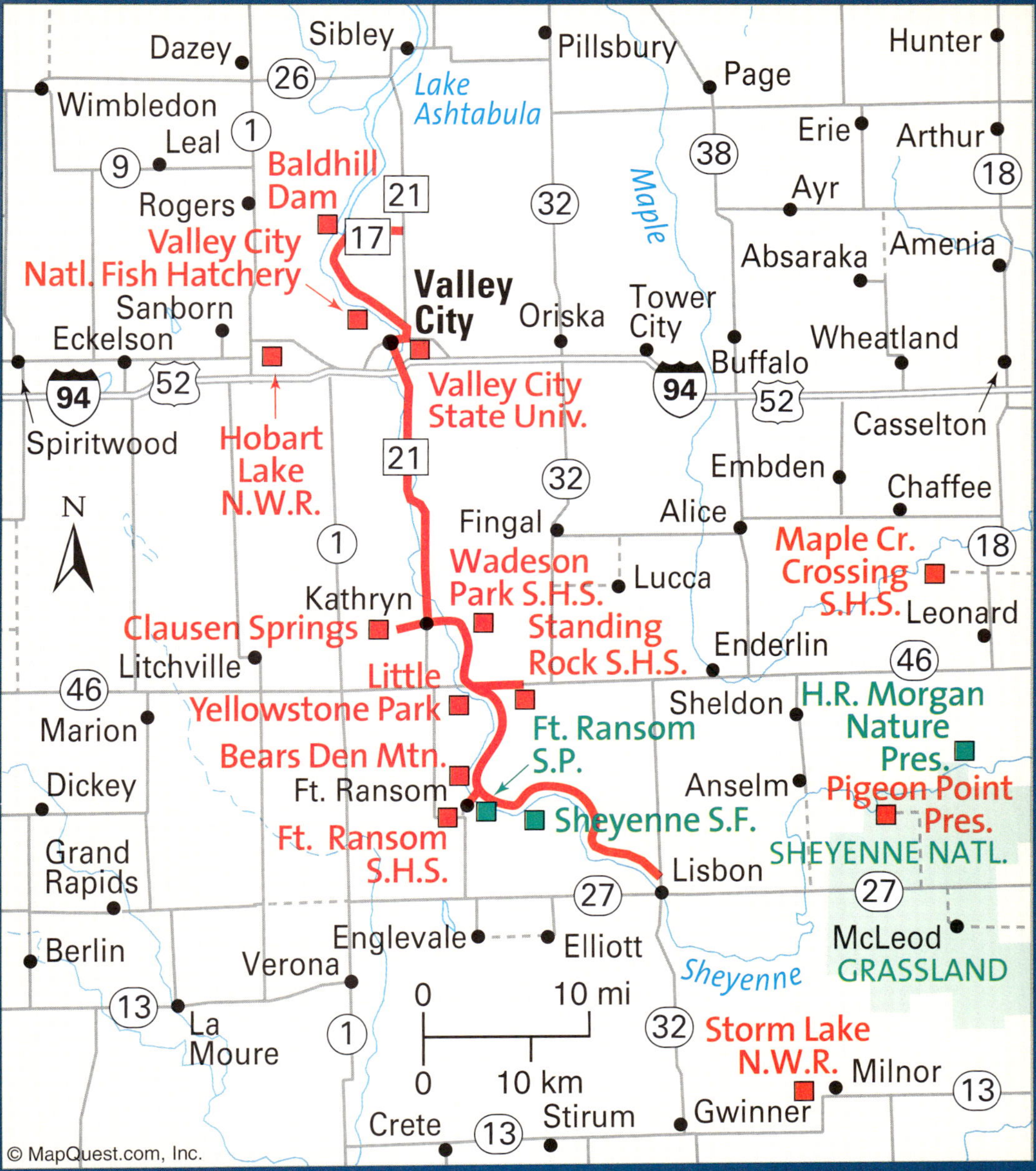

Sheyenne River Valley Scenic Byway

OH Amish Country Byway

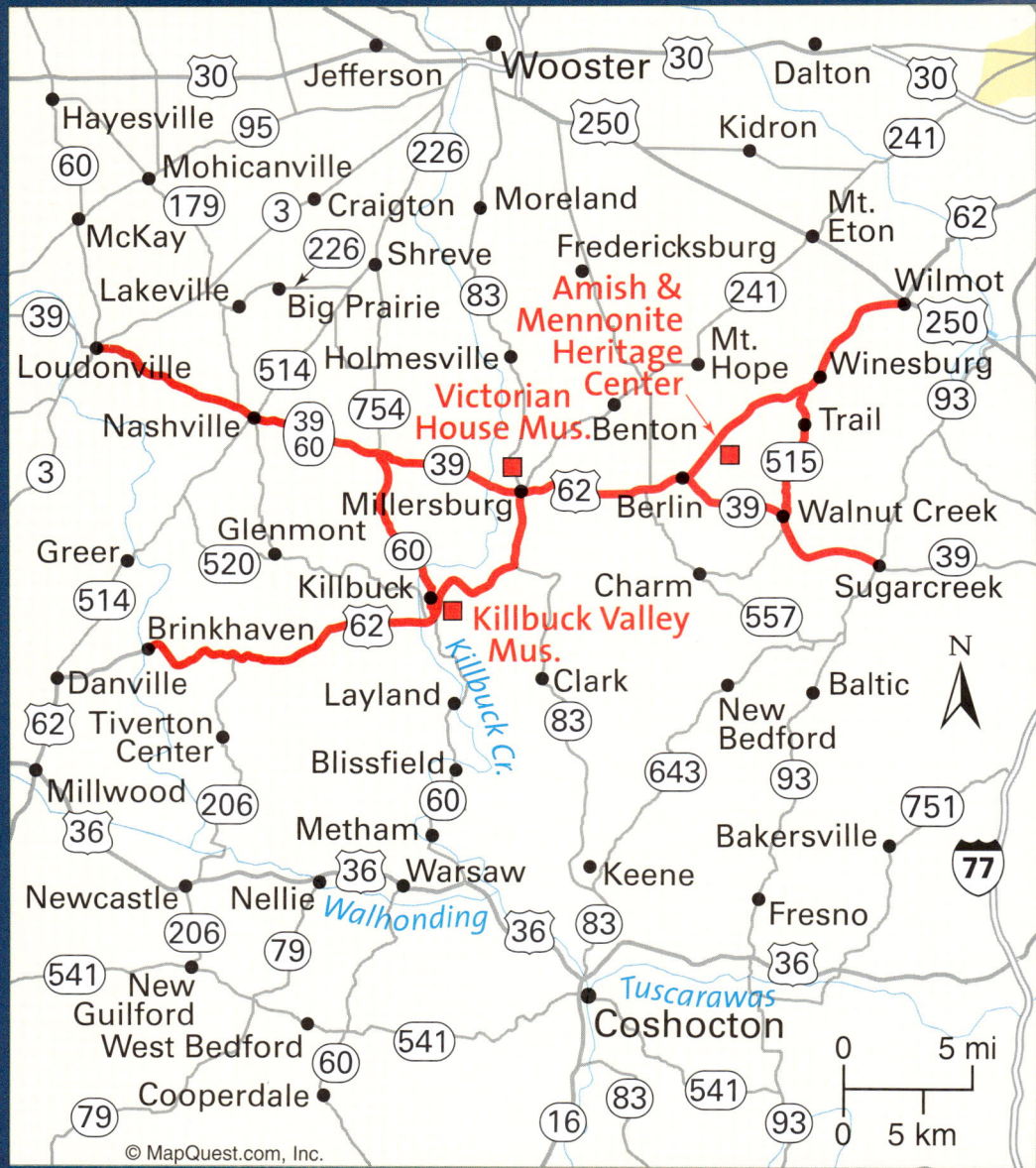

A20 AMERICA'S BYWAYS™

CanalWay Ohio Scenic Byway

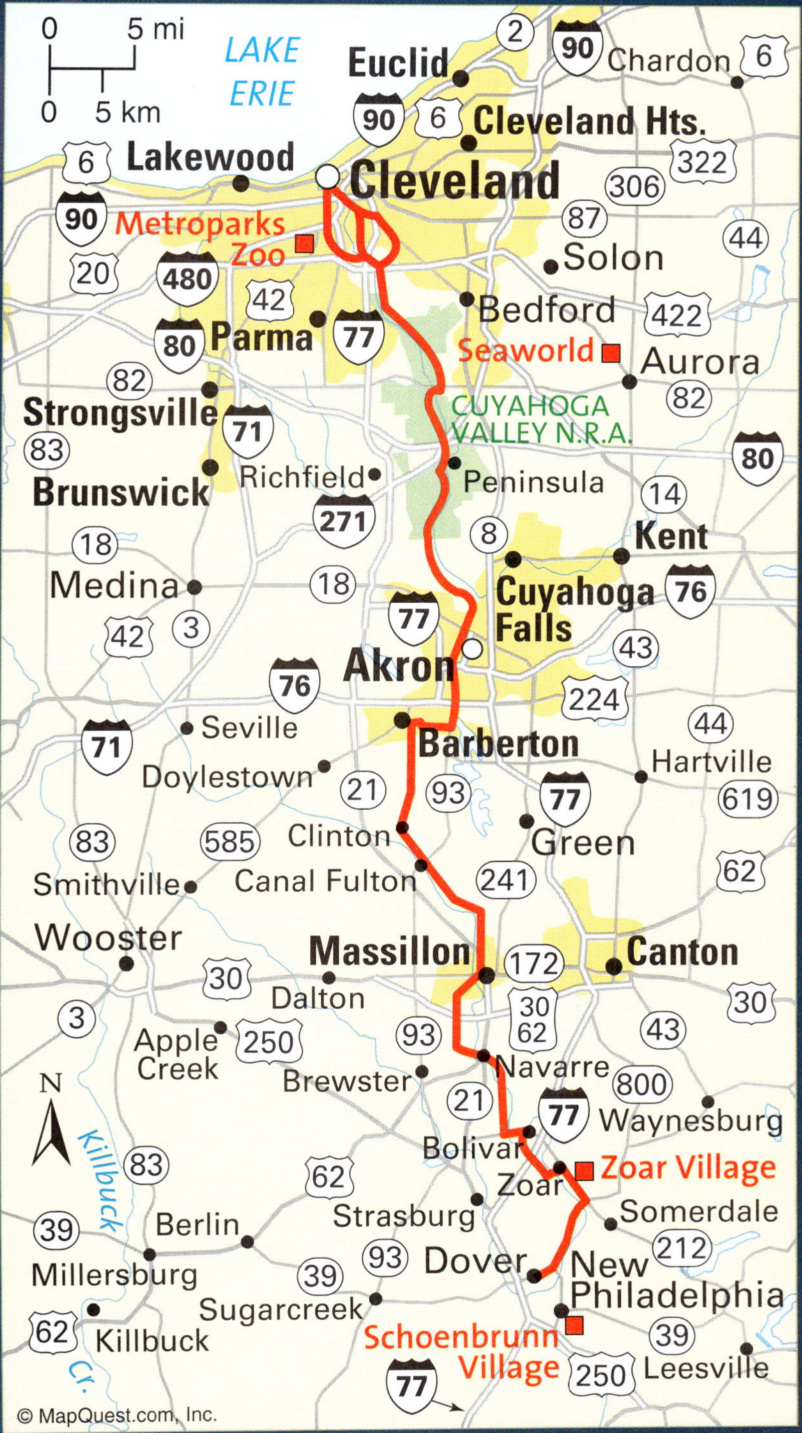

SD ❋ The Native American Scenic Byway

Peter Norbeck Scenic Byway — SD

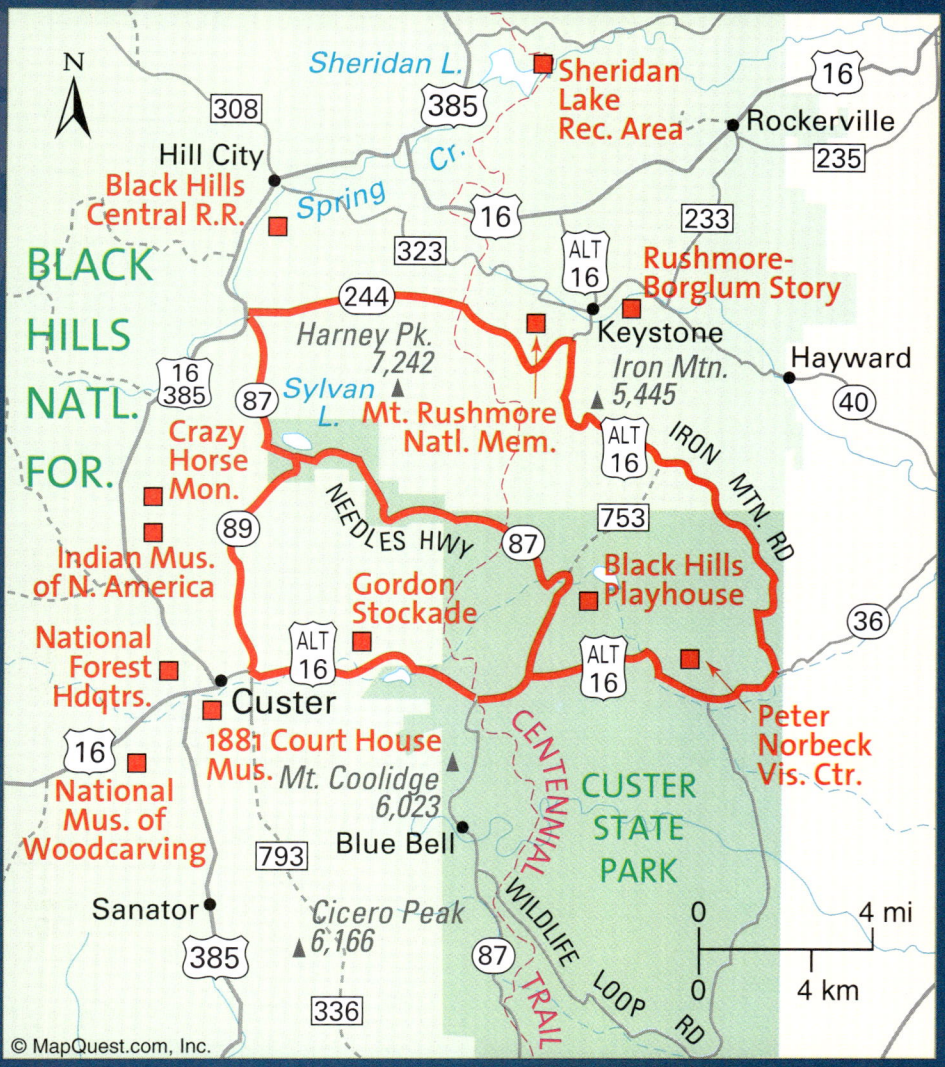

MIDWEST A23

come CLOSER to *the heart and soul of* your AMERICA

America's Byways are a distinctive collection of American roads, their stories and treasured places. They are roads to the heart and soul of America. Each and every Byway has unique qualities that make it special, whether it is a coastal highway known for its striking scenery or a historic route that takes travelers on a path through the country's past. The image on this page gives you only a taste of what the Byways in this region have to offer. For more photos, see the individual Byways.

North Shore Scenic Drive

Grand Portage National Monument

The Grand Portage National Monument, located along the North Shore Scenic Drive, preserves the history of both Ojibwe Indians and the fur trade that thrived here during the 18th, 19th, and 20th centuries.

Minnesota

America's Byways Series | MOBIL TRAVEL GUIDE

A Word to Our Readers

Travelers are on the roads in great numbers these days. They're exploring the country on day trips, weekend getaways, business trips, and extended family vacations, visiting major cities and small towns along the way. Because time is precious and the travel industry is ever-changing, having accurate, reliable travel information at your fingertips is critical. Mobil Travel Guide has been providing invaluable insight to travelers for more than 45 years, and we are committed to continuing this service well into the future.

The Mobil Corporation (known as Exxon Mobil Corporation since a 1999 merger) began producing the Mobil Travel Guide books in 1958, following the introduction of the US highway system in 1956. The first edition covered only five southwestern states. Since then, our books have become the premier travel guides in North America, covering the 48 contiguous states and Canada. Now, Mobil Travel Guide presents a brand-new series in partnership with the National Scenic Byways Program. We also recently introduced road atlases and specialty publications; a robust new Web site; as well as the first fully integrated, road-centric travel support program called MobilCompanion, the driving force in travel.

Since its founding, Mobil Travel Guide has served as an advocate for travelers seeking knowledge about hotels, restaurants, and places to visit. Based on an objective process, we make recommendations to our customers that we believe will enhance the quality and value of their travel experiences. Our trusted Mobil One- to Five-Star rating system is the oldest and most respected lodging and restaurant inspection and rating program in North America. Most hoteliers, restaurateurs, and industry observers favorably regard the rigor of our inspection program and understand the prestige and benefits that come with receiving a Mobil star rating.

The Mobil Travel Guide process of rating each establishment includes:

- Unannounced facility inspections
- Incognito service evaluations for Mobil Four- and Five-Star properties
- A review of unsolicited comments from the general public
- Senior management oversight

For each property, more than 450 attributes, including cleanliness, physical facilities, employee attitude, and courtesy, are measured and evaluated to produce a mathematically derived score, which is then blended with the other elements to form an overall score. These quantifiable scores allow comparative analysis among properties and form the basis that Mobil Travel Guide uses to assign its Mobil One- to Five-Star ratings.

This process focuses largely on guest expectations, guest experience, and consistency of service, not just physical facilities and amenities. It is fundamentally a relative rating system that rewards those properties that continually strive for and achieve excellence each year. Indeed, the very best properties are consistently raising the bar for those that wish to compete with them. These properties proactively respond to consumers' needs even in today's uncertain times.

Only facilities that meet Mobil Travel Guide's standards earn the privilege of being listed in our books. Deteriorating, poorly managed establishments are deleted. A Mobil Travel Guide listing constitutes a positive quality

A25

A WORD TO OUR READERS

recommendation; every listing is an accolade, a recognition of achievement. Our Mobil One- to Five-Star rating system highlights its level of service. Extensive in-house research is constantly underway to determine new additions to our lists.

- The **Mobil Five-Star Award** indicates that a property is one of the very best in the country and consistently provides gracious and courteous service, superlative quality in its facility, and a unique ambience. The lodgings and restaurants at the Mobil Five-Star level consistently and proactively respond to consumers' needs and continue their commitment to excellence, doing so with grace and perseverance.
- Also highly regarded is the **Mobil Four-Star Award,** which honors properties for outstanding achievement in overall facility and for providing very strong service levels in all areas. These award-winners provide a distinctive experience for the ever-demanding and sophisticated consumer.
- The **Mobil Three-Star Award** recognizes an excellent property that provides full services and amenities. This category ranges from exceptional hotels with limited services to elegant restaurants with a less-formal atmosphere.
- A **Mobil Two-Star property** is a clean and comfortable establishment that has expanded amenities or a distinctive environment. A Mobil Two-Star property is an excellent place to stay or dine.
- A **Mobil One-Star property** is limited in its amenities and services but focuses on providing a value experience while meeting travelers' expectations. The property can be expected to be clean, comfortable, and convenient.

Allow us to emphasize that we do not charge establishments for inclusion in our guides. We have no relationship with any of the businesses and attractions we list and act only as a consumer advocate. In essence, we do the investigative legwork so that you won't have to.

Keep in mind, too, that the hospitality business is ever-changing. Restaurants and lodgings—particularly small chains and standalone establishments—change management or even go out of business with surprising quickness. Although we make every effort to double-check information during our annual updates, we nevertheless recommend that you call ahead to make sure the place you've selected is still open and offers all the amenities you're looking for. We've provided phone numbers; when available, we also list Web site addresses.

We hope that your travels are enjoyable and relaxing and that our books help you get the most out of every trip you take. If any aspect of your accommodation, dining, or sightseeing experience motivates you to comment, please drop us a line. We depend a great deal on our readers' remarks, so you can be assured that we will read your comments and assimilate them into our research. General comments about our books are also welcome. You can write to us at Mobil Travel Guide, 1460 Renaissance Drive, Suite 401, Park Ridge, IL 60068, or send an e-mail to info@mobiltravelguide.com.

Take your Mobil Travel Guide books along on every trip you take. We're confident that you'll be pleased with their convenience, ease of use, and breadth of dependable coverage.

Happy travels!

America's Byways Series | MOBIL TRAVEL GUIDE

Overview of the National Scenic Byways Program

WHAT ARE AMERICA'S BYWAYS™?

Under the National Scenic Byways Program, the US Secretary of Transportation recognizes certain roads as National Scenic Byways or All-American Roads based on their archaeological, cultural, historic, natural, recreational, and scenic qualities. There are 96 such designated Byways in 39 states. The Federal Highway Administration promotes the collection as America's Byways™.

America's Byways™ are a distinctive collection of American roads, their stories and treasured places. They are roads to the heart and soul of America. Byways are exclusive because of their outstanding qualities, not because Byways are confined to a select group of people.

Managing the intrinsic qualities that shape the Byway's story and interpreting the story are equally important in improving the quality of the visitors' experience. The National Scenic Byways Program is founded upon the strength of the leaders for individual Byways. It is a voluntary, grassroots program. It recognizes and supports outstanding roads. It provides resources to help manage the intrinsic qualities within the broader Byway corridor to be treasured and shared. Perhaps one of the underlying principles for the program has been articulated best by the Byway leader who said, "The program is about recognition, not regulation."

WHAT DEFINES A NATIONAL SCENIC BYWAY AND AN ALL-AMERICAN ROAD?

To be designated as a National Scenic Byway, a road must possess at least one of the six intrinsic qualities described below. To receive an All-American Road designation, a road must possess multiple intrinsic qualities that are nationally significant and contain one-of-a-kind features that do not exist elsewhere. The road or highway must also be considered a destination unto itself. That is, the road must provide an exceptional traveling experience so recognized by travelers that they would make a drive along the highway a primary reason for their trip.

Anyone may nominate a road for possible designation by the US Secretary of Transportation, but the nomination must be submitted through a state's official scenic byway agency and include a corridor management plan designed to preserve and enhance the unique qualities of the Byway.

The Byways themselves typically are supported through a network of individuals who volunteer their time and effort. It is a bottom-up, grassroots-oriented program. Local citizens and communities create the vision for their Byway, identify the resources comprising the intrinsic qualities, and form the theme or story that stirs the interest and imagination of visitors about the Byway and its resources. Local citizens and communities decide how best to balance goals, strategies, and actions for promoting the Byway and preserving its intrinsic qualities. The vision, goals, strategies, and actions for the Byway are laid out in the required corridor management plan.

Nomination is not about filling out an application. It's all about telling the Byway's story. That's the premise that is driving the FHWA's work on requesting nominations for possible national designation. Nominees might want to think of their Byway's nomination as a combination of the community's guide and a visitor's guide for the Byway.

A27

OVERVIEW OF THE NATIONAL SCENIC BYWAYS PROGRAM

WHAT ARE INTRINSIC QUALITIES?

An intrinsic quality is a scenic, historic, recreational, cultural, archaeological, or natural feature that is considered representative, unique, irreplaceable, or distinctly characteristic of an area. The National Scenic Byways Program provides resources to the Byway community and enhances local quality of life through efforts to preserve, protect, interpret, and promote the intrinsic qualities of designated Byways.

- **Archaeological quality** involves those characteristics of the Byway corridor that are physical evidence of historic or prehistoric life that is visible and capable of being inventoried and interpreted.

- **Cultural quality** is evidence and expressions of the customs or traditions of a distinct group of people. Cultural features include, but are not limited to, crafts, music, dance, rituals, festivals, speech, food, special events, and vernacular architecture that are currently practiced.

- **Historic quality** encompasses legacies of the past that are distinctly associated with physical elements of the landscape, whether natural or man-made, that are of such historic significance that they educate the viewer and stir an appreciation of the past.

- **Natural quality** applies to those features in the visual environment that are in a relatively undisturbed state. These features predate the arrival of human populations and may include geological formations, fossils, landforms, water bodies, vegetation, and wildlife.

- **Recreational quality** involves outdoor recreational activities directly associated with, and dependent upon, the natural and cultural elements of the corridor's landscape.

- **Scenic quality** is the heightened visual experience derived from the view of natural and man-made elements of the visual environment.

For more information about the National Scenic Byways Program, call 800/4BYWAYS or visit the Web site www.byways.org.

America's Byways Series | MOBIL TRAVEL GUIDE

Introduction

America's Byways™ are a distinctive collection of American roads, their stories, and treasured places. They are the roads to the heart and soul of America. This book showcases a select group of nationally designated Byways and organizes them by state, and within each state, alphabetically by Byway. Information in this book is collected from two sources:

- The National Scenic Byways Program (NSBP) provides content about the Byways themselves—quick facts, the Byway story, and highlights. NSBP's information contributors include federal, regional, and state organizations, as well as private groups and individuals. These parties have been recognized as experts in Byways and are an authoritative source for the Byways information that appears in this book.

- Information in this book about lodgings, restaurants, and most sights and attractions along the Byways comes from Mobil Travel Guide, which has served as a trusted aid to auto travelers in search of value in lodging, dining, and destinations since its inception in 1958. The Mobil One- to Five-Star rating system is the oldest and most respected lodging and restaurant inspection and rating program in North America. This trusted, well-established tool directs you to satisfying places to eat and stay, as well as to interesting events and attractions in thousands of locations.

The following sections explain the wealth of information you'll find about the Byways that appear in this book: information about the Byway, things to see and do along the way, and places to stay and eat.

Quick Facts

This section gives you an overview of each Byway, including the following quick facts:

LENGTH: The number of miles from one end of the Byway to the other.

TIME TO ALLOW: How much time to allow to drive the entire length. For some Byways, the suggested time is several days because of the length or the number of attractions on or near the Byway; for others, the time is listed in hours.

BEST TIME TO DRIVE: The season(s) in which the Byway is most appealing. For some Byways, you also discover the peak season, which you may want to avoid if you're looking for a peaceful, uncrowded drive.

BYWAY TRAVEL INFORMATION: Telephone numbers and Web sites for the Byway organization and any local travel and tourism centers.

SPECIAL CONSIDERATIONS: Words of advice that range from the type of clothing you'll want to bring to winter-weather advisories.

RESTRICTIONS: Closings or other cautionary tips.

BICYCLE/PEDESTRIAN FACILITIES: Explains whether the Byway is safe and pleasant for bicycling and/or walking.

A29

INTRODUCTION

THE BYWAY STORY

As explained in the preceding section titled "Overview of the National Scenic Byways Program," a road must possess intrinsic qualities and one-of-a-kind features to receive a National Scenic Byway or All-American Road designation. An All-American Road must also be considered a destination unto itself. This section describes the unique qualities of each Byway, with a separate section for each of its intrinsic qualities. Here you'll find information about the history and culture of the roadway, the wildlife and other natural features found along the Byway, and the recreational opportunities that are available to visitors to the area.

HIGHLIGHTS

Some local Byway organizations suggest tours or itineraries that cover all or part of the Byway. Where these itineraries are available, they're included in this book under the heading "Highlights."

THINGS TO SEE AND DO

Mobil Travel Guide offers information about nearly 20,000 museums, art galleries, amusement parks, historic sites, national and state parks, ski areas, and many other types of attractions. A white star on a black background ★ signals that the attraction is a must-see—one of the best in the area. Because municipal parks, public tennis courts, swimming pools, and small educational institutions are common to most towns, they generally are not mentioned.

When a Byway goes through or comes quite close to a particular town, city, or national park, attractions in those towns or parks are included. Otherwise, attractions are limited to those along the Byway.

Attractions for the entire Byway are listed alphabetically by name. Following an attraction's description, you'll find the months, days, and, in some cases, hours of operation; the address/directions, telephone number, and Web site (if there is one); and the admission price category. The following are the ranges we use for admission fees:

- **FREE**
- **DONATION**
- **$** = Up to $5
- **$$** = $5.01-$10
- **$$$** = $10.01-$15
- **$$$$** = $15.01 and up

PLACES TO STAY

For each Byway, recommended lodgings are listed in alphabetical order, based on the cities in which they're located. In general, only lodgings that are close to or located right on the Byway are listed.

Each lodging listing gives the name, address/location (when no street address is available), neighborhood and/or directions from downtown (in major cities), phone number(s), Web site (if available), total number of guest rooms, and seasons open (if not year-round). Also included are details on business, luxury, recreational, and dining facilities on the property or nearby. A key to the symbols at the end of each listing can be found in the "Terms, Abbreviations, and Symbols in Listings" section of this Introduction.

Because most lodgings offer the following features and services, information about them does not appear in the listings unless exceptions exist:

- Year-round operation with a single rate structure
- Major credit cards accepted (note that Exxon or Mobil Corporation credit cards cannot be used to pay for room or other charges)
- Air-conditioning and heat, often with individual room controls
- Bathroom with tub and/or shower in each room
- Cots and cribs available

- Daily maid service
- Elevators
- In-room telephones

For every property, we also provide pricing information. Because lodging rates change frequently, we list a pricing category rather than specific prices. The pricing categories break down as follows:

- ¢ = Up to $90
- $ = $91-$150
- $$ = $151-$250
- $$$ = $251-$350
- $$$$ = $351 and up

All prices quoted are in effect at the time of publication; however, prices cannot be guaranteed. Note that in some locations, short-term price variations may exist because of special events or holidays. Certain resorts have complicated rate structures that vary with the time of year; always confirm rates when making your plans.

All listed establishments have been inspected by experienced field representatives and/or evaluated by a senior staff member. Our ratings are based on detailed inspection reports of the individual properties, on written evaluations of staff members who stay and dine anonymously, and on an extensive review of reader comments. Rating categories reflect both the features a property offers and its quality in relation to similar establishments.

Here are the definitions for the star ratings for lodgings:

- ★★★★★: A Mobil Five-Star lodging provides consistently superlative service in an exceptionally distinctive luxury environment, with expanded services. Attention to detail is evident throughout the hotel, resort, or inn, from bed linens to staff uniforms.
- ★★★★: A Mobil Four-Star lodging provides a luxury experience with expanded amenities in a distinctive environment. Services may include, but are not limited to, automatic turndown service, 24-hour room service, and valet parking.
- ★★★: A Mobil Three-Star lodging is well appointed, with a full-service restaurant and expanded amenities, such as a fitness center, golf course, tennis courts, 24-hour room service, and optional turndown service.
- ★★: A Mobil Two-Star lodging is considered a clean, comfortable, and reliable establishment that has expanded amenities, such as a full-service restaurant on the premises.
- ★: A Mobil One-Star lodging is a limited-service hotel, motel, or inn that is considered a clean, comfortable, and reliable establishment.

PLACES TO EAT

For each Byway, dining establishments are listed in alphabetical order, based on the cities in which they're located. These restaurants and other eateries are either right on or close to the Byway chapter in which they're listed. All establishments listed have a full kitchen and offer table service and a complete menu. Parking on or near the premises, in a lot or garage, is assumed.

Each listing also gives the cuisine type, address (or directions if no street address is available), neighborhood and/or directions from downtown (in major cities), phone number, Web site (if available), meals served, days of operation (if not open daily year-round), reservation policy, and pricing category. We also indicate whether a children's menu is offered. The pricing categories are defined as follows per diner and assume that you order an appetizer, entrée, and one drink:

- $ = Up to $15
- $$ = $16-$35
- $$$ = $36-$85
- $$$$ = $86 and up

All listed establishments have been inspected by experienced field representatives and/or evaluated by a senior staff member. Our ratings are based on detailed inspection reports of the individual properties, on written evaluations of staff members who stay and dine anonymously, and on an extensive review of reader comments. Rating categories reflect both the

INTRODUCTION

features a property offers and its quality in relation to similar establishments.

The Mobil star ratings for restaurants are defined as follows:

- ★★★★★: A Mobil Five-Star restaurant offers one of few flawless dining experiences in the country. These establishments consistently provide their guests with exceptional food, superlative service, elegant décor, and exquisite presentations of each detail surrounding a meal.
- ★★★★: A Mobil Four-Star restaurant provides professional service, distinctive presentations, and wonderful food.
- ★★★: A Mobil Three-Star restaurant has good food, warm and skillful service, and enjoyable décor.
- ★★: A Mobil Two-Star restaurant serves fresh food in a clean setting with efficient service. Value is considered in this category, as is family friendliness.
- ★: A Mobil One-Star restaurant provides a distinctive experience through culinary specialty, local flair, or individual atmosphere.

TERMS, ABBREVIATIONS, AND SYMBOLS IN LISTINGS

The following terms, abbreviations, and symbols are used throughout the Mobil Travel Guide lodging and restaurant listings to indicate which amenities and services are available at each establishment. We've done our best to provide accurate and up-to-date information, but things do change, so if a particular feature is essential to you, please contact the establishment directly to make sure that it is available.

Continental breakfast: Usually coffee and a roll or doughnut.

In-room modem link: Every guest room has a connection for a modem that's separate from the main phone line.

Laundry service: Either coin-operated laundry facilities or overnight valet service is available.

Luxury level: A special section of a lodging, spanning at least an entire floor, that offers

Byway experts from around the country recommend special restaurants and/or lodgings along their particular Byways that can make your trip even more pleasant. You'll see these special recommendations throughout this book. Look for this symbol next to the hotel or restaurant name:

increased luxury accommodations. Management must provide no less than three of these four services: separate check-in and check-out, concierge, private lounge, and private elevator service (with key access). Complimentary breakfast and snacks are commonly offered.

MAP: Modified American plan (lodging plus two meals).

Movies: Prerecorded videos are available for rental or check-out.

Prix fixe: A full, multicourse meal for a stated price; usually available at finer restaurants.

Valet parking: An attendant is available to park and retrieve your car.

VCR: VCRs are present in all guest rooms.

VCR available: VCRs are available for hookup in guest rooms.

- Pet allowed
- Fishing
- Horseback riding
- Snow skiing nearby
- Golf, nine-hole minimum, on premises
- Tennis court(s) on premises
- Swimming
- In-house fitness room
- Jogging
- Major commercial airport within 10 miles
- Nonsmoking guest rooms
- Senior citizen rates
- Business center

SPECIAL INFORMATION FOR TRAVELERS WITH DISABILITIES

The Mobil Travel Guide D symbol indicates establishments that are at least partially accessible to people with mobility problems. Our criteria for accessibility are unique to our publications. Please do not confuse them with the universal symbol for wheelchair accessibility.

When the D symbol follows a listing, the establishment is equipped with facilities to accommodate people using wheelchairs or crutches or otherwise needing easy access to doorways and rest rooms. Travelers with severe mobility problems or with hearing or visual impairments may or may not find the facilities they need. Always phone ahead to make sure that an establishment can meet your needs.

All lodgings bearing our D symbol have the following facilities:

- ISA-designated parking near access ramps
- Level or ramped entryways to buildings
- Swinging building entryway doors a minimum of 39 inches wide
- Public rest rooms on the main level with space to operate a wheelchair and handrails at commode areas
- Elevator(s) equipped with grab bars and lowered control buttons
- Restaurant(s) with accessible doorway(s), rest rooms with space to operate a wheelchair, and handrails at commode areas
- Guest room entryways that are at least 39 inches wide
- Low-pile carpet in rooms
- Telephones at bedside and in the bathroom
- Beds placed at wheelchair height
- Bathrooms with a minimum doorway width of 3 feet
- Bath with an open sink (no cabinet) and room to operate a wheelchair
- Handrails at commode areas and in the tub
- Wheelchair-accessible peepholes in room entry door
- Wheelchair-accessible closet rods and shelves

All restaurants bearing our D symbol offer the following facilities:

- ISA-designated parking beside access ramps
- Level or ramped front entryways to the building
- Tables that accommodate wheelchairs
- Main-floor rest rooms with an entryway that's at least 3 feet wide
- Rest rooms with space to operate a wheelchair and handrails at commode areas

The Great River Road
❋ ILLINOIS
Part of a multistate Byway; see also IA, MN, WI.

Quick Facts

LENGTH: 557 miles.

TIME TO ALLOW: 4 or 5 days.

BEST TIME TO DRIVE: Fall months are the best due to the beautiful colors of the foliage. High season is late spring to early fall.

BYWAY TRAVEL INFORMATION: Western Illinois Tourism Development Office: 309/837-7460, toll-free 877/GRR-7007.

SPECIAL CONSIDERATIONS: Gas, food, and lodging are available at various cities along the Byway.

RESTRICTIONS: Generally, the entire route of Illinois' portion of the Great River Road is within the 100-year flood plain. While flooding does not occur regularly, roads are closed and detours are marked when flooded.

BICYCLE/PEDESTRIAN FACILITIES: The Great River Road accommodates bicycle and pedestrian traffic. A 62.5-mile bike path is under construction along the route from Mississippi Palisades State Park in Savanna to Sunset Park in Rock Island, where it meets with the American Discovery Trail crossing. At this time, approximately 42 miles have been built. This portion of the bike path is primarily a two-lane, off-road trail winding through the trees and over specially constructed bridges following the route of the Great River Road in Illinois.

Experiencing the Mississippi River for the first time is a memory few can forget. The awe that many people feel toward this river may come from the power of a flood or the beauty of a golden sunset that reflects off the still winter waters and turns graceful steel bridges into shimmering lines of color.

Looking out over the river, it is almost impossible to comprehend the complex layers of history that have been acted out along its banks. From the large communities of the Hopewell Indian culture (the most complex society in North America that existed from approximately AD 700 to 1400) and early French colonial settlements and fortifications to the frightened, cautious, and optimistic eyes of slaves seeking freedom on the Underground Railroad, this corridor has played a role in many of this continent's most dramatic hours. Today, 15 percent of the nation's shipping passes through the river's complex system of locks and dams, yet such commercial activity occurs under the spreading wings of the newly thriving American bald eagle.

It is from the Great River Road that most visitors and residents understand and define their relationship with the Mississippi. It is from this road that the historic sites and cultural artifacts of the area can be accessed, from Native American mounds to the Mormon (Church of Jesus Christ of Latter-Day Saints) temple. The beautiful Mississippi bluffs tower over the Byway as permanent sentinels for the great river. Whether directly along the banks of the river or winding through the vast flood plain miles from the water, the Great River Road links resources, people, and history.

Illinois

❋ The Great River Road

THE BYWAY STORY

The Great River Road tells archaeological, cultural, historical, natural, recreational, and scenic stories that make it a unique and treasured Byway.

Archaeological

A little-known treasure trove of archaeological sites, the Illinois Great River Road has several places for visitors to discover pieces of the past. Among the archaeological qualities that can be found along this road are burial mounds of native tribes that lived along the river. The mounds, many of which were built more than 2,000 years ago, are representative of Native American religious practices and reverence for their ancestors. Cahokia Mounds near East St. Louis and Collinsville has been designated as a United Nations World Heritage Site. Among the most fascinating of the archaeological structures on the Great River Road is Monk's Mound, a 100-foot-tall, four-tiered platform that took 300 years to build.

In addition to Native American sites, many villages on the Byway offer a taste of archaeology in their preservation of the not-so-distant past. Many villages, like Maeystown, Galena, and Nauvoo, are listed on the National Register of Historic Places. These villages often re-create the lifestyles of the first settlers along the Great River Road for visitors who want to know more about the nation's past. With both Native American heritage sites and historic sites of the earliest European settlers, the Great River Road offers opportunities for you to discover America's archaeology all along the way.

Cultural

Some of the first people to settle along the banks of the Mississippi River were Native American tribes. These tribes were embedded in a culture that held the utmost respect for nature and the resources of the land. Their inextricable connection to the land can be seen in the burial mounds they left behind, as well as in museums and monuments.

Since the habitation of the first cultures in the area, several other cultures have passed through the Illinois Great River Road area, and some have stayed permanently. During the 1800s, the now-historic communities along the Great River Road were settled for reasons that ranged from gold rushes to religious freedom. The people who live in these communities maintain a distinct place on the Byway, with their styles of architecture and inventiveness. Today, the culture of the Great River Road embodies the relaxed hometown pace. The towns and villages along the Byway offer you a change of scenery and a chance to slow down. These towns are often small and full of rich historical detail that influences cultures even today.

Historical

As an area that has enraptured American Indians, explorers, and settlers, the Illinois Great River Road holds pieces of the past that are intriguing to today's visitors. Since 1938, the road has been protected and enhanced in order to preserve the scenic and historical qualities found along it. The heritage of the native tribes of the Sauk and Fox Indians remains prevalent in many places along the Byway.

You can find historic architecture in several of the historic towns along the road: Nauvoo, Quincy, Alton, Belleville, and Cairo allow you to experience the Great River Road as the settlers of nearly 200 years ago did. These cities all have their share of historic places and buildings that are full of Civil War tales and pioneer stories. As a passage on the Underground Railroad, the river represents a piece of African-American history as well. The river itself holds a story of steamboats chugging up the river. It represents the ingenuity of inventors and engineers in the earliest days of travel. The river is the lifeblood of the area that has drawn so many people to its shores.

Natural

Among the bluffs and rolling hills of the Illinois Great River Road area, you can observe wildlife and nature at its fullest. The lands surrounding the Byway are home to white-tailed deer, wild turkeys, ducks, and geese. Supported by the rich natural resources that abound in the river area, these creatures can be seen throughout the drive. During the fall, the trees along the Byway exhibit a beautiful spectrum of color, providing a fringe of brightness along the river. By the time winter sets in, there is a new visitor to the Great River Road. The American bald eagle arrives in late November, and by late December, hundreds of these magnificent birds are roosting in the rocky walls of the bluffs overlooking the river. Travelers come from miles around to watch them dive and soar in the air above the bluffs.

All along the banks and bluffs of the river, you will enjoy many interesting sights. At one point on the Illinois Great River Road, you will see a formation known as Tower Rock. This formation is an isolated mass of limestone that divides the river in half. In the areas surrounding the river, you'll also find lakes, wetlands, and swamps that provide their own style of natural beauty.

Recreational

On and around the river, you have places to go and different ways to get there. Hikers and bikers find riverside trails attractive, while other travelers may prefer to enjoy a pleasant afternoon on a riverboat. Ferries, canoes, and even old-fashioned steamboats give you a closer view of the greatest river in the nation. To see more of the communities on the Byway, you may enjoy a trolley tour or a park area, as well as museums and historical buildings. Museums and monuments to the past are sprinkled along the road to give you a sense of what came before on the Great River Road.

Other forms of fun can be found on the Byway as well. More than 75 golf courses help you track your progress along the Byway by greens. Travelers who would like to test their luck can try a riverboat casino. Communities all along the Byway offer numerous stops for antique shoppers who are looking for a piece of Illinois' past to take home with them, and if antiques aren't enough, plenty of novelty shops and gift shops abound. For the hungry traveler, many restaurants along the Byway are sure to suit your fancy. Entertainment is an element of the Byway's recreational offerings, too. Many towns host musicals, dinner shows, and old-fashioned theater experiences.

Chances to enjoy the outdoors along the Great River Road come often. In addition to the Shawnee National Forest, 29 state recreation and/or conservation areas are available along the route of the Great River Road. The Mississippi Palisades State Park and National Landmark offers phenomenal views to and from the bluffs (palisades) along the Mississippi River. The facilities for tent and trailer camping, fishing, cross-country skiing, and ice fishing are top notch.

The Big River State Forest is a 2,900-acre facility dedicated to demonstrating sound forestry practices. Fire breaks and a fire tower afford breathtaking views and hikes. Nearby, camping, hiking, and river and lake fishing are available at Delabar State Park. In the south, Horseshoe Lake Conservation Area is one of

see page A5 for color map

Illinois

✤ *The Great River Road*

the loveliest places to hike, camp, hunt, and boat. Horseshoe Lake is a quiet, shallow lake lined with cypress and tupelo gum and wild lotus. You can find places for bird-watching and exploring wetlands, and canoeing along the river is a widely recognized source of recreation all along the Byway.

Scenic

The Mississippi River itself is a natural phenomenon that few visitors will forget. This body of moving water presents a picture of the forces of nature at work with their surroundings. Perhaps one of the prettiest sights you will see along the Byway is the great waters of the Mississippi River flanked by the glacier-carved bluffs at the river's edge. Along the Byway, observe scenic vistas and bluffs that overlook the river: erosion from glacial movement has left unique formations of rock in the riverside topography. Feast your eyes on the rich architecture that has been a part of this area's history. From grand courthouses to historic bridges, sights all along the Byway complement the natural beauty of the Great River Road. In the summer, the fields along the Byway are adorned with wildflowers. During the fall, several communities host festivals celebrating the season, and the drive along the Byway becomes even more scenic with every leaf that dons its fall color. And keep in mind that a sunset on the Mississippi River is a sight not to be missed.

HIGHLIGHTS

When traveling the Moline-to-Nauvoo section of the Illinois Great River Road, consider using the following itinerary.

- Both the past and present of the world-famous John Deere & Company operations are centered in Moline, where you begin your tour. At the **John Deere Commons,** catch historic trolleys to other Deere sites, tour the John Deere Pavilion with interactive displays of historic and modern farm equipment, and visit the John Deere Store. The Deere Administrative Center, Deere corporate headquarters, lies on the outskirts of Moline. This building, designed by Eero Saarinen, and grounds are widely regarded as masterworks of architecture and landscape architecture. The Deere-Wiman House and Butterworth Center are mansions built in the late 1800s by Charles Deere. Guided tours of the homes and gardens are available.

- **Rock Island Arsenal** lies on spectacular Rock Island in the Mississippi River directly in front of the John Deere Commons. Visitors to the island can visit Historic Fort Armstrong (1812), the Rock Island Arsenal Museum (with exhibits of military equipment and small firearms), and other historic structures. The Rock Island Arsenal is the largest weapons manufacturing arsenal in the country. Located next to Lock and Dam 15, the largest roller dam in the world, the US Army Corps of Engineers **Mississippi River Visitor Center** features an observation deck for tow boats and birds. The visitor center includes displays about Upper Mississippi geography, ecology, and the lock-and-dam system. It is also a designated Great River Road interpretive center.

- Two miles south of Rock Island lies the next stop on the tour, **Black Hawk State Historic Site**—a wooded, steeply rolling 208-acre tract. Prehistoric Indians and 19th-century settlers made their homes here, but the area is most closely identified with the Sauk nation and the warrior-leader whose name it bears—Black

Hawk. The site, which is also noted for its many natural features, is managed by the Illinois Historic Preservation Agency. The **Hauberg Indian Museum,** located in the lodge constructed by the Civilian Conservation Corps in 1939, interprets the culture of the Sauk and the Mesquackie. Nearly 175 species of birds and 30 species of wildflowers, as well as a prairie restoration, can be observed here. Dickson Pioneer Cemetery is where many early settlers are buried. Picnicking and hiking are also available.

- Following the Byway along the Mississippi River for another 50 miles, you arrive at the 2,900-acre **Big River State Forest.** The forest lies in Henderson County, 6 miles north of Oquawka, where gas and food are available. The area's oldest pine plantation, the **Milroy Plantation,** with towering red, white, and jack pines lies within. The forest is a remnant of a vast prairie woodland border area that once covered much of Illinois. Two endangered plants, penstemon and Patterson's bindweed, are found here. A prominent landmark in the forest is its fire tower, located at the headquarters area and accessible to the public at non-emergency times. Sixty miles of firebreaks interlace Big River State Forest, which are used by hikers, horseback riders, and snowmobilers. Tent, trailer, and equestrian camping sites, boat launch, picnic areas, hunting, stables, and scenic drives are available.

- Located on the Mississippi River about 4.5 miles south of Big River State Forest and 1.5 miles north of Oquawka, the 89-acre **Delabar State Park** offers quality outdoor experiences for anglers, hikers, campers, and picnickers. More than 50 species of birds have been sighted in the park, making it a destination for bird-watching, too. Picnic areas, playground facilities, tent and trailer camping, trailer dumping, hiking trails, river and lake fishing, boat launching, ice fishing, and ice skating are available in the area.

- This tour of a short section of the Byway terminates about 45 miles south of Delbar

State Park in **Nauvoo.** The town is located at a picturesque bend in the river at Hancock County. Nauvoo was settled by Joseph Smith and members of the Church of Jesus Christ Latter-Day Saints (LDS) and served as the religious, governmental, and cultural center of the church from 1839 until Joseph Smith's death in 1846. Two visitor centers interpret the remaining town sites. The LDS Visitor Center features costumed hosts, interpretive displays, a sculpture garden, and tours of 25 Nauvoo town sites. The Joseph Smith Visitor Center, run by the Reorganized Church of Jesus Christ of Latter-Day Saints (RLDS), features displays, an informative video, and access to the gravesite and homes of Joseph Smith and family. In late 1999, the LDS church began rebuilding the historic limestone temple destroyed in the late 19th century. Nearby Nauvoo State Park features recreational opportunities. The wine and cheese traditions of the French Icarians, who came to Nauvoo after the LDS church, are still pursued.

THINGS TO SEE AND DO

Driving along the Great River Road will certainly keep your senses engaged, but if you yearn to get out of the car and stretch your legs, or if you'd like to make a mini-vacation out of your trip, check out these attractions along the route.

ALTON BELLE RIVERBOAT CASINO. *1 Front St, Alton (62002). Phone toll-free 800/711-GAME. www.argosycasinos.com.* Entertainment complex includes slots, showrooms, lounges, and restaurants. Open daily 8-6 am. **FREE**

BELVEDERE MANSION. *1008 Park Ave, Galena (61036). Phone 815/777-0747.* This Italianate/Steamboat Gothic mansion (1857) with 22 rooms is restored and furnished with antiques, including pieces used on the set of *Gone With the Wind.* Open Memorial Day-Oct, daily. **$$**

Illinois

❈ *The Great River Road*

BLACK HAWK STATE HISTORIC SITE. *1510 46th Ave, Rock Island (61201). Phone 309/788-0177. www.state.il.us/hpa/Blackhawk.htm.* These wooded, steeply rolling hills provided the site on which the westernmost battle of the Revolutionary War was fought. The area was occupied for nearly a century by the capital villages of the Sauk and Fox nations. The Watch Tower, on a promontory 150 feet above the Rock River, provides a view of the river valley and surrounding countryside. The Hauberg Indian Museum contains an outstanding collection of Native American artifacts, paintings, and relics; dioramas of Sauk and Fox daily life, prehistoric display; also changing displays. Fishing; hiking, picnicking. Open 8:30 am-noon and 1-4:30 pm; closed Jan 1, Thanksgiving, Dec 25. **DONATION**

BRUSSELS FERRY. *Hwy 100 at Hwy 3, Brussels (62013). Phone 618/786-3636.* Ferry boat navigates the Illinois River at the confluence of the Mississippi River. Open daily. **FREE**

CHESTNUT MOUNTAIN RESORT. *Blackjack Rd, Galena (61036). Phone 815/777-1320; toll-free 800/397-1320. www.chestnutmtn.com.* Two quad, three triple chairlifts; three surface lifts; patrol, school, rentals, snowmaking; restaurants, cafeteria, bars, lodge, nursery. Longest run 3,500 feet; vertical drop 475 feet. Open mid-Nov to mid-Mar, daily. **$$$$**

CUSTOM HOUSE MUSEUM. *1400 Washington Ave, Cairo (62914). Phone 618/734-1019.* This 19th-century Federal-style building contains artifacts and replicas from Cairo's past. Open Mon-Fri. **FREE**

FAIRMOUNT PARK. *2 miles W on US 40 jct I-255, Collinsville (62234). Phone 618/345-4300. www.fairmountpark.com.* Thoroughbred racing Sun-Tues 1 pm; Fri, Sat 7:30 pm from mid-Apr to Oct. **$**

FORT DEFIANCE STATE PARK. *Rtes 3 and 51, Cairo (62914). Phone 618/734-3015.* Splendid view of the confluence of the Ohio and Mississippi rivers on 39 acres; site of a Civil War post commanded by General Ulysses S. Grant.

GALENA POST OFFICE & CUSTOMS HOUSE. *110 Green St, Galena (61036). Phone 815/777-0225.* Named a Great American Post Office by the Smithsonian Institution, this was the first post office to receive the honor; built in 1859, it is the second oldest post office in the United States still in use. Open Mon-Sat. **FREE**

HORSESHOE LAKE STATE CONSERVATION AREA. *Rte 3, Olive Branch (62969). Phone 618/776-5689.* Large flocks of Canadian geese migrate to these 10,336 acres in winter. Fishing, boating (ramp; 10-hp motor limit mid-Mar-mid-Nov); hunting, hiking, picnicking, concession, camping.

JOHN DEERE COMMONS. *1400 River Dr, Moline (61265). Phone 309/765-1000. www.johndeerecommons.com.* On the banks of the Mississippi River, this complex is home to the John Deere Pavilion, a visitor center with interactive displays about agriculture and vintage and modern John Deere equipment. Also here are a John Deere Store; a restaurant; hotels; The MARK of the Quad Cities, a 12,000-seat arena hosting high-profile events; and Centre Station (the transportation hub and information center for the Quad Cities). Open daily. **FREE**

JOSEPH SMITH HISTORIC CENTER. *149 Water St, Nauvoo (62354). Phone 217/453-2246. www.cofchrist.org/js/.* A guided tour begins in the visitor center and includes a video about the Latter-Day Saint movement in this area in the 1840s. Open Mon-Sat 9 am-5 pm, Sun 1-5 pm; closed holidays. **FREE**

MAGNOLIA MANOR. *2700 Washington Ave, Cairo (62914). Phone 618/734-0201.* Italianate Victorian mansion (1869) with 14 rooms was built for a wealthy flour merchant and contains period furnishings and items of local historical interest. Enjoy views of the Mississippi and Ohio rivers from the tower. Open daily; closed holidays. **$$**

NAUVOO RESTORATION, INC., VISITOR CENTER. *Young and Main sts, Nauvoo (62354). Phone 217/453-2237; toll-free 800/453-0022. www.beautifulnauvoo.com/site/nri.cfm.* The center has a 20-minute movie and exhibits on Nauvoo history. Open daily. Also here are Seventy Hall, an 1840s meetinghouse; Lyon Drug Store; the Nauvoo Temple site; Montrose Crossing Monument; Sarah Kimball Home; William Weeks Home; Noble-Smith Home; Pendleton log house; Webb blacksmith and wagon shop; Stoddard tin shop; Riser Cobbler shop; 1840 theater; brick kiln; Clark store; Old Post Office and Merryweather Mercantile; Family Living Center with barrel-, candle-, and pottery-making; Jonathan Browning Gunshop; Scovil Bakery; and other significant structures. **FREE**

NAUVOO STATE PARK. *980 S Bluff St, Nauvoo (62354). Phone 217/453-2512. www.dnr.state.il.us.* On 148 acres. Restored house with wine cellar and century-old vineyard adjoining; museum open May-Sept. Fishing, boating (ramp, electric motors only); hiking, picnic area (shelter), playgrounds, camping. Open daily. **FREE**

OLD CARTHAGE JAIL. *307 Walnut St, Carthage (62321). Phone 217/357-2989.* Restored jail (1839-1841) where Joseph Smith and his brother were killed; ten-minute tour; visitor center has an 18-minute film presentation and exhibits. Open daily. **FREE**

PERE MARQUETTE STATE PARK. *100 Great River Rd, Grafton (62037). Phone 618/786-3323. www.dnr.state.il.us.* Pere Marquette State Park is one of Illinois' largest state parks. Nestled along the banks of the Illinois River on the Byway near Grafton, a myriad of trails take you within the wild forests and up to spectacular viewing areas along the bluff line above. Rangers can assist you with hiking plans and provide valuable insight on how to best explore this giant park. Don't miss the visitor center, where you can learn about the natural treasures in and around the park. A lodge and cabins are available for overnight visitors, but plan far ahead, as this is one of the most popular attractions in the Midwest.

PIASA BIRD PAINTING REPRODUCTION. *Alton (62002). Phone toll-free 800/258-6645. www.altonweb.com/history/piasabird/.* According to Native American legend, a monster bird frequented these bluffs and preyed on all who came near. When Marquette sailed down the Mississippi in 1673, he spotted "high rocks with hideous monsters painted on them" at this spot. The paintings, destroyed by quarrying in the 19th century, were reproduced in 1934. These reproductions, in turn, were destroyed by the construction of Great River Road. They were reproduced a second time on a bluff farther up the river. **FREE**

QUAD CITY BOTANICAL CENTER. *2525 4th Ave, Rock Island (61201). Phone 309/794-0991. www.qcgardens.com.* Sun garden conservatory features more than 100 tropical plants and trees, a 14-foot waterfall over reflecting pools, and a horticulture resource center. Open Mon-Sat 10 am-5 pm, Sun noon-5 pm; closed Jan 1, Thanksgiving, Dec 25. **$**

ROCK ISLAND ARSENAL AND US ARMY ARMAMENT, MUNITIONS AND CHEMICAL COMMAND. *Rock Island (61201). Phone 309/782-6001. www.ria.army.mil.* On Arsenal Island, between Rock Island, Illinois, and Davenport, Iowa, is the Rock Island Arsenal Museum, which contains an extensive firearms collection and Court of Patriots memorial (open daily 10 am-4 pm; closed holidays; free; phone 309/782-5021); a replica of Fort Armstrong blockhouse; the Colonel Davenport house (open May-mid-Oct, tours Thurs-Sun, free; phone 309/786-7336); a Confederate soldiers cemetery; and the Rock Island National Cemetery, at the center of the island, which has approximately 1,300 interments. It was the site of the first railroad bridge to span the Mississippi; lock and dam with visitor center (open daily; phone before visiting for site restrictions). Bicycle trail (7 miles) around the island.

Illinois

❋ The Great River Road

ULYSSES S. GRANT HOME STATE HISTORIC SITE. *500 Bouthillier St, Galena (61036). Phone 815/777-3310. www.granthome.com.* This Italianate house was given to General Grant upon his return from the Civil War (1865); original furnishings and Grant family items. Interpretive tours. Picnicking. Open Wed-Sun 9 am-5 pm; closed holidays. **DONATION**

VILLAGE OF ELSAH. *Elsah (62028). Phone 618/374-1568. www.elsah.org.* Many buildings are more than 100 years old. Museum open Apr-Nov, Thurs-Sun afternoons. **FREE**

VINEGAR HILL LEAD MINE & MUSEUM. *8885 N Three Pines Rd, Galena (61036). Phone 815/777-0855. www.enjoyillinois.com.* Guided tour of an old mine showing early mining techniques. Open June-Aug, daily 9 am-5 pm; May, Sept, Oct, Sat-Sun 9 am-5 pm. **$**

PLACES TO STAY

If you choose to include an overnight stay in your trip along this Byway, Mobil Travel Guide recommends the following lodgings.

Alton

★★ **HOLIDAY INN.** *3800 Homer Adams Pkwy, Alton (62002). Phone 618/462-1220; toll-free 800/465-4329. www.holiday-inn.com.* 137 rooms, 4 story. Complimentary breakfast buffet. Check-out noon. TV; cable (premium); VCR (movies). In-room modem link. Restaurant, bar; entertainment. Room service. In-house fitness room, sauna. Health club privileges. Game room. Indoor pool, whirlpool. Free airport transportation. **$**

★ **SUPER 8.** *1800 Homer Adams Pkwy, Alton (62002). Phone 618/465-8885; toll-free 800/800-8000. www.super8.com.* 61 rooms, 3 story. Pet accepted; fee. Complimentary continental breakfast. Check-out 11 am. TV; cable (premium). In-room modem link. **¢**

Collinsville

★ **BEST WESTERN HERITAGE INN.** *2003 Mall Rd, Collinsville (62234). Phone 618/345-5660; toll-free 800/780-7234. www.bestwestern.com.* 80 rooms, 2 story. Complimentary continental breakfast. Check-out noon. TV; cable (premium), VCR available (movies). In-house fitness room. Indoor pool, whirlpool. **¢**

★★ **HOLIDAY INN.** *1000 Eastport Plaza Dr, Collinsville (62234). Phone 618/345-2800; toll-free 800/551-5133. www.holiday-inn.com.* 229 rooms, 5 story. Pet accepted; fee. Complimentary full breakfast. Check-out noon. TV; cable (premium). In-room modem link. Restaurant, bar. Room service 24 hours. In-house fitness room, sauna. Health club privileges. Game room. Indoor pool, whirlpool. Free airport transportation. **$**

★ **MAGGIE'S BED & BREAKFAST.** *2102 N Keebler Ave, Collinsville (62234). Phone 618/344-8283.* 5 rooms, 3 story. No room phones. Pet accepted, some restrictions; fee. Complimentary full breakfast. Check-out noon, check-in 4-6 pm. TV; VCR (movies). Game room. Whirlpool. Built in 1900; former boarding house. Totally nonsmoking. Credit cards not accepted. **$**

East Dubuque

★★ **TIMMERMAN'S HOTEL & RESORT.** *7777 Timmerman Dr, East Dubuque (61025). Phone 815/747-3181.* 74 rooms, 3 story. Pet accepted, some restrictions; fee. Complimentary continental breakfast. Check-out noon. TV; VCR (movies). In-room modem link. Restaurant, bar. Room service. Sauna. Game room. Indoor pool, whirlpool. Downhill ski 10 miles, cross-country ski 1/4 mile. **¢**

Galena

★★ **ALDRICH GUEST HOUSE.** *900 3rd St, Galena (61036). Phone 815/777-3323. www.galena.com/aldrich/.* 12 rooms, 2 story. No room phones. Children over 12 years only. Complimentary full breakfast. Check-out 11 am, check-in 4 pm. TV in sitting room; cable (premium), VCR. Library/sitting room, fireplace. Downhill, cross-country ski 8 miles. Street parking. Built in 1845. Screened porch. Totally nonsmoking. ¢

★ **ALLEN'S VICTORIAN PINES LODGING.** *11383 US 20 W, Galena (61036). Phone 815/777-2043; toll-free 866/847-4637. www.victorianpineslodging.com.* 51 rooms, 1-2 story. Some room phones. Complimentary continental breakfast. Check-out 11 am. TV. Massage. Sauna. Downhill ski 9 miles, cross-country ski 6 miles. Complex consists of a motel; Ryan House (1876), a 24-room Italianate/Victorian mansion with antiques; and Bedford House (1850), an Italianate structure with original chandeliers, leaded glass, and walnut staircase. ¢

★★★ **EAGLE RIDGE INN & RESORT.** *444 Eagle Ridge Dr, Galena (61036). Phone 815/777-2444; toll-free 800/892-2269. www.eagleridge.com.* A 6,800-acre resort, Eagle Ridge features 63 holes of championship golf, as well as a complete line of spa amenities. 80 rooms, 2 story. Check-out noon, check-in 4 pm. TV; cable (premium), VCR (movies). Fireplaces. Dining room, bar; entertainment Tues-Sun. Room service. Supervised children's activities, ages 4-12. In-house fitness room, sauna. Massage. Game room. Indoor pool, whirlpool. Golf on premise, greens fee $77. Tennis. Downhill ski 10 miles, cross-country ski on site. Bike rental. Sand beach, private lake, marina. Canoes, pontoon boats. Sleigh rides. Trail rides. Free airport transportation. $$

★★ **PARK AVENUE GUEST HOUSE.** *208 Park Ave, Galena (61036). Phone 815/777-1075; toll-free 800/359-0743. www.galena.com/parkave/.* 4 rooms, 3 with shower only, 3 story. No room phones. Children over 12 years only. Complimentary full breakfast. Check-out 11 am, check-in 4 pm. TV; cable (premium), VCR available (movies). Some fireplaces. Downhill, cross-country ski 7 miles. Built in 1893. Totally nonsmoking. $

Grafton

★ **PERE MARQUETTE LODGE.** *100 Great River Rd, Grafton (62037). Phone 618/786-2331. www.pml.ilresorts.com.* 72 rooms, 2 story. Check-out noon. TV; VCR available (movies). Restaurant, bar. In-house fitness room, sauna. Game room. Indoor pool, whirlpool. Lighted tennis. Lawn games. Hiking, bike trails. On the Mississippi River. ¢

Moline

★ **BEST WESTERN AIRPORT INN.** *2550 52nd Ave, Moline (61265). Phone 309/762-9191; toll-free 800/780-7234. www.bestwestern.com.* 50 rooms, 2 story. Complimentary continental breakfast. Check-out 11 am. TV; cable (premium). Indoor pool, whirlpool. Downhill, cross-country ski 15 miles. Near the Quad City Airport. ¢

Illinois

❋ *The Great River Road*

★ **HAMPTON INN.** *6920 27th St, Moline (61265). Phone 309/762-1711; toll-free 800/426-7866. www.hamptoninn.com.* 138 rooms, 2 story. Pet accepted; fee. Complimentary continental breakfast. Check-out noon. TV; cable (premium). Free airport transportation. ¢

Nauvoo

★★ **HOTEL NAUVOO.** *1290 Mulholland St (IL96), Nauvoo (62354). Phone 217/453-2211. www.hotelnauvoo.com.* 8 rooms. Check-out 11 am, check-in 4 pm. TV; cable (premium). Dining room, bar. Restored historic inn (1840), originally a private residence. Totally nonsmoking. $

★★ **MISSISSIPPI MEMORIES BED & BREAKFAST.** *1 Riverview Terrace, Nauvoo (62354). Phone 217/453-2771.* 4 rooms, 3 story. Closed the week of Dec 25. Complimentary full breakfast. Check-out 10 am. TV available. Two fireplaces. Totally nonsmoking. $

★ **MOTEL NAUVOO.** *1610 Mulholland St (Hwy 96), Nauvoo (62354). Phone 217/453-2219.* 11 rooms. Check-out 11 am. TV; cable (premium). Totally nonsmoking. ¢

★★ **NAUVOO FAMILY INN & SUITES.** *1875 Mulholland St, Nauvoo (62354). Phone 217/453-6527; toll-free 800/416-4470. www.nauvoonet.com.* 104 rooms, 2 story. Check-out 11 am. TV; cable (premium). Indoor pool. $

Quincy

★ **TRAVELODGE.** *200 S 3rd St, Quincy (62301). Phone 217/222-5620; toll-free 800/578-7878. www.travelodge.com.* 68 rooms, 2 story. Pet accepted; fee. Complimentary continental breakfast. Check-out noon. TV; cable (premium). Laundry services. Restaurant, bar. Room service. Pool. ¢

PLACES TO EAT

A long day of driving is sure to make you hungry. At the end of your journey, take a table at one of the following restaurants.

Alton

★★ **TONY'S.** *312 Piasa St, Alton (62002). Phone 618/462-8384. www.tonyssteaks.com.* Italian menu. Closed holidays. Dinner. Bar. Children's menu. Outdoor seating. $$

East Dubuque

★★ **BITTERSWEET ON THE BLUFF.** *7010 Donna's Dr, East Dubuque (61025). Phone 815/747-2360. www.bittersweetonthebluff.com.* American menu. Dinner. Family-owned and -operated business overlooking the Mississippi River with a 20-year history. $$

Galena

★ **BACKSTREET STEAK & CHOPHOUSE.** *216 S Commerce St, Galena (61036). Phone 815/777-4800. www.backstreetgalena.com.* Steak menu. Dinner. Offering a traditional steakhouse experience in beautiful downtown Galena. $$$

★★ **BERNADINE'S TEA ROOM.** *513 Bouthiller St, Galena (61036). Phone 815/777-0557. www.stillmaninn.com.* American menu. Open Fri, Sat. Dinner. Located in historic downtown Galena in a restored bed-and-breakfast. $$

★★ **CAFÉ ITALIA/TWISTED TACO CAFÉ.** *301 N Main St, Galena (61036). Phone 815/777-0033.* Italian, American menu. Dinner. Bar. Children's menu. Built in 1855. $$

★★ **FRIED GREEN TOMATOES.** *1301 Irish Hollow Rd, Galena (61036). Phone 815/777-3938. www.friedgreen.com.* Italian, American menu. Closed Thanksgiving, Dec 25. Dinner. Bar. Three dining levels in a historic (1851) rural setting. $$

★ **LOG CABIN.** *201 N Main St, Galena (61036). Phone 815/777-0393.* Steak, seafood menu. Closed Mon; Thanksgiving, Dec 24-25. Lunch, dinner. Bar. Children's menu. **$$**
[D]

★★ **WOODLANDS RESTAURANT.** *444 Eagle Ridge Dr, Galena (61036). Phone toll-free 800/998-6338. www.eagleridge.com.* American menu. Breakfast, lunch, dinner. Serving Midwest regional cuisine in a refined setting with beautiful lake views. **$$**

Nauvoo

★ **GRANDPA JOHN'S.** *1255 Mulholland St (IL 96), Nauvoo (62354). Phone 217/453-2310.* Closed Jan-Feb. Lunch. Children's menu. Soda fountain. Established in 1912. Totally nonsmoking. **$**
[D]

★★★ **HOTEL NAUVOO.** *1290 Mulholland St (IL 96), Nauvoo (62354). Phone 217/453-2211. www.hotelnauvoo.com.* Built in 1840 for Mormon founder Joseph Smith, this inn and restaurant resides in west-central Illinois along the Mississippi River. From mid-Apr to mid-Nov, visitors can sample the all-American, fixed-price buffet including such favorites as fried chicken and apple pie. Closed Mon. Dinner, Sun brunch. Bar. Reservations recommended. **$$**
[D]

The Historic National Road

ILLINOIS AN ALL-AMERICAN ROAD
Part of a multistate Byway; see also IN, OH.

Quick Facts

LENGTH: 165 miles.

TIME TO ALLOW: 12 hours.

BEST TIME TO DRIVE: All seasons have their unique attractions on this Byway. High season is during the fall.

BYWAY TRAVEL INFORMATION: Effingham Convention & Visitors Bureau: 800/772-0750; Collinsville Convention & Visitors Bureau: 618/345-4999; Cahokia Mounds State Historic Site: 618/346-5160; Byway local Web site: www.nationalroad.org.

RESTRICTIONS: Some delays may be experienced during severe weather, and seasonal storms may increase driving times.

BICYCLE/PEDESTRIAN FACILITIES: All counties, except for Madison (which is on the western end of the Byway), rate the route as suitable for bicycles. In Madison County, bicyclists are urged to be cautious; bicycling is not advisable due to high traffic.

The Historic National Road crosses the state of Illinois from near the Wabash River to the great Mississippi River. The rolling countryside, prairie fields, and small towns along the old trail whisper of an earlier time. Each of the seven counties of the old trail weaves its own story.

The route of the Historic National Road is a road of history. Nineteenth-century river transportation and commerce, along with historic cemeteries, tell of the struggles of the early settlers on the western frontier. County fairs and main-street storefronts speak of small towns where you can still find soda fountains, one-room schools, and old hotels where travelers stopped to rest.

Small and large museums, a National Register Historic District, and National Register Historic Sites are found all along the Byway. Prehistoric Native American life is evident here as well, along with giant earthwork mounds that took 300 years to build. This old trail still beckons as it did more than a century and a half ago, with lakes, streams, wildlife refuges, nature preserves, and trails where white-tailed deer play. The atmosphere of old-fashioned travel is stored in the little shops and towns along the way. The western end of the Byway takes you to Eads Bridge and the gateway to the West.

THE BYWAY STORY

The Historic National Road tells archaeological, historical, natural, recreational, and scenic stories that make it a unique and treasured Byway.

Archaeological

The Cahokia Mounds State Historic Site, an archaeological site with worldwide recognition,

Illinois

The Historic National Road

bisects the Historic National Road. This remarkable World Heritage Archaeological Site consists of the largest mound buildings built by Native Americans on the North American continent. As you pass the site while driving the Byway, you see Monks Mound rising out of the ground on the north side of the Byway, covering 14 acres and rising 100 feet into the air. You may also notice a large circle of wooden posts, known as Woodhenge, next to the road. The visitor center is the pride of the Illinois Historic Preservation Agency and provides more information on the site and the people who created it.

Historical

In 1806, Congress appropriated funds to construct a National Road that would run westward from Cumberland, Maryland, to the Mississippi River. It was the first federally funded road system in this great new country. The Illinois section was surveyed in 1828 by Joseph Schriver, and construction was started in 1831 under the supervision of William C. Greenup. The section to Vandalia was completed in 1836. However, the western section was never funded due to high costs and waning interest in road building. With the coming of the Terre Haute-Vandalia-St. Louis Railroad that paralleled the road, the National Road fell into disrepair, only to be resurrected in the early 1920s when it was hard surfaced and designated US 40. Today, most of the original alignment of the 1828 surveyed National Road is still in place and is in public hands.

The route passes through many historic towns and villages that were established in the mid-1800s along this great road. On the eastern end of the Byway is Marshall, a town that sports the oldest continuously operating hotel. Also on the eastern end is the village of Greenup, with its unique business section decorated with original overhanging porches. This village was designated as a Historic Business District on the National Register of Historic Places. In the central section is Vandalia, the second capital of Illinois, its original business district intact. The Capital Building, in which Abraham Lincoln passed his test to practice law, is now a State Historic Site and sits on the Historic National Road.

The Illinois Historic National Road ends at the historic Eads Bridge, just across the river from the Jefferson Memorial Expansion National Park and the St. Louis Arch, the gateway to the West.

Natural

The Historic National Road in Illinois is dominated at each end of the Byway by rivers. The Wabash River is on the east end, and the mighty Mississippi River lies on the west end. Rivers and lakes are interspersed throughout the middle area, with flat prairies and hilly landscapes combining to create many natural features along the Byway. Different species of wildlife make their homes all along the route, and fish are plentiful in the many lakes and rivers.

The topography along the Byway was created by glaciers that advanced and retreated over the land during the Pleistocene Period, leaving behind moraines and glacial deposits that created regions of undulating landscape in some areas and flat prairies in other areas. The landscape of the Byway is defined by three major areas: the Embarras River Basin on the east side, the Wabash River Basin in the central area, and the Sinkhole Plain on the west side (which is contained in the Mississippi River Basin).

The Mississippi River is the third largest river in the world, and as a result of its size, this river has played an important part in the lives of those who have called this area home. From flooding to fertile land, the river has shaped the lives of Native Americans, pioneers, settlers, and current residents of the many cities that dot the banks of the river. The river, through irrigation, has been the lifeblood of farmers, and also has been a recreational destination for many. In addition, over 400 species of wildlife—including ancient lineages of fish—live on and near the Mississippi River. In fact, 40 percent of North America's duck, goose, swan, and wading bird populations use the Mississippi as a migration corridor.

see page A6 for color map

Recreational

The Historic National Road in Illinois offers many recreational opportunities. You can bike or hike on the various trails that are accessible from the Byway. Numerous state parks allow you to enjoy the natural characteristics of the Byway; many of the towns have city parks and recreational facilities. From camping to bird-watching, there is something for everyone along this Byway.

At Lincoln Trail Lake State Park, you can travel in the route Abraham Lincoln took from Kentucky to Illinois. This park is on part of that route, and today, it is a place where you can enjoy hiking, fishing, boating, or camping. Summer is not the only time to enjoy this area, however; wintertime sports include ice fishing, ice skating (when the lake allows it), and cross-country skiing.

Many lakes, rivers, and streams provide recreational opportunities. Carlyle Lake is a 26,000-acre multipurpose lake known for its great fishing and waterfowl hunting. You can catch bass, bluegill, crappie, catfish, walleye, and sauger fish. At Eldon Hazlet State Park, controlled pheasant hunting is available, and bird-watching is also a popular activity at the lakes. Carlyle Lake is well known among sailors, and you can rent a houseboat at the park. Camping and golf courses are available here as well.

Numerous small towns are spread across the Byway, providing a variety of activities. If more adventurous activities are what you are looking for, you can board the *Casino Queen*, a riverboat casino that is docked in East St. Louis. At Collinsville, the Gateway International Raceway hosts motor sports in a state-of-the-art facility.

Scenic

Diverse and changing, the landscape of the Historic National Road in Illinois offers many scenic views to the Byway traveler. The route is dotted with towns and rural communities, interspersed with rural lands and farms. The large metropolitan area of the western edge of the Byway, in Collinsville and East St. Louis, provides a different kind of scene. From historic buildings and bridges to gently rolling hills, this Byway exemplifies a scenic drive.

The natural layout of the land is one of variation. In the east, the rolling hills and interspersed forests provide a different view than the flat, unbroken views presented on the western edge of the Byway. In between, cultivated fields, distant barns, farmhouses, and grazing livestock all speak of the nature of the land. Small communities were developed around the agriculture of the area, and now these towns beckon visitors with historic buildings and one-of-a-kind features, such as Greenup's historic porches.

Many features of the Byway revolve around transportation because of the importance of the National Road. Bridges, such as the S-bridge, remain to provide visitors a chance to glimpse these engineering feats. Picturesque stone bridges, as well as covered bridges, may be seen on the Byway. In addition, the Eads Bridge stands in East St. Louis on the Mississippi, giving the metropolitan skyline a distinct look.

HIGHLIGHTS

The Illinois National Road Museum Tour takes you through some of the museums located on the National Road in Illinois. The tour begins in Martinsville and goes to Collinsville (east of St. Louis).

- **Lincoln School Museum:** This quality museum is located in Martinsville, about 18 miles west of the Indiana state line. The building itself was built in 1888, and the school is open to groups for an interpretation of early pioneer days.

- **Franciscan Monastery Museum:** Dating to 1858, this historic monastery has a wonderful museum that displays artifacts from early

Illinois

❋ The Historic National Road

settlers as well as the Franciscan Fathers. Visitors can view pioneer items such as toys and kitchen utensils, and religious items such as Bibles and vestments. There are also antique legal documents on display, such as marriage licenses. The monastery is located about 5 miles east of Effingham.

- **My Garage Corvette Museum:** This museum, located in Effingham, about 30 miles west of Martinsville, is a must for automobile lovers. On display are vintage Corvettes from the 1950s and 1960s—a perfect museum for the National Road.

- **Collinsville Historical Museum:** Located about 90 miles west of Effingham in Collinsville, this museum offers visitors a unique glance into the region's residents all the way back to John Cook, the first settler in 1810. Many interesting artifacts are on display, including a variety of Civil War objects and miners' tools. No museum tour would be complete without a visit to this high-quality museum.

THINGS TO SEE AND DO

Driving along the Historic National Road will certainly keep your senses engaged, but if you yearn to get out of the car and stretch your legs, or if you'd like to make a mini-vacation out of your trip, check out these attractions along the route.

LITTLE BRICK HOUSE MUSEUM. *621 St. Clair St, Vandalia (62471). Phone 618/283-0667. www.vandaliaillinois.com/tourism/.* In this museum, you'll find simple Italianate architecture with six restored rooms furnished primarily in the 1820-1839 period, including antique wallpapers, china, wooden utensils, dolls, doll carriages, pipes, parasols, powder horns, oil portraits, and engravings. The Berry-Hall Room contains memorabilia of James Berry, artist, and James Hall, writer. The house also pays tribute to members of the Tenth General Assembly of Illinois. There are outbuildings and a period garden with original brick pathways around the house. Guided tours by appointment only.

RAMSEY LAKE STATE PARK. *850 N and 700 E, Ramsey (62080). Phone 618/423-2215. www.dnr.state.il.us.* At approximately 1,960 acres, Ramsey Lake State Park offers a 47-acre lake stocked with bass, bluegill, and red ear sunfish. Fishing, boating (ramp, rentals, electric motors only), hunting, hiking, horseback riding, picnicking (shelters), concession, and camping. **FREE**

RICHARD W. BOCK SCULPTURE MUSEUM. *315 E College Ave, Greenville (62446). Phone 618/664-2800 or 618/664-6724. www.greenville.edu/campus/bock/.* Inside the original Almira College house (1855) is a collection of works by sculptor Richard W. Bock (1865-1949), who, between 1895 and 1915, executed a number of works for Frank Lloyd Wright-designed buildings. Also on display are Wright-designed prototypes of leaded-glass windows and lamps for Wright's Dana House in Springfield. Open Wed, Fri 1-5 pm; Sat 10 am-2 pm; closed during the summer; call for an appointment. **FREE**

✪ **VANDALIA STATEHOUSE STATE HISTORIC SITE.** *315 W Gallatin St, Vandalia (62471). Phone 618/283-1161. www.state.il.us.HPA/hs/Vandalia.htm.* President Lincoln and Stephen Douglas served in the House of Representatives in this two-story, Classical Revival building built by townspeople in 1836 in an effort to keep the state capital in Vandalia. Many antiques and period furnishings. Open daily 8:30 am-5 pm (until 4 pm Nov-Feb); closed holidays. **FREE**

PLACES TO STAY

If you choose to include an overnight stay in your trip along this All-American Road, Mobil Travel Guide recommends the following lodgings.

Collinsville

★★ **HOLIDAY INN.** *1000 Eastport Plaza Dr, Collinsville (62234). Phone 618/345-2800; toll-free 800/551-5133. www.holiday-inn.com.* 229 rooms, 5 story. Pet accepted, some restrictions. Complimentary full breakfast. Check-out noon. TV; cable (premium). In-room modem link. Restaurant, bar. Room service 24 hours. In-house fitness room, sauna. Health club privileges. Game room. Indoor pool, whirlpool. Free airport transportation. **$**

★ **MAGGIE'S BED & BREAKFAST.** *2102 N Keebler Rd, Collinsville (62234). Phone 618/344-8283.* 5 rooms, 3 story. No room phones. Pet accepted, some restrictions. Complimentary full breakfast. Check-out noon, check-in 4-6 pm. TV; VCR (movies). Game room. Whirlpool. Built in 1900; former boarding house. Totally nonsmoking. Credit cards not accepted. **$**

Effingham

★ **BEST INN.** *1209 N Keller Dr, Effingham (62401). Phone 217/347-5141; toll-free 800/237-8466. www.bestinn.com.* 83 rooms, 2 story. Pet accepted, some restrictions. Complimentary breakfast. Check-out 1 pm. TV. Pool. **¢**

★★ **BEST WESTERN RAINTREE INN.** *1809 W Fayette Ave, Effingham (62401). Phone 217/342-4121; toll-free 800/780-7234. www.bestwestern.com.* 65 rooms, 2 story. Pet accepted, some restrictions. Complimentary continental breakfast. Check-out 11 am. TV; cable (premium). Restaurant, bar. Pool. **¢**

★ **SUPER 8.** *1400 Thelma Keller Ave, Effingham (62401). Phone 217/342-6888; toll-free 800/800-8000. www.super8.com.* 49 rooms, 2 story. Pet accepted, some restrictions. Complimentary breakfast. Check-out 11 am. TV; cable (premium). **¢**

★★ **THE DAISY INN BED & BREAKFAST.** *315 E Illinois St, Greenup (62428). Phone 217/923-3050. www.bbonline.com/il/daisy.* 5 rooms, 2 story. A restored Victorian inn with comfortable rooms and private baths. **$**

Greenville

★ **BEST WESTERN COUNTRY VIEW INN.** *I-70 and IL 127, Greenville (62246). Phone 618/664-3030; toll-free 800/780-7234. www.bestwestern.com.* 83 rooms, 2 story. Pet accepted, some restrictions; fee. Complimentary continental breakfast. Check-out 11 am. TV. In-house fitness room. Pool. **¢**

Illinois

✸ *The Historic National Road*

★ **THE RENCH HOUSE.** 316 W Main St, Greenville (62246). Phone 618/664-9698. www.greenville-chamber.com/renchhouse. 3 rooms, 2 story. Built in 1852, this Victorian-style house is within walking distance of downtown Greenville. $

Marshall

★★ **THE ARCHER HOUSE.** 717 Archer Ave, Marshall (62441). Phone 217/826-8023. www.thearcherhouse.com. 8 rooms, 2 story. Built in 1841 as a stagecoach stop, the Archer House is the oldest operating hotel in Illinois. $$

Vandalia

★ **JAY'S INN.** 720 W Gochenour St, Vandalia (62471). Phone 618/283-1200. 21 rooms, 2 story. Pet accepted. Check-out noon. TV; cable (premium). Restaurant, bar. ¢
SC

PLACES TO EAT

A long day of driving is sure to make you hungry. At the end of your journey, take a table at one of the following restaurants.

★ **RICHARD'S FARM RESTAURANT.** 607 NE 13th St, Casey (62420). Phone 217/932-5300. American menu. Lunch, dinner. Located in a converted barn; home of the 1-pound pork chop. $

★ **EL RANCHERITO.** 1313 Keller Dr, Effingham (62401). Phone 217/342-4753. www.elrancheritocantina.com. Mexican menu. Closed Thanksgiving. Lunch, dinner. Children's menu. $
SC

★ **NIEMERG'S STEAK HOUSE.** 1410 W Fayette Ave, Effingham (62401). Phone 217/342-3921. American menu. Closed Dec 25. Breakfast, lunch, dinner. Bar. Children's menu. Casual, family-style dining. $
D

★ **NUBY'S STEAKHOUSE.** 679 Hwy 40 W, Pocahontas (62275). Phone 618/669-2737. www.greenville-chamber.com/nubys. American menu. Closed Mon. Lunch, dinner. $

★ **DEPOT OF VANDALIA.** 107 S 6th St, Vandalia (62471). Phone 618/283-1918. French, Cajun menu. Lunch, dinner. Restaurant and lounge located in a turn-of-the-century train depot. $

Lincoln Highway
✵ ILLINOIS

Quick Facts

LENGTH: 179 miles.

TIME TO ALLOW: 5 to 6 hours.

BEST TIME TO DRIVE: Early summer.

BYWAY TRAVEL INFORMATION: Illinois Lincoln Highway Coalition: 866/455-4249; Byway local Web site: www.lincolnhwyil.com.

SPECIAL CONSIDERATIONS: This Byway is long and may be difficult to follow as it winds its way through and around the Illinois interstate system and countryside. To experience all that the Lincoln Highway has to offer, you may want to spend two days driving the route.

BICYCLE/PEDESTRIAN FACILITIES: Pedestrian travel is accommodated by sidewalks along the majority of the Byway. Although bicycle travel along the Lincoln Highway is not advised, many biking and hiking trails/routes can be found intersecting the highway or in the communities along the route.

The innovative development of the Lincoln Highway was the prototype of roadways as we know them today. This historic Byway follows the original alignment of the Lincoln Highway, the first paved transcontinental highway in the United States and the forerunner of the modern interstate transportation system as it was originally conceived in 1913.

The 179-mile route crosses the width of northern Illinois, starting in Lynwood on the Indiana border. The route travels through eight counties and 31 cities to Fulton at the Iowa border. The Illinois portion of the Lincoln Highway, located near the center of the 3,389-mile transcontinental route, was the site of the first seedling mile of paved roadway constructed to demonstrate the superiority of pavement over dirt roads. The Lincoln Highway was also the first instance in which transportation principles such as directional signs and urban bypasses were employed.

Not only does the Lincoln Highway tell the story of early automobile travel and highway design, but it also ties together the histories, economies, and identities of the cities and towns of northern Illinois. Be sure to stop at a few of the Byway communities to find out what makes each one unique. The highway winds its way through the downtowns of many of the towns along its length, and you may visit a newly constructed windmill, go on a scenic river walk, or visit a bustling downtown area near Chicago.

Illinois

✻ Lincoln Highway

THE BYWAY STORY

The Lincoln Highway tells cultural, historical, recreational, and scenic stories that make it a unique and treasured Byway.

Cultural

The culture of the Lincoln Highway centers around its communities, and these communities all share a pride in being part of the history and future of transportation. In most of the Lincoln Highway communities, you find at least one defining characteristic—one aspect that makes the community unique. In Dixon, it is the Victory Arch that spans the highway; in Chicago Heights, it is the Arche Fountain commemorating Abraham Lincoln; in Batavia, it is the unique architecture.

The people of the Lincoln Highway allow this historic road to string them all together and bring them to the threshold of history. But with this historical remembrance, the communities of the Lincoln Highway continue to grow and change.

Historical

In 1913, a core group of automobile industrialists and enthusiasts established an organization to promote the development of "good roads" and conceived a route for a paved, transcontinental road. This group sought, initially without government assistance, to secure private funding to build a road that would serve the needs of industry, particularly the automobile industry. They had the notion of paving 1-mile stretches of road along the Lincoln Highway route with concrete to lay the groundwork for the future. They called these stretches seedling miles. They purposely placed them away from major cities so that motorists drawn by publicity would have to struggle over unimproved dirt roads to reach them. The first such seedling mile was completed in Malta, Illinois, just west of DeKalb, in September of 1914. Four more seedling miles were constructed in 1915. These prototypes immediately became popular motorist destinations. The stark contrast between these smooth patches of pavement and the bumpy or muddy roads leading up to them created a groundswell of public opinion in favor of good roads.

This clamor for action was directed at local, state, and federal officials and resulted in the passage of the Federal Aid Road Act of 1916, which authorized and appropriated $75 million for the construction of what were called post roads. This amount was to be matched by the states seeking to build the roads, thus starting the practice of federal-state grant matching for road construction. Additionally, many segments were constructed by volunteer labor, such as the Mooseheart segment of the Lincoln Highway, which was built by area businessmen, manual laborers, and others to demonstrate their support of their community and of local businesses.

Shortly thereafter, America became involved in World War I, shifting national attention onto the war effort and away from the road-building effort. However, interest in good roads resumed in earnest shortly after the war ended in November 1918. In 1919, Lincoln Highway Association leader Harry Ostermann persuaded the War Department to conduct a transcontinental motor convoy trip from the East Coast to San Francisco on the marked route of the Lincoln Highway. A 76-vehicle convoy combining public and private vehicles took off from the White House on July 7, 1919. The convoy, primarily following the Lincoln Highway route, finally arrived in San Francisco, but not after considerable difficulty on the dirt roads traveled en route. The seedling miles of concrete made a strong impression. Among those participating in the convoy was Lt. Colonel Dwight D. Eisenhower, who much later applied his experiences on the Lincoln Highway, along with his experiences with World War II and the German Autobahn system, to conceive of an interstate road system to aid the movement of troops, goods, and people across the country.

Various aspects of the Lincoln Highway's early development predated and predicted some of the technical and fundamental elements of

see page A7 for color map

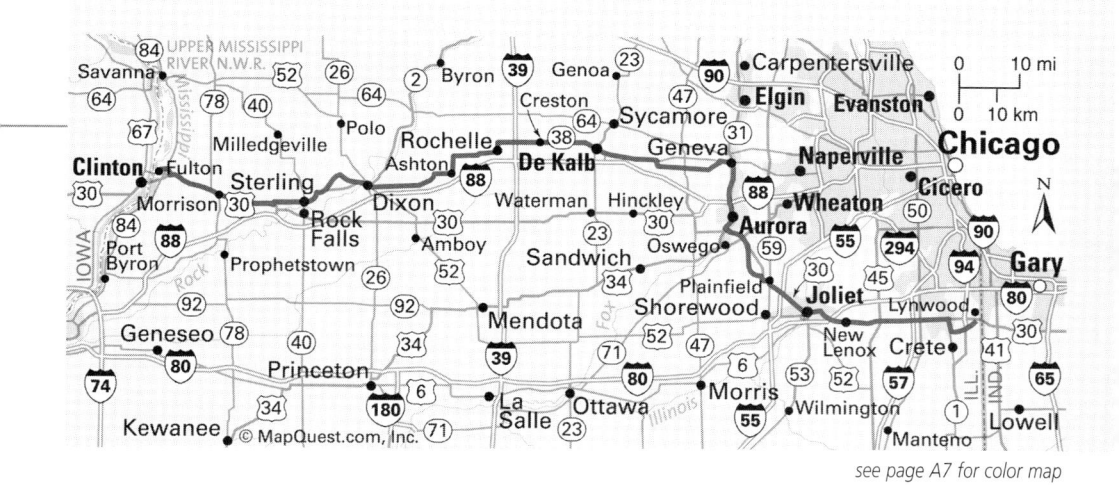

current US transportation policy. These aspects went beyond those of paved roadways and transcontinental travel. They included directional signage, a system of concrete markers designed to assist travelers determining their location along a given roadway. The markers included a small coin-like bust of Lincoln with the inscription, "This highway dedicated to Abraham Lincoln." Never before had a consistent road signage system been employed. Another new concept was the urban bypass: the Lincoln Highway was purposely routed 25-30 miles south and west of Chicago to avoid the congestion and time delays associated with traveling through the city.

With much of its original (and modified) objectives achieved, the Lincoln Highway Association was dissolved in 1935. Its legacy for America has been the prominent role it played in the development of the roadway system in place today in the United States. Since 1935, much of the original Lincoln Highway has been paved over, bypassed, or converted to numbered US, state, and county highways or municipal streets. Very few of the 1928 cement markers still exist. However, the name Lincoln is still attached to much of the route in the form of roadway and street names, local Lincoln businesses and brochures, articles, and artifacts preserved in museums and historical societies along the route.

The Lincoln Highway was designated a National Scenic Byway in 2000 and presents many opportunities to enjoy the culture and history of Illinois. Most of all, the Lincoln Highway offers a unique opportunity to relive the days of the cross-country road trip and the sights of a developing nation.

Recreational

Recapture the adventure of the open road, just as avid motorists of nearly a century ago braved the elements to heed the call of wanderlust to travel the Lincoln Highway. From the rolling hills of western Illinois and the Mississippi River Valley to the sights and sounds of the Chicago metropolitan area, the Lincoln Highway includes an impressive collection of diverse recreational opportunities.

Near Franklin Grove, stop at Franklin Creek State Natural Area to enjoy a picnic by the edge of Franklin Creek. Hiking, skiing, horseback riding, and snowmobiling trails are available there. As you near Chicago, you'll find increased shopping opportunities in places like Chicago Heights and Joliet. In Geneva, tour the Japanese Gardens or go biking on the Riverwalk. While touring each community, you are sure to come across several enticing activities.

Whether for an overnight getaway trip or for an extended stay to enjoy northern Illinois' hospitality and charm, the Lincoln Highway beckons travelers of all ages to experience firsthand the history, sights, and stories of the highway.

Scenic

Although the scenic qualities of the Lincoln Highway may differ from those of other Byways, the highway is scenic nonetheless. Several architectural treats await you. Often, this architecture is combined with elements of nature for a scenic effect. Many of the most interesting sites on the Byway can be seen from the car, but you will likely want to get out and try some of the biking trails or river walks.

Lincoln Highway

HIGHLIGHTS

This tour begins in North Aurora, Illinois, and travels west across the state. If you wish to take the tour traveling east, read this list from the bottom and work your way up.

- **Mooseheart:** Just north of North Aurora on Illinois Route 31, you'll find Mooseheart. This lodge is an important piece of history for the Lincoln Highway, because members of the Moose Lodge raised $12,000 to fund the initial paving of the Lincoln Highway. Members of the lodge from all over the country then traveled to Illinois and helped grade the road using picks and shovels. In appreciation for the efforts of the lodge, the state later paved an extra 10-foot strip, which is still visible today, in front of Mooseheart.

- **DeKalb Memorial Park:** Traveling west from Mooseheart on SR 60, you come to the city of DeKalb. The Memorial Park in DeKalb is famous for its memorial clock called Soldiers and Sailors, which was originally dedicated in 1913. The clock was restored completely in 1996, and in 1999, a community mural was painted on the side of the old Chronicle building, located behind the park. The mural highlights the history of DeKalb and can be seen from the Byway.

- **First Seedling Mile:** Cement companies donated the cement to pave seedling miles on the highway. These paved miles of road were meant to show travelers what they had to look forward to when highways would be paved with a hard surface. Seedling miles were set in rural areas so that travelers wanting to see them would have to drive over a length of unpaved road and would quickly understand the advantages of the hard surface. The first seedling mile in the country is located west of Malta on SR 38.

- **Railroad Park:** Located in Rochelle, west of Malta and the first seedling mile, is Railroad Park, one of two X rail crossings in the country. People come from all over the world to watch the rail traffic of the Union Pacific and the Burlington Northern Santa Fe Railroads, the two major rail carriers in the western United States.

THINGS TO SEE AND DO

Driving along the Lincoln Highway will certainly keep your senses engaged, but if you yearn to get out of the car and stretch your legs, or if you'd like to make a mini-vacation out of your trip, check out these attractions along the route.

AURORA HISTORICAL MUSEUM. *20 E Downer Pl, Aurora (60505). Phone 630/906-0650. www.aurorahistoricalsociety.org/museums.html.* In the restored Ginsberg Building, this museum contains displays of 19th-century life, a collection of mastodon bones, a history center and research library, and Native American artifacts. Open Wed-Sun afternoons. **$**

BLACKBERRY FARM AND PIONEER VILLAGE. *100 S Barnes Rd, Aurora (60506). Phone 630/892-1550. www.foxvalleyparkdistrict.org/pages/facility/blackfrm.html.* An 1840s to 1920 living-history museum and working farm; children's animal farm; craft demonstrations; wagon rides, discovery barn, pony rides, peddle tracker, farm play area, and train. Open late Apr-Labor Day, daily 10 am-4:30 pm; Labor Day-Oct, Fri-Sun. **$**

BICENTENNIAL PARK THEATER/BAND SHELL COMPLEX. *201 W Jefferson St, Joliet (60432). Phone 815/724-3760.* Joliet Drama Guild and other productions (fee). Also outdoor concerts in the band shell (June-Aug, Thurs evenings). Historic walks. **FREE**

FERMI NATIONAL ACCELERATOR LABORATORY. *Wilson and Kirk rds, Batavia (60510). Phone 630/840-3351. www.fnal.gov.* The world's highest energy particle accelerator is on a 6,800-acre site. Also on the grounds are hiking trails and a buffalo herd. Obtain brochures for self-guided tours in the atrium of 15-story Wilson Hall. Art and cultural events, films in auditorium. Open daily 6 am-8 pm. **FREE**

GARFIELD FARM AND INN MUSEUM. *3N016 Garfield Rd, Geneva (60134). Phone 630/584-8485. www.garfieldfarm.org.* This 281-acre living-history farm includes an 1846 brick tavern, 1842 haybarn, and 1849 horse barn; poultry, oxen, and sheep; gardens and prairie. Special events throughout the year. Grounds open June-Sept, Sun, Wed; other times by appointment. **$$**

HARRAH'S JOLIET CASINO. *151 N Joliet St, Joliet (60432). Phone 815/740-7800. www.harrahs.com.* Harrah's features a Las Vegas-style casino, entertainment complex, hotel, and buffet. Games include more than 1,100 slots (from 5-cent to $100 machines), blackjack, Caribbean stud poker, mini-baccarat, craps, roulette, video poker, and video keno. Open daily.

JOHN DEERE HISTORIC SITE. *8393 S Main, Dixon (61021). Phone 815/652-4551. www.deere.com.* Site where the first self-scouring steel plow was made in 1837; reconstructed blacksmith shop (demonstrations Wed-Sun); restored house and gardens; 2-acre natural prairie. Open Apr-Oct, daily 9 am-5 pm; group tours available in winter by appointment. **$**

LINCOLN STATUE PARK. *100 Lincoln Statue Dr, Dixon (61021). Phone 815/288-3404.* Includes the site of Fort Dixon, around which the town was built, and a statue of Lincoln as a young captain in the Black Hawk War of 1832. **FREE**

RIALTO SQUARE THEATRE. *15 E Van Buren St, Joliet (60432). Phone 815/726-7171; box office 815/726-6600. www.rialtosquare.com.* This performing arts center, designed by the Rapp brothers and built in 1926, is considered one of the most elaborate and beautiful of old 1920s movie palaces. Tours (Tues and by appointment). **$**

RONALD REAGAN'S BOYHOOD HOME. *816 S Hennepin Ave, Dixon (61021). Phone 815/288-3404.* Two-story, three-bedroom house with 1920s furnishings; memorabilia connected with the former president's childhood and acting and political careers. Visitor center adjacent. Open Mar-Nov, daily; rest of year, Sat-Sun. **FREE**

PLACES TO STAY

If you choose to include an overnight stay in your trip along this Byway, Mobil Travel Guide recommends the following lodgings.

★ **COMFORT SUITES.** *111 N Broadway, Aurora (60505). Phone 630/896-2800; toll-free 877/424-6423. www.comfortsuitesaurora.com.* 82 suites, 3 story. Complimentary continental breakfast. Check-out 11 am, check-in 4 pm. TV; cable (premium), VCR available. In-room modem link. Fireplaces. In-house fitness room. Game room. Indoor pool. Business center. **$$**

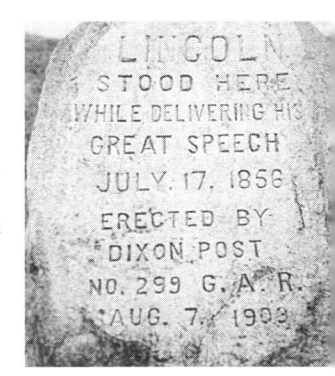

Illinois

✵ Lincoln Highway

★ **HOWARD JOHNSON EXPRESS INN.** *1321 W Lincoln Hwy, DeKalb (60115). Phone 815/756-1451; toll-free 800/446-4656. www.hojo.com.* 60 rooms, 2 story. Complimentary continental breakfast. Check-out noon. TV; cable (premium). ¢

★ **SUPER 8.** *800 Fairview Dr, DeKalb (60115). Phone 815/748-4688; toll-free 800/800-8000. www.super8.com.* 44 rooms, 2 story. Complimentary continental breakfast. Check-out 11 am. TV; cable (premium). Indoor pool, whirlpool. ¢

★★ **BEST WESTERN REAGAN HOTEL.** *443 IL Rte 2, Dixon (61021). Phone 815/284-1890; toll-free 800/780-7234. www.bestwestern.com.* 91 rooms, 2 story. Pet accepted; fee. Complimentary continental breakfast (Mon-Fri). Check-out noon. TV; VCR available (movies). Restaurant, bar. Room service. In-house fitness room. Outdoor pool, whirlpool. ¢

★★ **THE HERRINGTON INN.** *15 S River Ln, Geneva (60134). Phone 630/208-7433; toll-free 800/216-2466. www.herringtoninn.com.* 63 rooms, 3 story. Complimentary continental breakfast. Check-out noon, check-in 4 pm. TV; cable (premium), VCR available (movies). Some stereos, DVDs. Fireplaces. Restaurant. Room service. In-house fitness room. Health club privileges. Golf on premise, greens fee $22-$44. Bicycle rentals. Airport transportation. Business center. Concierge service. Built in the late 1800s. On the river. $$

★★★ **HILLENDALE BED & BREAKFAST.** *600 W Lincolnway, Morrison (61270). Phone 815/772-3454. www.hillend.com.* 29 rooms, 8 with private bath, 3 with shower only, 3 story. Children over 12 years only. Complimentary full breakfast. Check-out 11 am, check-in 3 pm. TV in some rooms; VCR available (movies). Some whirlpools. Fireplaces. In-house fitness room. Built in 1891. Totally nonsmoking. $$

★★★ **PINEHILL INN BED & BREAKFAST.** *400 Mix St, Oregon (61061). Phone 815/732-2067; toll-free 800/851-0131. www.pinehillbb.com.* 4 rooms, 3 story. This Italianate country estate was built in 1874 and is listed on the National Historic Register. $$

PLACES TO EAT

A long day of driving is sure to make you hungry. At the end of your journey, take a table at one of the following restaurants.

★ **WALTER PAYTON'S ROUNDHOUSE.** *205 N Broadway Ave, Aurora (60506). Phone 630/264-2739. www.eagleridge.com.* Lunch, dinner. Located in the former railroad roundhouse and originally owned by the late football star Walter Payton. $

★ **MERICHKA'S RESTAURANT.** *604 Theodore St, Crest Hill (60435). Phone 815/723-9371.* American menu. Lunch, dinner. This classic American roadside diner with a 70-year history is still run by the same family. $

★ **THE HILLSIDE RESTAURANT.** *121 N 2nd St, DeKalb (60115). Phone 815/756-4749. www.hillsiderestaurant.com.* American menu. Closed Tues. Lunch, dinner. The oldest restaurant in DeKalb, The Hillside Restaurant has hosted politicians and celebrities since 1955. $

★★ **ATWATER'S.** *15 S River Ln, Geneva (60134). Phone 630/208-8920. www.herringtoninn.com.* International menu. Breakfast, lunch, dinner, Sun brunch. Bar. Adjacent to river. Casual attire. Outdoor seating. **$$$**

★★ **MILL RACE INN.** *4 E State St, Geneva (60134). Phone 630/232-2030. www.themillraceinn.com.* American menu. Closed Dec 25. Lunch, dinner, Sun brunch. Bar. Entertainment Wed, Fri, Sat. Children's menu. Casual attire. Valet parking available. Gazebo dining. On the Fox River. **$$**

★★ **BAROLO RISTORANTE.** *158 N Chicago St, Joliet (60432). Phone 815/722-1744.* Italian menu. Closed Sun. Lunch, dinner. Located in historic downtown Joliet with a convivial European atmosphere and progressive Italian cuisine. **$$**

★★ **TRUTH RESTAURANT.** *808 W Jefferson St, Joliet (60435). Phone 815/744-5901.* American menu. Closed Sun, Mon. Dinner. Serving contemporary American regional cuisine in an inviting neighborhood atmosphere. **$$**

★ **WHITE FENCE FARM.** *Joliet Rd, Lemont (60439). Phone 630/739-1720. www.whitefencefarm.com.* American menu; specializes in fried chicken. Closed Mon; Thanksgiving, Dec 24-25; also the month of Jan. Lunch, dinner. Children's menu. Casual attire. Children's zoo adjacent. **$$** **D**

★★ **SYL'S RESTAURANT.** *829 Moen Ave, Rockdale (60436). Phone 815/725-1977. www.sylsrestaurant.com.* American menu. Lunch, dinner. A Joliet-area landmark since 1946, Syl's offers classic American cuisine in a stately environment. **$$**

Meeting of the Great Rivers Scenic Route
❋ ILLINOIS

Quick Facts

LENGTH: 57 miles.

TIME TO ALLOW: 2 hours.

BEST TIME TO DRIVE: In the fall, people come from miles around to see the colors of the leaves and to enjoy harvest in the orchards and vineyards. During the spring, the dogwoods and redwoods are in bloom. In the summer, recreational activities are abundant on the Byway. In the winter, look for eagles swooping near the bluffs.

BYWAY TRAVEL INFORMATION: Greater Alton/Twin Rivers Convention and Visitors Bureau: 618/465-6676.

SPECIAL CONSIDERATIONS: One of the only gas stations along the entire Byway is located in Grafton.

RESTRICTIONS: Roads can be closed during the spring due to flooding and during the winter due to ice. When this happens, alternate routes are provided. Also, ferries cannot operate when the river freezes. During these times, use the Joe Paige Bridge at Hardin to connect with Kampsville in Calhoun County.

BICYCLE/PEDESTRIAN FACILITIES: Touted as the longest and perhaps the most picturesque bicycle path in the region, the 25-mile Vandalabene (Great River Road) Bicycle Trail runs parallel to the Meeting of the Great Rivers Scenic Route from Alton to the north of Pere Marquette State Park. The trail provides walking and bicycling opportunities, and it is the non-motorized, interconnecting link to 80 percent of the attractions along the Byway. There are also pedestrian trails within Pere Marquette State Park.

Within a 25-mile expanse, the Mississippi, Missouri, and Illinois rivers meet to form a 35,000-acre floodplain. This confluence is the backdrop for the Meeting of the Great Rivers Scenic Route. The river systems have been vital transportation routes as long as there has been human habitation, moving people and goods to world markets.

The Meeting of the Great Rivers Scenic Route offers a dramatic composite of the Mississippi River. Beneath white cliffs, the Byway runs next to the Mississippi, beginning in an industrial, urban setting and changing to a scenic, natural area. The area is so magnificent that the Illinois legislature called it "the most beautiful stretch of the entire Mississippi River." As though moving back through time, expanses of pastoral countryside and stone houses are reminders of a time long ago in the Lower Illinois River Valley.

The Meeting of the Great Rivers Scenic Route crosses the Illinois River on a ferry to the Kamp Store Museum in Kampsville. Artifacts of the earliest aboriginal people in America reside here, and the Byway's rich historical and archaeological qualities unfold. The road then travels to Alton, where history converges with present developments in the engineering wonder of the Clark Bridge and the memory of the last Lincoln-Douglas debate. Little towns along the Byway seem almost forgotten by time, giving travelers a look at historic architecture and small-town life along the Mississippi River.

Illinois

✳ Meeting of the Great Rivers Scenic Route

THE BYWAY STORY

The Meeting of the Great Rivers Scenic Route tells archaeological, cultural, historical, natural, recreational, and scenic stories that make it a unique and treasured Byway.

Archaeological

Despite present-day development, archaeological remains are largely intact along the Meeting of the Great Rivers Scenic Route. For example, the Koster Site, located south of Eldred, is world renowned because of the evidence found that shows that humans lived on the site 8,000 years ago. Structures dating back to 4200 BC are considered to be the oldest such habitations found in North America, and villages flourished here circa 3300 BC, 5000 BC, and 6000 BC. More than 800 archaeological sites have been inventoried along the route. Experts believe that the Mississippi and Illinois rivers, known as the Nile of North America, nourished the development of complex and sophisticated Native American cultures. So complete are cultural records that archaeologists term the area "the crossroads of prehistoric America."

Cultural

Visitors discover real river towns along this Byway. For example, the historic town of Alton has a solid Midwestern appeal. Legends and folklore of the past have formed the communities along the Byway into the unique places they are today. From the Piasa Bird to riverside amusements, things to do and see are plentiful along the Meeting of the Great Rivers Scenic Route. The people of the Byway are aware of the area's role in the past and have restored and preserved many of the places and sights along the Byway that represent the evolving culture of the Meeting of the Great Rivers. Monuments and museums stand as a tribute to the past that present visitors can enjoy.

Many important cultural events have occurred along this Byway. For example, Lewis and Clark trained the Corps of Discovery, Lincoln and Douglas had their final debate, and Elijah Lovejoy was martyred here while defending the freedom of the press and fighting slavery. The tallest man in history, Robert Wadlow, called Alton home. Also, spanning the Mississippi River at Alton is the famous Clark Bridge, a suspension marvel that was featured in a two-hour PBS *Nova* documentary entitled "Super Bridge." It is a beautiful structure that proves that intelligent and compassionate engineers can marry function and form.

To celebrate each unique aspect of their culture, many of the communities on the Byway have established their own museums. In addition, the region displays an appreciation of high culture through orchestras, theaters, galleries, institutions of higher learning, and many diverse festivals that celebrate the arts. Throughout the year, more than 50 festivals and fairs celebrate the history, art, music, and crafts of this region. As you explore these cultural venues, the unique flavor of the communities along the Byway will permeate your senses.

Historical

The Mississippi River is internationally famous. Father Marquette and Louis Joliet first made their expedition down the Mississippi in 1673. Later, when the Illinois Territory was formed, the Missouri River was the gateway to the unexplored West, and the Illinois River led to the Great Lakes and was also a connection to the East. Early American explorers began in the confluence area. Lewis and Clark, for example, embarked from Fort Dubois near the mouth of the Missouri. Eventually, towns were settled on the shores of the rivers, providing a secure way for travel and commerce using the rivers. The buildings that these towns were composed of still stand, many of them dating to the early 1800s.

As the nation grew and developed, many of the towns along the Byway were growing and developing as well. Although many of the towns that stand today seem to be nestled somewhere in history, some of the Byway's communities have been at the edge of new ideas. During pre-Civil War times, the Underground Railroad ran through this area, bringing escaped slaves to the safety of the north. The city of Alton was also the site of the last Lincoln-Douglas debate. Confederate prison ruins found on the Byway are another testament to this corner of Illinois' involvement in the Civil War.

By the 19th century, Mark Twain's Mississippi River stories had inspired an ideal of Mississippi legends, history, and culture in the minds of Americans. Meanwhile, as the river and its uses were also evolving, paddleboats gave way to barges and tows. River traffic increased as industries grew, and Lock and Dam 26 was built. Today, historic 18th-century river towns, islands, bars, points, and bends create beautiful scenery beneath limestone bluffs, which are covered by forests that extend nearly 20,000 acres. Historical and cultural features in the 50-mile corridor have received national recognition, with seven sites presently registered on the National Register of Historic Places.

Natural

Nature abounds along the Meeting of the Great Rivers Scenic Route. The wetlands from three different waterways, the rock bluffs, and the stately trees all harbor native creatures and provide lovely views along the Byway. The palisade cliffs and towering bluffs provide a characteristic drive along the riverside where visitors can see the results of this great channel of water carving its way through post-glacial terrain. You may want to enjoy the nature of the Byway from the car, look for hikes along the way, or get out and explore a wildlife refuge.

Located right in the middle of the US is Piasa country, a bird-watcher's heaven. Migratory flyways using the Mississippi, Illinois, and Missouri rivers converge within a 25-mile zone from Alton to Grafton. This offers amazing opportunities to see many species of birds that pass through this chokepoint region, from the American bald eagle to the white pelican. Deer, otters, and beavers are present, as well as raccoons, opossums, and squirrels. Fishing enthusiasts will discover many species in the local waters.

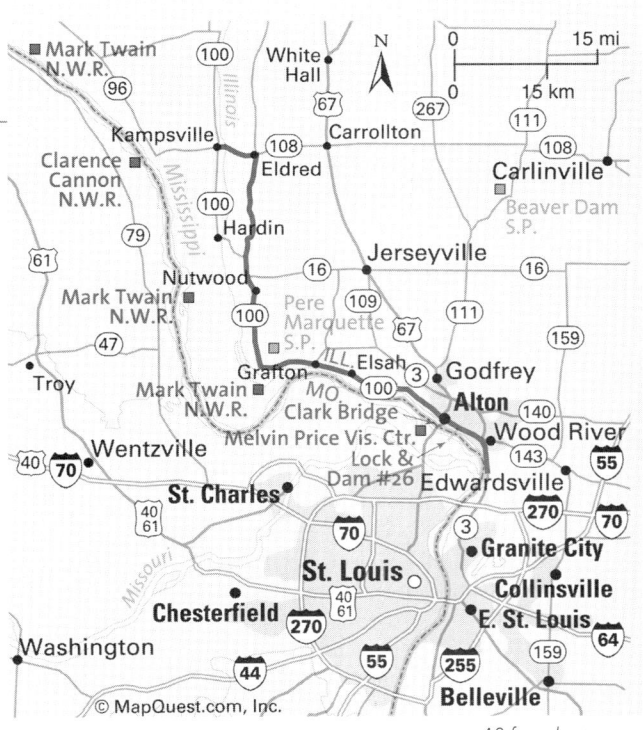

Many natural points of interest dot this Byway. For example, Pere Marquette State Park is one of Illinois' largest state parks. It is nestled along the banks of the Illinois River on the Byway near Grafton. Here, a myriad of trails take you within the wild forests and up to spectacular viewing areas along the bluff line above.

The Riverlands Environmental Wetlands Area is another natural point of interest. Located near Alton, this US Army Corps of Engineers site provides a fertile wetland that attracts all types of wildlife. Early-morning travelers frequently see wildlife making their way to the river. The Mark Twain Refuge is located near Pere Marquette State Park and is often open to the public. The preserve offers sanctuary to rare and endangered migratory birds on their long flights up and down the Mississippi and Illinois rivers.

Illinois

✻ Meeting of the Great Rivers Scenic Route

All the stunning views that can be enjoyed from a vehicle can also be enjoyed on foot or by bike. The Sam Vadalabene Trail, a bicycle and walking trail, winds more than 25 miles from Alton to Pere Marquette State Park on the Byway, making this a Byway that encourages and accommodates hikers and bikers.

Recreational

After you have seen the sights on the Meeting of the Great Rivers Scenic Route, you may decide to enjoy the surroundings on a closer level. Trails and paths along the Byway offer excitement for hikers and bikers. Also, forests that line the roadsides are perfect places for camping, picnicking, or simply enjoying the peaceful solitude that nature affords. Be sure to tour the historic districts of the Byway communities and stop at the museums and visitor centers that provide a closer look at the Byway and its characteristics.

There is always fun to be found on the Mississippi River. Visitors enjoy the water in every way, from parasailing to jet skiing. Sailboats and riverboats keep the river alive with movement year-round. During the summer, families stop at one of the two water parks along the Byway or travel on one of the four free river ferries located in the area. The Meeting of the Great Rivers Scenic Route is one of the most accommodating to bikers, with a bicycle path that goes directly along the Byway.

Shopping for crafts and antiques in the historic riverside towns along the Byway is a pleasant pastime, and golfers enjoy the ten courses in the region. There's a theater in Wood River and an amphitheater in Grafton for musical productions, stage productions, and other kinds of entertainment. It is hard to miss the *Alton Belle* Riverboat Casino on a leisurely cruise down the river. In addition to the attractions along the Byway, festivals, fairs, and events are always occurring in its communities.

Scenic

The Mississippi River is like a chameleon. Depending on weather conditions, sun angles, and the color of the sky, the waters can turn from serene pale blue to dark navy to muddy brown. Insiders' favorite time for viewing the river is early in the morning as the sun is rising. Often, the river is glass-like, creating a mirror of the sky above. The blue is sweet and clear, and the reflections of the bluffs and trees are remarkable. Majestic bluffs tower above the Byway, creating a stunning wall of trees, rocky cliffs, and soaring birds. The meandering curves of the river provide amazing views, and you can see up and down the river for miles. The bluffs, which are imposing when immediately adjacent to the road, diminish into the far horizon at several viewing areas.

Note the unusual sunsets along these parts. Most think of the Mississippi as a southbound river that cuts up and down the center of the nation. This is not true here. In Piasa country, the Mississippi River makes a distinct turn and the current flows from west to east. In Alton, Grafton, and Elsah, the sun rises and sets in the long stretch of the water. On many evenings at dusk, the fiery reds, yellows, and oranges run nearly the entire length of the river. One of the great pastimes along this Byway is celebrating these glorious and unique sunsets.

Along the road to Eldred, the bluffs give way to rolling hills, farms, and forests. Depending on the season, roadside stands with fruits and vegetables may entice you to stop. The apples and peaches in Jersey, Calhoun, and Greene counties are legendary. In Eldred, an old-fashioned Illinois town full of Americana, most travelers stop for a slice of pie and get out to smell the crisp, fresh air. Eldred is a small town, but the smiles are big. Moving northward, you see the great Illinois farmlands that bring the bounty of food to both America's and the world's dinner tables. Soon the road branches westward, and the journey ends with another free ferry over the Illinois River into Kampsville.

This Byway is a must-see destination during all four seasons. In the spring, the trees and shrubs turn the bluffs and countryside into a wonderful tapestry of colorful buds and blossoms. Summer brings festivals, fairs, and river recreation. Autumn hosts the Fall Color Caravan and some of America's most amazing foliage, accented by the nearby rivers. Finally, the winter brings the American bald eagle by the hundreds to winter along the bluffs and feed along the banks of the rivers. The rivers, majestic bluffs, fantastic trees and wildlife, quaint villages, and rolling farmlands all make this Byway a wonderful adventure.

HIGHLIGHTS

This must-see tour of the Meeting of the Great Rivers Scenic Route begins at the northernmost point (Kampsville) and concludes at the southernmost point (Alton). If you're traveling in the other direction, simply read this itinerary from the bottom and work your way up.

- **Kampsville:** In Calhoun County on the Illinois River. Free ferry ride (drive east on Highway 108 approximately 5 miles to Eldred). Home of the **American Center for Archaeology;** site of **Old Settlers Days** with Lewis and Clark, Civil War, and other reenactments. Gas, food, shops.

- **Eldred:** In Greene County at Highway 108 and Blacktop Road. Wonderful Illinois village with gas, food, shops. The **Eldred Home** shows a glimpse of life in the 1800-1900s. Turn south (right) onto Blacktop Road. Drive approximately 15 miles to the intersection of Blacktop Road, Highway 100, and Highway 16; continue straight ahead and onto Highway 100 southbound. Continue approximately 10 miles to Pere Marquette State Park.

- **Pere Marquette State Park:** In Jersey County on Route 100. This 7,895-acre preserve overlooks the Mississippi and Illinois rivers. Nature trails, prehistoric sites, horseback riding, camping, fishing, boating, and hiking. Wonderful lodge built in the 1930s by the Civilian Conservation Corps. The fireplace alone soars 50 feet into the grand hall, and the great room is rich with massive timber beams and stone. Continue southward (left) out of Pere Marquette State Park onto Highway 100. Continue approximately 3 miles to Brussels Ferry.

- **Brussells Ferry:** On the Illinois River. Take a free ride across the Illinois River (it's okay to turn around and come right back!) and get a feel of the river under the wheels of your vehicle. Nearby is the **Mark Twain Wildlife Refuge,** the seasonal home for hundreds of thousands of migratory birds (American bald eagles, herons, owls, pelicans, geese, ducks, and many rare species) on the Mississippi Flyway. Return to Highway 100 (where you boarded the ferry) and turn eastbound (right) for approximately 2 miles.

- **Grafton:** On Highway 100 in Jersey County. All but wiped out by the Great Flood of 1993, this amazing river town bounced back and is now considered one of the most important stops on the Byway. Bed-and-breakfast inns, antiques and specialty shops, casual family dining, riverside entertainment, summer outdoor family amphitheater, a small museum, a visitor center, parasailing, jet skiing, pontoon boats, fishing, hunting, hiking, bike trails, cottages, horseback riding, a mystery dinner theater, and much more.

Illinois

❋ Meeting of the Great Rivers Scenic Route

Festivals abound throughout the spring, summer, and fall. Continue southbound approximately 1/2 mile to the bluffs running along the Mississippi River.

- **Scenic bluffs:** Without question, the most spectacular view anywhere along this route is from just outside Grafton, approximately 15 miles southeast to Alton. The bluffs tower above the river with the Byway road surface immediately between the peaks and the riverbank. The ever-flowing Mississippi is alive with commercial traffic, sailboats, and wildlife, in contrast to the majestic bluffs soaring overhead. Any time is good viewing, but late afternoon and sunset are very rewarding. Because the river runs west to east, the sun illuminates the geologic structures, creating a vista unlike anywhere else on the Mississippi. Be careful: many people stop along the highway to take pictures of the bluffs and river. Try to remain in your car to photograph the scenery. Also exercise caution because of many bicyclists and fast-moving traffic at all times. Continue eastbound on Highway 100 about 5 miles from Grafton to Elsah. Be prepared to make an abrupt northward turn (left).

- Village of **Elsah:** In Jersey County. Considered by many national travel writers as the river town that time forgot. This adorable village contains more than two dozen homes built in the 1800s, when Elsah was an important riverboat stop. Because the town has almost no contemporary structures, you immediately feel like you have been transported back into the mid-1800s. Bed-and-breakfasts and small shops abound. Continue eastbound (left from Elsah) onto Highway 100 about 10 miles to the Cliffton Terrace Park.

- **Cliffton Terrace Park:** In Madison County in Godfrey. Pleasant roadside park with facilities for picnics, seasonal wildlife viewing, comfort station, and playground. Continue eastbound (left from Cliffton Terrace Park) onto Highway 100 about 5 miles to the legendary Piasa Bird. Be alert for an abrupt turn northward (left) as you begin seeing riverside barges tied up along the banks.

- **Piasa Bird:** In Madison County in Alton. This mythical creature was seen by American Indian tribes and early European explorers. Today, a gigantic bluff painting depicts the half-dragon, half-cat creature. Restored from early sketches and photography of the 1800s, the site is being developed into an interpretive park and wetlands area. Continue southeasterly (left from Piasa Bird) about 1 mile on Highway 100 and enter Alton.

- **Alton:** In Madison County. The community dates to the early 1800s as a major river port just north of St. Louis and can best be summed up by the word "historic." The Reverend Elijah Lovejoy was martyred here in his stand against slavery and for freedom of the press. It was also the site of the last Lincoln-Douglas debate. Nearby in Hartford-Wood River, Lewis and Clark built Camp Dubois, assembled the Corps of Discovery, and set off on their monumental expedition. The world's tallest man, Robert Wadlow, called Alton home. During the Civil War, thousands of Confederate soldiers were held at the Federal Penitentiary; today, a solemn monument and cemetery honors the dead. Alton has fantastic recreational facilities, including golf courses and ball fields that welcome national championship tournaments. The city has unique casual and fine restaurants, bed-and-breakfasts, inns, hotels, an antique shopping district, a shopping mall, parks, riverboat gaming, and other leisure activities throughout the year. Continue eastbound on Highway 100, going approximately 3 miles to the stoplight. Turn north (right) into the Melvin Price Locks and Dam Complex and National Great Rivers Museum site.

- **Melvin Price Locks and Dam, National Great Rivers Museum:** About 2 miles past the Clark Bridge, on Highway 100, is the Melvin Price Locks and Dam and the site of the **National Great Rivers Museum.** This colossal structure tames the mighty river and aids in flood control and navigation. A wonderful riverfront walkway surrounds the dam and museum. Watch long strings of barges full of fuel and grain pass through the locks to be lowered or raised as the river winds down to the delta. It offers a wonderful view of the Alton skyline and Clark Bridge.

THINGS TO SEE AND DO

Driving along the Meeting of the Great Rivers Scenic Route will certainly keep your senses engaged, but if you yearn to get out of the car and stretch your legs, or if you'd like to make a mini-vacation out of your trip, check out these attractions along the route.

ALTON BELLE RIVERBOAT CASINO. *1 Front St, Alton (62002). Phone toll-free 800/711-GAME. www.argosycasinos.com.* Entertainment complex includes slot machines, showrooms, lounges, and restaurants. Open daily 8-6 am. **FREE**

BRUSSELS FERRY. *Hwy 100 at Hwy 3, Brussels (62013). Phone 618/786-3636.* Ferry boat navigates the Illinois River at the confluence of the Mississippi River. Open daily. **FREE**

★ **PERE MARQUETTE STATE PARK.** *100 Great River Rd, Grafton (62037). Phone 618/786-3323. www.dnr.state.il.us.* Pere Marquette State Park is one of Illinois' largest state parks. Nestled along the banks of the Illinois River on the Byway near Grafton, a myriad of trails take you within the wild forests and up to spectacular viewing areas along the bluff line above. Rangers can assist you with hiking plans and provide valuable insights on how to best explore this giant park. Don't miss the visitor center, where you can learn firsthand about the natural treasures in and around the park. A lodge and cabins are available for the overnight visitors, but plan far ahead, as this is one of the most popular attractions in the Midwest. Campgrounds available; boat launch.

PIASA BIRD PAINTING REPRODUCTION. *Alton (62002). Phone toll-free 800/258-6645. www.altonweb.com/history/piasabird/.* According to Native American legend, a monster bird frequented these bluffs and preyed on all who came near. When Marquette sailed down the Mississippi in 1673, he spotted "high rocks with hideous monsters painted on them" at this spot. The paintings, destroyed by quarrying in the 19th century, were reproduced in 1934. These reproductions, in turn, were destroyed by the construction of Great River Road. They were reproduced a second time on a bluff farther up the river.

VILLAGE OF ELSAH. *Elsah (62028). Phone 618/374-1568. www.elsah.org.* Many buildings are more than 100 years old. Museum (open Apr-Nov, Thurs-Sun afternoons). **FREE**

PLACES TO STAY

If you choose to include an overnight stay in your trip along this Byway, Mobil Travel Guide recommends the following lodgings.

★★ **HOLIDAY INN.** *3800 Homer Adams Pkwy, Alton (62002). Phone 618/462-1220; toll-free 800/465-4329. www.holiday-inn.com.* 137 rooms, 4 story. Complimentary breakfast buffet. Check-out noon. TV; cable (premium), VCR (movies). In-room modem link. Restaurant, bar; entertainment. Room service. In-house fitness room, sauna. Health club privileges. Game room. Indoor pool, whirlpool. Free airport transportation. **$**

★ **SUPER 8.** *1800 Homer Adams Pkwy, Alton (62002). Phone 618/465-8885; toll-free 800/800-8000. www.super8.com.* 61 rooms, 3 story. Pet accepted; fee. Complimentary continental breakfast. Check-out 11 am. TV; cable (premium). In-room modem link. **¢**

Illinois

❋ *Meeting of the Great Rivers Scenic Route*

★ **PERE MARQUETTE LODGE.** *100 Great River Rd, Grafton (62037). Phone 618/786-2331.* 72 rooms, 2 story. Check-out noon. TV; VCR available (movies). Restaurant, bar. In-house fitness room, sauna. Game room. Indoor pool, whirlpool. Lighted tennis. Lawn games. On the Mississippi River. ¢

★ **THE HOMERIDGE BED AND BREAKFAST.** *1470 N State St, Jerseyville (62052). Phone 618/498-3442.* Built in 1867; Italianate Victorian décor. Previous home of Senator Theodore S. Chapman. 5 rooms, 3 story. No room phones. Complimentary full breakfast. Check-out 11 am, check-in 4-6 pm. Game room, pool, lawn games. Totally nonsmoking. $

PLACES TO EAT

A long day of driving is sure to make you hungry. At the end of your journey, try the following restaurant.

★★ **TONY'S.** *312 Piasa St, Alton (62002). Phone 618/462-8384.* Italian menu. Closed holidays. Dinner. Bar. Children's menu. Outdoor seating. $$$

Ohio River Scenic Byway

❋ ILLINOIS
Part of a multistate Byway; see also IN, OH.

Quick Facts

LENGTH: 188 miles.

TIME TO ALLOW: 8 to 10 hours.

BEST TIME TO DRIVE: Year-round; high season is summer.

BYWAY TRAVEL INFORMATION: Southernmost Illinois Tourism Bureau: 618/845-3777.

SPECIAL CONSIDERATIONS: There are no regular seasonal limitations on the Ohio River Scenic Byway in Illinois. In general, the route location and flood control systems allow the route to be traveled during ordinary periods of high water. Also, passenger vehicles are easily accommodated on the Byway in southern Illinois. The roads are well surfaced and in good condition. A narrow and hilly segment of the Byway near Tower Rock in Hardin County is not advised for RVs or tour buses. An alternate route has been identified for these vehicles along the roadside.

RESTRICTIONS: During periods of significant flooding, segments of the route may be closed, especially the terminus of the route in Cairo.

BICYCLE/PEDESTRIAN FACILITIES: The Ohio River Scenic Byway in Illinois offers hiking in the Shawnee National Forest, along the River to River Trail and along the proposed southern route of the American Discovery Trail—sharing several segments of the eastern Byway route. Officially designated bike routes exist along the Byway within the Shawnee National Forest and along Illinois State Route 146 between Cave-in-Rock and Golconda. Pedestrians are easily accommodated at all identified sites in the Byway corridor and in all communities along the route.

This Byway's history is closely tied to the Ohio River, which it follows. For example, many forts from the Civil War and the French and Indian War were placed strategically along this route, and the Underground Railroad had many stops along this Byway. This is where you can find the Cave-in-Rock, an enormous cave that was once home to river pirates. It is now a great vantage point from which to watch today's river traffic.

The outstanding scenery found along this Byway is largely due to the river, which offers winter homes to thousands of Canadian geese. Also, because of the river and the Shawnee National Forest (which the Byway goes through), you'll find plenty of places to picnic, camp, boat, fish, hike, and hunt.

THE BYWAY STORY

The Ohio River Scenic Byway tells archaeological, cultural, historical, natural, and recreational stories that make it a unique and treasured Byway.

Archaeological

Archaeological digs were conducted at Fort Massac State Park in Metropolis from 1939 to 1942, and again in 1966 and 1970. From these digs, reconstruction of the 1970s fort was started. A museum houses artifacts from the original excavations.

Cultural

The spirit of those who live along the Ohio River can be seen at Fort Defiance Park in the town of Cairo. This park, located south of Cairo on Route 51, was once called "probably the ugliest park in America, the park that no one wants," by the *Chicago Tribune*.

Illinois

Ohio River Scenic Byway

A grassroots organization of Cairo citizens, Operation Enterprise, leased the park from the state to renovate and maintain it. The park is now a beautiful place to watch the constant meeting of the Ohio and Mississippi rivers that refuse to merge. The Ohio River waters become a blue ribbon, rippling far down the brown Mississippi currents. Gulls wheel above the placid barges and tugs navigate the point.

Because all this magnificent scenery is found along the Ohio River, there is much evidence in Cairo of the people's desire for beauty. For example, located in the Halliday Park at Washington and Poplar is "The Hewer," a statue that was made in 1906. Sculpted by George Grey Barnard and exhibited at the St. Louis World's Fair, this original bronze statue was declared by Laredo Taft to be one of the finest nudes in America.

You can also experience many annual cultural events and celebrations along this Byway. The Superman Celebration happens the second weekend in June. The Massac County Youth Fair takes place each year in July. The Superman Jet Rally is held the first weekend in October, and the Fort Massac Encampment occurs the third weekend in October.

The Living History Weekends in Fort Massac State Park happen throughout the year. Also consider visiting during the Labor Day Celebration or the Home Town Christmas Light Display from Thanksgiving through New Year's Eve.

Historical

Many historical sites are situated along the Ohio River Scenic Byway. These sites cover everything from early Native American cultures to Civil War sites. Kinkaid Mounds, a designated Byway site, is a monument to the Native American people who once inhabited this region.

The cities of Old Shawneetown and Golconda have numerous historic structures from the 1820s and 1830s. Cairo, positioned at the confluence of the Ohio and Mississippi rivers, was a thriving port city and an important strategic site during the Civil War. Fort Massac State Park in Metropolis is a reconstructed fort from the French and Indian War.

The Ohio River has been a river of both opportunity and tragedy. While the state of Illinois possesses many Underground Railroad sites, the Byway also passes near the Slave House, where captured fugitive slaves and even freed blacks were incarcerated before being returned to the South. This constituted the little-known Reverse Underground Railroad.

In addition, the Trail of Tears crossed the Ohio River in Golconda. At the Buel House (certified by the National Park Service as an official Trail of Tears designated site), the Cherokee were offered food and hospitality.

You'll find many historic buildings and structures along this Byway, starting in Cairo. For example, St. Patrick Catholic Church was built in 1894 and is located on 9th and Washington. The present church is a stately Romanesque structure of Bedford stone. It has three bells and numerous stained-glass windows. It is the oldest church in Cairo, and parishioners celebrated the church's 160th anniversary in 1998.

St. Mary's Park Pavilion was built in 1876 and is located on 28th Street between Magnolia Drive and Park Place West in Cairo. It was built as a memorial to the centennial. This pavilion was the site of President Roosevelt's address during his visit to Cairo in 1907. The pavilion was also the center for numerous Victorian celebrations.

The US Custom House was built in 1872 and is located on 14th and Washington in Cairo. This building is a rare example of Romanesque architecture. The monumental limestone structure was completed for a costly $225,000 at the time. It is one of the few remaining works of noted government architect A. B. Mullet and is currently under renovation. Within the Custom House is the flagpole from General Grant's flagship, the *Tigress*. As a river packet commandeered by the Union Army, the *Tigress* carried

Grant up the Tennessee to the battle of Shiloh on April 6, 1862. A year later, the *Tigress* was sunk while running the shore batteries at Vicksburg. Her crew survived and returned the flag staff to Cairo. A Civil War cannon on the outside of the building complements Grant's standard.

Natural

The natural features of the Ohio Valley in Illinois represent some of the most dramatic features along the entire Ohio River. For example, the Garden of the Gods in the Shawnee National Forest preserves a grouping of unique limestone features with such names as Camel Rock and the Devil's Smokestack. Nearby Rim Rock and Pounds Hollow are also linked by the Byway route. Along the Byway, the Spur to Cave-in-Rock State Park is a great limestone cavern on the Ohio River. The enormous cave, once home to river pirates, provides a vantage point from which you can watch today's river traffic. Farther west, the designated Cache River Spur takes visitors to a rare wetland ecosystem, designated by the United Nations as a Wetland of International Importance.

The Horseshoe Lake Conservation Area in Miller City, 7 miles north of Cairo, has 10,645 acres, including a 2,400-acre lake. The first 49 acres of the park were purchased by the Department of Natural Resources in 1927 for development as a Canadian goose sanctuary. Additional tracts of land, including Horseshoe Island, continued to be purchased in order to create the conservation area that greets visitors today. Canadian geese began wintering at the site in 1928. The original 1,000 birds increased to a population of more than 40,000 by 1944, but this number dropped to 22,000 by 1947. Today, however, more than 250,000 Canadian geese winter at this site, thanks to improved refuge management and harvest controls.

Recreational

Horseshoe Lake Conservation Area resembles Louisiana bayous with its swamp cypress and wading herons. As the winter home of thousands of Canadian geese, Horseshoe Lake is a biologist's and birder's paradise throughout the year; with its large stands of trees around its 20-mile shoreline, it is a beautiful body of water. Since 1930, when a concrete, fixed spillway was constructed, the lake has maintained a fairly constant 4-foot depth. Four picnic areas are found in the park, and each site includes tables, park stoves, and parking. Children appreciate the playground that is located at the picnic area near the spillway. Drinking water is available at hydrants and fountains located conveniently throughout the park. For more information, call 618/776-5689 or 618/776-5215.

You can enjoy camping in both the Fort Massac State Park and throughout the Shawnee National Forest. The Shawnee National Forest in Illinois has 338 miles of equestrian/hiking trails, 454 campsites, 16 designated campgrounds, and 27 designated picnic areas. Recreational opportunities range from primitive make-your-own campsites and trails to developed campsites with beaches, showers, and electricity. For more information, call the Shawnee National Forest toll-free at 800/699-6636. Seven wilderness areas in the Shawnee National Forest are available for wilderness study, including Bald Knob, Burden Falls, Garden of the Gods, Panther Den, Bay Creek, Clear Springs, Lust Creek, and Ripple Hollow (conditionally).

In the tradition of the Mississippi steamships that featured gambling, drinking, and Old West living, a casino steamship still leaves the port of Metropolis. It takes visitors for two-hour cruises on the river during the hours of 9 am to 11 pm. For further information, call toll-free 800/929-5905 or 800/935-1111. Along

Illinois

❋ Ohio River Scenic Byway

the Metropolis riverfront is the Merv Griffin Theater, which presents a range of entertainers on special dates throughout the year (phone toll-free 800/949-5740).

HIGHLIGHTS

The Ohio River Scenic Byway must-see tour begins in Cairo, Illinois, and runs to the Indiana border. If you're traveling the opposite way, simply read this list from the bottom and work your way up.

- **Fort Defiance Park** is right on the Byway. Stop and view the confluence of the Ohio and Mississippi rivers, where you can see tugs working. There are picnic tables and the Boatmens Memorial, which affords great photo opportunities at the spot where the nation's largest rivers meet.

- Taking Highway 51 right through Cairo (on Washington Avenue), travel 1 1/2 miles to **The Hewer** statue in Halliday Park on the corner of Washington and Poplar avenues. About 1/2 mile later, you can see the **Safford Memorial Library,** where you can pick up a 50-cent book out of the "treasure bin" to read in bed that night. Across the street is the **US Custom House Museum.**

- Where Highway 51 and Washington Avenue split, you can take a short detour following Washington Avenue along **Millionaire's Row** to see turn-of-the-century mansions such as River Lore and Magnolia Manor, which is open for tours. Then turn right on 28th Street to rejoin Byway.

- Heading north on Highway 51 out of Cairo, about 5 miles at the junction of Highway 51 and Highway 37, is the **Mound City National Cemetery.** From Mound City, the Byway route goes onto Highway 37.

- Traveling north on Highway 37, about 5 miles later is the **Olmstead Lock and Dam** project, which offers a lookout pavilion and rest rooms. The route then goes up Highway 37 about 5 miles to the Grand Chain Joppa Blacktop, and then it runs into Highway 45.

- Highway 45 goes south into Metropolis, where you can see **Superman Square** in the center of town. A casino is located in Metropolis along the Ohio River.

- **Fort Massac State Park** is the next stop (2 miles south of Superman Square), located on Highway 45, with picnic tables, a museum, and a variety of events that occur throughout the year. Follow the Byway out through Brookport, Unionville, and Liberty up through Baycity to Golconda (about 45 miles from Metropolis), where you will find the **Golconda Marina,** the **Buel House,** and many antique and novelty shops.

- Continue on Highway 46 and go through Elizabethtown and on to **Cave-in-Rock State Park,** with a lodge, a restaurant, picnic tables, trails, and much more.

- North of Cave-in-Rock, you can stop by **Garden of the Gods Wilderness Area** in the Shawnee National Forest to enjoy the hiking trails and picnic tables. Then continue on the Byway until it joins Indiana's portion of the Ohio River Scenic Byway.

THINGS TO SEE AND DO

Driving along the Ohio River Scenic Byway will certainly keep your senses engaged, but if you yearn to get out of the car and stretch your legs, or if you'd like to make a mini-vacation out of your trip, check out these attractions along the route.

CAVE-IN-ROCK STATE PARK. *#1 New State Park Rd, Cave-In-Rock (62919). Phone 618/289-4325.* This park offers a stunning view of the Ohio River from high bluffs sets in a heavily wooded area. The central feature of the park is a 55-foot-wide limestone cave. Numerous hiking trails and picnic areas. Boating, fishing, camping. Marina. Restaurant, lodge (phone 618/289-4545).

CUSTOM HOUSE MUSEUM. *1400 Washington Ave, Cairo (62914). Phone 618/734-1019.* This 19th-century Federal-style building contains artifacts and replicas from Cairo's past. Open Mon-Fri. **FREE**

FORT DEFIANCE STATE PARK. *Rtes 3 and 51, Cairo (62914). Phone 618/734-3015.* Splendid view of the confluence of Ohio and Mississippi rivers on 39 acres; site of Civil War post commanded by General Ulysses S. Grant.

GOLCONDA MARINA and **SMITHLAND POOL.** *RR 2, Golconda (62958). Phone 618/683-5875 or 618/949-3394.* The marina is the gateway to Smithland Pool, a 23,000-acre recreational area off the Ohio River that offers excellent fishing and boating opportunities.

MAGNOLIA MANOR. *2700 Washington Ave, Cairo (62914). Phone 618/734-0201.* This Italianate Victorian mansion (1869) with 14 rooms, built for a wealthy flour merchant, contains period furnishings and items of local historical interest. View of Mississippi and Ohio rivers from the tower. Open daily; closed holidays. **$$**

SHAWNEE NATIONAL FOREST. *50 Hwy 145 S, Marion (62946). Phone 618/253-7114; toll-free 800/699-6636. www.fs.fed.us/r9/Shawnee/.* Approximately 278,000 acres, bordered on the east by the Ohio River and on the west by the Mississippi River; unusual rock formations, varied wildlife. Swimming, fishing, boating; hunting, hiking and bridle trails, picnicking, camping on a first-come, first-served basis. **$$**

PLACES TO STAY

If you choose to include an overnight stay in your trip along this Byway, Mobil Travel Guide recommends the following lodgings.

★★ **RIVER ROSE INN.** *1 Main St, Elizabethtown (62931). Phone 618/287-8811.* 5 rooms, 3 story. Located in a Greek Gothic mansion on the banks of the Ohio River in the Shawnee National Forest. **$**

★★★ **MANSION OF GOLCONDA.** *515 Columbus St, Golconda (62938). Phone 618/683-4400.* 4 rooms, 3 story. Built in 1894, this gabled 21-room Victorian mansion is a full-service inn listed on the National Historic Register. **$**

★★ **ISLE OF VIEW BED & BREAKFAST.** *205 Metropolis St, Metropolis (62960). Phone 618/524-5838; toll-free 800/566-7491. www.isle-of-view.net.* 5 rooms, 3 story. Built in 1889 and renowned as one of the area's finest bed-and-breakfasts, this Victorian inn lies one block from the Ohio River. **$**

Illinois

❈ *Ohio River Scenic Byway*

PLACES TO EAT
A long day of driving is sure to make you hungry. At the end of your journey, take a table at one of the following restaurants.

Golconda
★★ **MANSION OF GOLCONDA.** 515 Columbus St, Golconda (62938). Phone 618/683-4400. American menu. Closed Sun, Mon. Lunch, dinner. The main-floor restaurant of this inn is known throughout the region for fine dining with superb regional cuisine in an elegant setting. **$$**

Paducah, KY
Paducah isn't far south of the Byway and offers additional dining options.

★ **C. C. COHEN.** 103 S 2nd St, Paducah (42001). Phone 270/442-6391. Steak, seafood menu. Closed Sun; Thanksgiving, Dec 25. Lunch, dinner. Bar. Entertainment Fri-Sat. Children's menu. In the Cohen Building (circa 1870); early decorative metalwork. Many antiques. **$$**

★ **JEREMIAH'S.** 225 Broadway, Paducah (42001). Phone 270/443-3991. American menu. Closed Sun; Dec 25. Dinner. Bar. Own brewery. Former bank (1800s); rustic décor. **$$**

★★ **WHALER'S CATCH.** 123 N 2nd St, Paducah (42001). Phone 270/444-7701. Specializes in Southern-style seafood. Closed Sun; major holidays. Lunch, dinner. Bar. Children's menu. Outdoor seating. **$$$**

The Historic National Road

❋ INDIANA   AN ALL-AMERICAN ROAD

Part of a multistate Byway; see also IL, OH.

Quick Facts

LENGTH: 156 miles.

TIME TO ALLOW: 3 to 9 hours.

BEST TIME TO DRIVE: High seasons are spring and fall. In the spring, poppies, irises, and wildflowers are spread along the roadside. Black locusts and redbuds are also in bloom. During September and October, community festivals occur all along the Byway.

BYWAY TRAVEL INFORMATION: Indiana National Road Association: 765/478-3172.

BICYCLE/PEDESTRIAN FACILITIES: In rural portions of the Historic National Road, you will not find sidewalks for pedestrians or shoulders specifically designated for bicycles. However, in rural portions of the Byway where traffic is often light, bicyclists and pedestrians can travel many stretches of the route in relative safety and comfort.

One of America's earliest roads, the National Road was built between 1828 and 1834 and established a settlement pattern and infrastructure that is still visible today. Nine National Register Districts are found along the route, as are 32 individually designated National Register Sites offering education and entertainment. As you travel Indiana's Historic National Road, you find a landscape that has changed little since the route's heyday in the 1940s.

Historic villages with traditional main streets and leafy residential districts still give way to the productive fields and tranquil pastures that brought Indiana prosperity. From the Federal-style architecture of an early pike town (a town that offered traveling accommodations and little else) to the drive-ins and stainless-steel diners of the 1940s, you can literally track the westward migration of the nation in the buildings and landscapes that previous generations have left behind.

Along the way, you will find many of the same buildings and towns that were here during the earliest days of westward expansion. A visit to Antique Alley gives you a chance to do some antique shopping and exploring along this historic road. The Indiana Historic National Road is a unique way to experience the preserved pike towns along the route, such as Centerville and Knightstown.

THE BYWAY STORY

Indiana's portion of the Historic National Road tells archaeological, cultural, historical, natural, recreational, and scenic stories that make it a unique and treasured Byway.

Indiana

❋ *The Historic National Road*

Archaeological

Eastern Indiana was the home of two groups of Native Americans identified by scholars as the Eastern Woodland Societies, who made their homes in the area following the retreat of the glaciers. One group occupied the area around 7000 to 1000 BC, the other from approximately 1000 to 700 BC. Many of their campsites have been found in the area of the Whitewater River Gorge. The Whitewater River Gorge was an important area after glacier movement and activity had stopped in the area. The area was excellent for hunting and fishing, with flowing streams and an abundance of resources.

Cultural

The National Road brought the nation to Indiana. The lure of limitless opportunities and the romance of the West drew tens of thousands of pioneers through Indiana between 1834 and 1848. Many stayed and settled in the Hoosier State, thus creating a new culture—the foundation for our national culture. This is because religious and economic groups left the distinctive colonial societies of the eastern seaboard and merged in the Midwest. Settlers to Indiana brought with them their own particular mix of customs, languages, building styles, religions, and farming practices. Quakers, European immigrants, and African Americans looking for new opportunities all traveled the National Road. Evidence of this mix of cultural influences can be seen along the corridor today in the buildings and landscapes. It can also be learned at the Indiana State Museum's National Road exhibit, and it can be experienced on a Conestoga Wagon at Conner Prairie or at a Civil War encampment along the route.

As the region matured, the culture continued to evolve under the influence of the nation's primary east-west route. Richmond was home of the Starr Piano Company, and later the Starr-Gennet recording studios, where jazz greats like Hoagy Carmichael and Louis Armstrong made recordings in an early jazz center. The Overbeck sisters, noted for their Arts and Crafts pottery, lived and worked in Cambridge City. The poet James Whitcomb Riley, author of "Little Orphant Annie" and "Raggedy Man," lived in Greenfield. Also, Indianapolis, the largest city on the entire Historic National Road, became an early center for automobile manufacturing. Today, visitors experience such attractions as the Children's Museum of Indianapolis (the largest in the world) and the Eiteljorg Museum of American Indian and Western Art, as well as a variety of other museums and cultural institutions.

The Historic National Road in Indiana represents one segment of the historic National Road corridor from Maryland to Illinois. The historic and cultural resources within Indiana are intimately tied to traditions and customs from the eastern terminus of the road in Cumberland, Maryland, and are built on goals and expectations of a nation looking west.

Historical

The National Road was the first federally funded highway in the United States. Authorized by Thomas Jefferson in 1803, the road ran from Cumberland, Maryland, west to Vandalia, Illinois. Designed to connect with the terminus of the C&O Canal in Cumberland, the National Road gave agricultural goods and raw materials from the interior direct access to the eastern seaboard. It also encouraged Americans to settle in the fertile plains west of the Appalachians. For the first time in the United States, a coordinated interstate effort was organized and financed to survey and construct a road for both transportation purposes and economic development.

Built in Indiana between 1828 and 1834, the National Road established a settlement pattern and infrastructure that is still visible today. The historic structures along the National Road illustrate the transference of ideas and culture from the east as the road brought settlement and

see page A6 for color map

commerce to Indiana. The National Road still passes through well-preserved, Federal-style pike towns and Victorian streetcar neighborhoods, and it is lined with early automobile-era structures, such as gas stations, diners, and motels.

Natural

The topography of Indiana was created by glaciers that advanced and retreated over the land during the Pleistocene Period. Leaving behind moraines and an undulating landscape, the glaciers also helped to create the Whitewater River Gorge, where fragments of limestone, clay, and shale bedrock can be seen. The gorge and surrounding region is known internationally among geologists for its high concentration of Ordovician Period fossils.

Recreational

You can find many opportunities for recreation along the Historic National Road in Indiana, as well as in nearby cities. Golf is a popular sport along the highway, as evidenced by the many golf courses, such as the Glen Miller Public Golf Course, the Hartley Hills Country Club, the Highland Lake Public Golf Course, and the Winding Branch Public Golf Course. Biking and hiking are other extremely popular sports along the Byway. Local park and recreation facilities are often directly accessible from the Byway or can be found nearby.

Professional sports can be enjoyed along the Byway as well. White River State Park in Indianapolis offers you an opportunity to enjoy a Triple-A baseball game at Victory Field (or to visit the Indianapolis Zoo). Just off the Historic National Road in downtown Indianapolis are the RCA Dome, home of the Indianapolis Colts, and Conseco Fieldhouse, home of the Indiana Pacers.

Scenic

The Historic National Road is a combination of scenes from rural communities, small towns, and a metropolitan city. This combination makes the Byway a scenic tour along one of the most historically important roads in America. Small-town antique shops and old-fashioned gas pumps dot the Byway, making the Historic National Road a relaxing and peaceful journey. Broad views of cultivated fields, distant barns and farmhouses, and grazing livestock dominate the landscape. In other areas, courthouse towers, church steeples, and water towers signal approaching communities that draw you from the open areas into historic settlements. The topography of the land affords vistas down the corridor and glimpses into natural areas that sit mostly hidden in the rural landscape. This repeating pattern of towns and rural landscapes is broken only by metropolitan Indianapolis.

HIGHLIGHTS

The following are just some of the points of interest available to you when traveling west across the Indiana portion of this Byway from the western border of Ohio. If you're traveling east, read this list from the bottom up.

- **Historic Richmond:** As one of Indiana's oldest historic towns (founded in 1806), Richmond has one of the Hoosier State's largest intact collections of 19th-century architecture. You can visit four National Register Historic Districts; **Hayes Regional Arboretum;** a bustling historic downtown full of unique shops and restaurants; and a fascinating collection of local museums, including the **Wayne County Museum,** the **Richmond Art Museum,** the **Gaar Mansion,** the **Indiana Football Hall of Fame,** the **Joseph Moore Museum** at Earlham College, and the **Rose Gardens** located along the road on the city's east side.

- **Centerville:** One of the historic highway's most intact and quaint National Road-era pike towns is listed in the National Register of Historic Places for its fine collection of architecture. Centerville also has a noteworthy

Indiana

The Historic National Road

collection of small antique and specialty shops and is home to the world's largest antique mall, just several blocks north of the National Road.

- **Pike towns and Antique Alley (Richmond to Knightstown):** You can meander along the National Road and enjoy the tranquil agricultural landscape interspersed with pike towns that recall the early years, when travelers needed a place to rest every 5 miles or so. This route is also heralded as Antique Alley, with more than 900 antique dealers plying their trade in and between every community along the route.
- **Huddleston Farmhouse Inn Museum, Cambridge:** A restored National Road-era inn and farm tells the story of the historic highway and the people who formed communities along its length. The museum is owned and operated by the Historic Landmarks Foundation of Indiana and is the home office of the Indiana National Road Association, the National Scenic Byway management nonprofit group. The museum displays the way of life of an early Hoosier farm family and the experience of westward travelers who stopped for food and shelter. Cambridge City is also listed in the National Register of Historic Places and has unique historic buildings that are home to diverse shops and local eateries.
- **Knightstown:** Knightstown grew because of its location on the National Road between Richmond and Indianapolis. The town has retained its significant collection of 19th- and 20th-century architecture; a large section of the town is listed in the National Register of Historic Places. Today, you can visit four antique malls, watch a nationally known coppersmith, and stay in one of two bed-and-breakfasts. Also available is the **Big Four Railroad Scenic Tour.**
- **Greenfield: James Whitcomb Riley's Old Home and Museum** on the National Road in Greenfield tells the story of the Hoosier poet and allows you to experience his life and community with guided tours. The town also is rich in small-town local flavor, with many shops and restaurants to satisfy you.
- **Irvington:** A classic 1870s Indianapolis suburb was developed as a getaway on the city's east side. Irvington has since been swallowed by the city but retains its stately architecture and peaceful winding cul-de-sacs. Listed in the National Register of Historic Places, Irvington recalls turn-of-the-century progressive design principles and allows the modern visitor a glimpse into the city's 19th-century development.
- **Indianapolis:** The center of Indiana's National Road is also its state capital. Downtown Indianapolis offers a growing array of activities and amenities, from the state's best shopping at **Circle Centre Mall,** an unprecedented historic preservation development that

incorporates building facades from the city's past into a state-of-the-art mall experience, to gourmet dining and an active nightlife and sports scene. Along Washington Street just east of downtown, you can visit the **Indiana Statehouse,** the **Indianapolis Zoo,** the **Eiteljorg Museum of American and Indian Art,** and **White River State Park.** The **Indianapolis Colts** play at the RCA Dome, and the **Indiana Pacers** continue the ritual of Hoosier Hysteria at Conseco Fieldhouse downtown.

- **Plainfield:** Twentieth-century automobile culture dominates this area. Motels and gas stations remain from the early days and are interspersed with the sprawl and development of the modern city. The Diner, on the east side of Plainfield, is a remnant from the early days of travel, a stainless-steel café with an atmosphere reminiscent of the 1940s, an atmosphere that is quickly disappearing. From Plainfield to Brazil, look for roadside farmers' markets.

- **Brazil:** The western extension of the National Road was surveyed through what is now Brazil in 1825; today, its National Register-listed Meridian Street remains a classic example of how the historic highway promoted the growth of communities along its length. The village is also full of curios and collectibles.

- **Terre Haute:** The western edge of Indiana's National Road is anchored by Terre Haute, a community offering historic points of interest and cultural experiences of various kinds. The **Rose-Hulman Institute of Technology** on the city's east side was founded in 1874 and is an exceptionally beautiful college campus; just west of the city on State Road 150, the **St. Mary-of-the-Woods College** campus offers a touch of European elegance in the Indiana forest. Its campus is in a beautiful wooded setting and has several buildings dating to its 1841 founding. You can choose National Road restaurants along the city's **Wabash Avenue,** located in historic buildings.

Dobbs Park, 1/2 mile south of the highway at the intersection of Highways 46 and 42, is home to a nature center and **Native American Museum. The Sheldon Swope Art Museum** at 25 South 7th Street features 19th- and 20th-century artworks in a 1901 Renaissance Revival-style building with an Art Deco interior. The **Children's Science and Technology Museum** at 523 Wabash Avenue houses rooms full of hands-on learning displays and special exhibits. Larry Bird fans can see memorabilia at **Larry Bird's Boston Connection** (55 South 3rd Street) and view a museum of his career keepsakes, including his Olympic medal, MVP trophies, photographs, and other mementos. Continue on Indiana's Old National Road to the Illinois portion of the road.

THINGS TO SEE AND DO

Driving along the Historic National Road will certainly keep your senses engaged, but if you yearn to get out of the car and stretch your legs, or if you'd like to make a mini-vacation out of your trip, check out these attractions along the route.

ANTIQUE ALLEY. *5701 National Rd E, Richmond (47374). Phone 765/935-8687. www.antiquealley.us.* More than 900 dealers display their treasures within a 33-mile loop. Hours vary. **FREE**

✪ **THE CHILDREN'S MUSEUM OF INDIANAPOLIS.** *3000 N Meridian St, Indianapolis (46208). Phone 317/334-3322. www.childrensmuseum.org.* The largest museum of its kind in the nation, with ten major galleries. Exhibits cover science, social cultures, space, history, and exploration and include the Welcome Center, SpaceQuest Planetarium (fee), 30-foot-high Water Clock, Playscape gallery for preschoolers, Computer Discovery Center, hands-on science exhibits, simulated limestone cave, carousel rides (fee), and a performing arts theater. The largest gallery, Center for Exploration, is designed for ages

Indiana

✤ *The Historic National Road*

12 and up. Open Mar-Labor Day, daily; rest of year, Tues-Sun; closed Easter, Thanksgiving, Dec 25. **$$**

CIRCLE CENTRE MALL. *49 W Maryland St, Indianapolis (46204). Phone 317/681-8000. www.simon.com.* Indianapolis' newest mall in the heart of downtown offers more than 100 shopping, dining, and entertainment options. Anchor stores Nordstrom and Parisian are flanked by national chains such as Gymboree, Banana Republic, and Williams-Sonoma in this four-story structure that spans two city blocks. Open Mon-Sat 10 am-9 pm, Sun noon-6 pm; closed Thanksgiving, Dec 25. **FREE**

CITY MARKET. *222 E Market St, Indianapolis (46204). Phone 317/634-9266. www.indianapoliscitymarket.com.* This renovated marketplace was constructed in 1886 and includes this building and two adjacent areas. The market features smoked meats, dairy, specialty bakery and fruit stands, and ethnic foods. Open Mon-Fri 6 am-6 pm, Sat 8 am-4 pm; closed holidays. **FREE**

✪ **CONNER PRAIRIE.** *13400 Allisonville Rd, Indianapolis (46038). Phone 317/776-6000; toll-free 800/966-1836. www.connerprairie.org.* This 250-acre nationally acclaimed living-history museum offers costumed interpreters who depict the life and times of early settlement in this 1836 village. The area contains 39 buildings, including a Federal-style brick mansion (1823) built by fur trader William Conner (self-guided tours). Working blacksmith, weaving, and pottery shops; woodworkers complex. Visitor center with changing exhibits. Hands-on activities at Pioneer Adventure Area; games, toys. Picnic area, restaurant, gift shop. Special events throughout year. Open Tues-Sun until 5 pm; closed holidays, Tues in Nov. **$$$**

CRISPUS ATTUCKS MUSEUM. *1140 Martin Luther King Jr. St, Indianapolis (46202) Phone 317/226-4613.* Four galleries have been established to recognize, honor, and celebrate the contributions made by African Americans. Open Mon-Fri 10 am-2 pm; closed holidays. **FREE**

CROWN HILL CEMETERY. *700 W 38th St, Indianapolis (46208). Phone 317/925-8231; toll-free 800/809-3366. www.crownhill.org.* This is the third largest cemetery in the nation. President Benjamin Harrison, poet James Whitcomb Riley, novelist Booth Tarkington, and gangster John Dillinger are among the notables buried here.

EAGLE CREEK PARK. *7840 W 56th St, Indianapolis (46254). Phone 317/327-7110. www.indy.gov/indyparks/.* Approximately 3,800 acres of wooded terrain with a 1,300-acre reservoir make up the nation's second largest city park. Fishing, boat ramps, rentals, swimming beach (open Memorial Day-Labor Day), bathhouse, water sports center; shelters, golf course, cross-country skiing, hiking trails, playgrounds, picnicking. Open daily; some facilities closed in winter. **$**

✪ **EITELJORG MUSEUM OF AMERICAN INDIAN AND WESTERN ART.** *500 W Washington St, Indianapolis (46204). Phone 317/636-9378. www.eiteljorg.org.* The museum features collections of Native American and American Western art, considered one of the finest collections of its kind. Open Memorial Day-Labor Day, Mon-Sat, also Sun afternoons; rest of year, Tues-Sat, also Sun afternoons; closed Easter, Thanksgiving, Dec 25. **$$**

FOWLER PARK PIONEER VILLAGE. *3000 E Oregon Church Rd, Terre Haute (47802). Phone 812/462-3391.* An 1840s pioneer village with 12 log cabins, a general store, schoolhouse, and gristmill. Open summer weekends; also by appointment. **FREE**

THE GERMAN FRIENDSHIP GARDEN. *2500 National Rd E, Richmond (47374).* Features 200 German hybridized roses sent by the German city of Zweibrücken from its own rose garden. In bloom May-Oct. **FREE**

GLEN MILLER PARK. *2514 E Main St, Richmond (47374). Phone 765/983-7285.* A 194-acre park; E. G. Hill Memorial Rose Garden, nine-hole golf (fee), natural springs, picnic shelters, concessions, fishing, paddleboats, playground, tennis courts, and outdoor amphitheater (summer concerts). **FREE**

HAYES REGIONAL ARBORETUM. *801 Elks Rd, Richmond (47374). Phone 765/962-3745. www.hayesarboretum.org.* A 355-acre site with trees, shrubs, and vines native to this region; 40-acre beech-maple forest; auto tour (3 1/2 miles) of the site. Fern garden; spring house. Hiking trails; bird sanctuary; nature center with exhibits. Open Tues-Sun 9 am-5 pm; closed holidays. **FREE**

HISTORIC LOCKERBIE SQUARE. *528 Lockerbie St, Indianapolis (46202). Phone 317/631-5885.* Late 19th-century private houses have been restored in this six-block area. Cobblestone streets, brick sidewalks, and fine architecture make this an interesting area for sightseeing.

HOLCOMB OBSERVATORY & PLANETARIUM. *4600 Sunset Ave, Indianapolis (46208). Phone 317/940-9333. www.butler.edu/holcomb/.* Features the largest telescope in Indiana, a 38-inch Cassegrain reflector. Planetarium shows (call for schedule). **$$**

HUDDLESTON FARMHOUSE INN MUSEUM. *838 National Rd, Cambridge City (47327). Phone 765/478-3172.* This restored 1840s farmhouse/inn complex with outbuildings once served National Road travelers. Open May-Aug, Tues-Sat, also Sun afternoons; rest of year, Tues-Sat only; closed holidays and the month of Jan. **$**

INDIANA STATE MUSEUM. *202 N Alabama St, Indianapolis (46204). Phone 317/232-1637. www.in.gov/ism/.* Depicts Indiana's history, art, science, and popular culture with five floors of displays. Exhibits include the Indiana Museum of Sports, Indiana radio, forests of 200 years ago, a small-town community at the turn of the century, and paintings by Indiana artists. Changing exhibits. Open Mon-Sat 9 am-5 pm, Sun 11 am-5 pm; closed holidays. **$$**

★ INDIANAPOLIS MOTOR SPEEDWAY AND HALL OF FAME MUSEUM. *4790 W 16th St, Indianapolis (46222). Phone 317/484-6747. www.brickyard.com.* Site of the famous 500-mile automobile classic held each year the Sun before Memorial Day. Many innovations in modern cars have been tested at races here. The oval track is 2 1/2 miles long, lined by grandstands, paddocks, and bleachers. Hall of Fame Museum (fee) has exhibits of antique and classic passenger cars, many built in Indiana; more than 30 Indianapolis-winning race cars. Open daily; closed Dec 25. **$$**

INDIANAPOLIS ZOO. *1200 W Washington St, Indianapolis (46222). Phone 317/630-2001. www.indyzoo.com.* This 64-acre facility includes the state's largest aquarium, an enclosed whale and dolphin pavilion, and more than 3,000 animals from around the world. Sea lions, penguins, sharks, polar bears; daily whale and dolphin shows; camels and reptiles of the deserts; lions, giraffes, and elephants in the Plains; tigers, bears, and snow monkeys in the Forests. Encounters features domesticated animals from around the world, and a 600-seat outside arena offers daily programs and demonstrations. Living Deserts of the World is a conservatory covered by an 80-foot-diameter transparent dome. New **White River Gardens** is a conservatory and gardens. Commons Plaza include a restaurant and snack bar; additional animal exhibits; and an amphitheater for shows and concerts. Horse-drawn streetcar, elephant, camel, carousel, and miniature train rides. Open daily. **$$$**

Indiana

❋ *The Historic National Road*

JAMES WHITCOMB RILEY HOME. *250 W Main St, Greenfield (46140). Phone 317/462-8539.* Boyhood home of the poet from 1850 to 1869. Riley wrote "When the Frost is on the Punkin" and many other verses in Hoosier dialect. Tours. Museum adjacent. Open late Mar-early Nov, Mon-Sat 10 am-4 pm; closed Sun; major holidays. **$**

JOSEPH MOORE MUSEUM OF NATURAL SCIENCE. *801 National Rd W, Richmond (47374). Phone 765/983-1303.* On the Earlham College campus. Birds and mammals in natural settings, fossils, mastodon and allosaurus skeletons. Open mid-Sept-mid-Dec and mid-Jan-early May, Mon, Wed, Fri, Sun 1-5 pm; rest of year, Sun 1-5 pm. **FREE**

LEVI COFFIN HOUSE STATE HISTORIC SITE. *113 US 27 N, Fountain City (47341). Phone 765/847-2432.* Federal-style brick home of the Quaker abolitionist who helped 2,000 fugitive slaves escape to Canada; period furnishings. Tours. Open June-late Aug, Tues-Sat 1-4 pm; Sept-late Oct, Sat 1-4 pm; closed July 4. **$**

LIEBER STATE RECREATION AREA. *1317 W Lieber Rd, Greencastle (46120). Phone 765/795-4576.* This area contains approximately 775 acres on Cataract Lake (1,500 acres). Swimming, lifeguard, bathhouse, water-skiing, fishing, boating (dock, rentals); picnicking, concession, camping. Activity center. Adjacent are 342 acres of state forest and 7,300 acres of federal land, part of Cagles Mill Flood Control Reservoir Project. Open Memorial Day-Labor Day. **$**

MADAME WALKER THEATRE CENTER. *617 Indiana Ave, Indianapolis (46202). Phone 317/236-2099. www.walkertheatre.org.* The Walker Theatre, erected and embellished in an African and Egyptian motif, was built in 1927 as a tribute to Madame C. J. Walker, America's first self-made female millionaire. The renovated theater now features theatrical productions, concerts, and other cultural events. The center serves as an educational and cultural center for the city's African-American community. Tours. Open Mon-Fri. **$**

MADONNA OF THE TRAILS. *22nd and E Main sts, Richmond (47374). Phone 765/983-7200.* One of 12 monuments erected along the National Road (US 40) in honor of pioneer women. **FREE**

MIDDLEFORK RESERVOIR. *IN 27 and Sylvan Nook Dr, Richmond (47374). Phone 765/983-7293.* A 405-acre park with a 175-acre stream and a spring-fed lake. Fishing, boating (dock rental), bait and tackle supplies; hiking trails, picnicking, playground. **FREE**

NATIVE AMERICAN MUSEUM. *5170 E Poplar St, Terre Haute (47803). Phone 812/877-6007.* Exhibits include dwellings, clothing, weapons, and music of Eastern Woodland Native American cultures. Hands-on activities. Open Mon-Sat 9 am-5 pm, Sun noon-5 pm; closed major holidays. **FREE**

OLD LOG JAIL AND CHAPEL-IN-THE-PARK MUSEUMS. *US 40 and Apple St, Greenfield (46140). Phone 317/462-7780.* Historical displays include arrowheads, clothing, china, and local memorabilia. Open Apr-Nov, Sat and Sun 1-5 pm. **$**

PRESIDENT BENJAMIN HARRISON HOME. *1230 N Delaware St, Indianapolis (46202). Phone 317/631-1888. www.presidentbenjaminharrison.org.* This residence of 23rd president of the United States includes 16 rooms with original furniture, paintings, and the family's personal effects. Herb garden. Guided tours (every 30 minutes). Open Mon-Sat, 10 am-3:30 pm; July, Aug, Dec, Sun 12:30-3:30 pm; closed holidays, the first three weeks in Jan, and 500 Race day. **$$**

SHAKAMAK STATE PARK. *6265 W State Rd 48, Jasonville (47438). Phone 812/665-2158. www.state.in.us/dnr/parklake/parks/shakamak.html.* More than 1,766 acres with three artificial lakes stocked with game fish. Swimming pool, lifeguard, bathhouse; boating (rentals, no gasoline motors). Picnicking, playground, hiking, camping, trailer facilities; cabins. Naturalist service. nature center. Open May-Aug.

SHELDON SWOPE ART MUSEUM. *25 S 7th St, Terre Haute (47807). Phone 812/238-1676. www.swope.org.* The museum's permanent collections include 19th- and 20th-century American art. Special exhibits, films, lectures, classes, and performing arts events. Open Tues-Fri, 10 am-5 pm; Sat and Sun, noon-5 pm; closed Mon, holidays. **FREE**

VIGO COUNTY HISTORICAL MUSEUM. *S 6th St, at Washington Ave, Terre Haute (47802) Phone 812/235-9717.* Local exhibits in 12 rooms of an 1868 house; one-room school, country store, military room, and dressmaker's shop. Open Tues-Sun; closed holidays.

WAYNE COUNTY HISTORICAL MUSEUM. *1150 N A St, Richmond (47374). Phone 765/962-5756. www.wchm.org.* Pioneer rooms include a general store; bakery, cobbler, print, bicycle, blacksmith, and apothecary shops; log cabin (1823); loom house; agricultural hall; decorative arts gallery; antique cars and old carriages; Egyptian mummy; and collections of the Mediterranean world. Open Tues-Fri, 9 am-4 pm, Sat and Sun, 1-4 pm; closed Mon, holidays. **$**

PLACES TO STAY

If you choose to include an overnight stay in your trip along this All-American Road, Mobil Travel Guide recommends the following lodgings.

Brazil

★ **MCKINLEY HOUSE.** *3273 E US Hwy 40, Brazil (47834). Phone 812/442-5308; toll-free 866/442-5308.* This nicely restored inn was built in 1872 by Green McKinley, one of the original contractors of Highway 40, and is constructed of red brick made from clay dug at nearby Croy Creek. It has been operated as an inn continuously since it was built. 4 rooms, 2 story. Complimentary full breakfast. Check-in 2 pm, check-out 11 am. **$**

Centerville

★ **LANTZ HOUSE INN B&B.** *214 W Main St, Centerville (47330). Phone 765/855-2936; toll-free 800/495-2689.* This inn is located in a historical commercial building in the heart of old Centerville, within walking distance of shops and restaurants. All rooms have private baths. 5 rooms, 2 story. **$**

Greencastle

★★★ **THE WALDEN INN.** *2 W Seminary St, Greencastle (46135). Phone 765/653-2761; toll-free 800/225-8655. www.waldeninn.com.* This beautiful country inn on the campus of DePauw University has warmth and charm. Guests can stroll through town and check out the area's covered bridges. 55 rooms, 2 story. Check-out 1 pm, check-in 4 pm. TV. Restaurant. Amish furniture. **$**

Greenfield

★ **LEES INN.** *2270 N State St, Greenfield (46140). Phone 317/462-7112; toll-free 800/733-5337. www.leesinn.com.* 100 rooms, 2 story. Pet accepted. Complimentary continental breakfast. Check-out noon. TV. **¢**

Indianapolis

★★★ **THE CANTERBURY HOTEL.** *123 S Illinois St, Indianapolis (46225). Phone 317/634-3000; toll-free 800/538-8186. www.canterburyhotel.com.* The Canterbury Hotel transplants the charm and elegance of England to downtown Indianapolis. Since the 1850s, it has enjoyed a proud history as the city's leading hotel. This intimate hotel provides visitors with a convenient city location and private access to the adjacent shopping mall, filled with upscale stores and restaurants. Mahogany furniture and traditional artwork complete the classic décor in the guest rooms. Spacious and inviting, the accommodations are well suited for modern

Indiana

The Historic National Road

travelers. The lovely restaurant dishes up American and continental favorites for breakfast, lunch, and dinner, while the traditional afternoon tea is a local institution. 99 rooms, 12 story. Complimentary continental breakfast. Check-out noon. TV, VCR available. In-room modem link. Restaurant. Concierge. Formal décor; four-poster beds, Chippendale-style furniture. Historic landmark. $$

★★★ **EMBASSY SUITES.** *110 W Washington St, Indianapolis (46204). Phone 317/236-1800; toll-free 800/EMBASSY. www.embassysuites.com.* Within walking distance of the business district, shopping, and local attractions, this hotel is conveniently located. Guest suites are large, comfortable, and affordable. 360 rooms, 18 story. Complimentary full breakfast. Check-out noon. TV, VCR available. In-room modem link. Restaurant, bar. In-house fitness room. Indoor pool, whirlpool. $$

★ **HAMPTON INN.** *105 S Meridian St, Indianapolis (46225). Phone 317/261-1200; toll-free 800/426-7866. www.hamptoninn.com.* 180 rooms, 9 story. Complimentary continental breakfast. Check-out noon. TV; cable (premium). In-room modem link. Restaurant, bar. Room service. In-house fitness room. Valet parking available. $

Richmond

★ **COMFORT INN.** *912 Mendelson Dr, Richmond (47374). Phone 765/935-4766; toll-free 800/228-5150. www.choicehotels.com.* 52 rooms, 2 story. Pet accepted. Check-out 11 am. Game room. Indoor pool, whirlpool. ¢

Terre Haute

★★ **THE FARRINGTON BED & BREAKFAST.** *931 S 7th St, Terre Haute (47807). Phone 812/238-0524.* 5 rooms, 3 story. Complimentary full breakfast. Located in the historical district of Terre Haute and built in 1898. $

★★ **HOLIDAY INN.** *3300 US 41 S, Terre Haute (47802). Phone 812/232-6081; toll-free 800/465-4329. www.holiday-inn.com.* 230 rooms, 2-5 story. Pet accepted. Check-out noon. TV; cable (premium). In-room modem link. Guest laundry. Restaurant, bar. Room service. In-house fitness room. Indoor pool, whirlpool. ¢

PLACES TO EAT

A long day of driving is sure to make you hungry. At the end of your journey, take a table at one of the following restaurants.

Centerville

★★ **PALAIS ROYAL CAFÉ.** *822 E Main St, Centerville (47330). Phone 765/939-1199. www.palaisroyalcafe.com.* American menu. Closed Sun-Wed. Dinner. Elegant dining in a turn-of-the-century farmhouse. $$

Greencastle

★★★ **A DIFFERENT DRUMMER.** *2 W Seminary St, Greencastle (46135). Phone 765/653-2761. www.waldeninn.com.* Inside the Walden Inn, diners will find conservative fare such as chicken cordon bleu, roasted eggplant with spinach, lamb chops, and veal medallions. Closed Dec 25. Breakfast, lunch, dinner. Bar. $$

Greenfield

★ **CARNEGIE'S.** *100 W North St, Greenfield (46140). Phone 317/462-8480.* American menu. Closed Sun, Mon. Lunch, dinner. This chef-owned restaurant features regional cuisine and is located in the town's first library. $

Indianapolis

★★ **THE MAJESTIC.** *47 S Pennsylvania, Indianapolis (46204). Phone 317/636-5418. www.majesticrestaurant.com.* Continental menu. Closed Sun; some major holidays. Lunch, dinner. Bar. Children's menu. $$$
D

★★★ **RESTAURANT AT THE CANTERBURY.** *123 S Illinois, Indianapolis (46225). Phone 317/634-3000. www.canterburyhotel.com.* Decorated more like an English club than a restaurant, this elegant, tranquil hotel dining room serves American continental cuisine, focusing on game dishes. Lunch prices can be very reasonable. Don't miss the afternoon tea service with live piano music. Continental menu. Breakfast, lunch, dinner, Sun brunch. Bar. Jacket required. Valet parking. $$$
D

★★ **ST. ELMO STEAK HOUSE.** *127 S Illinois, Indianapolis (46225). Phone 317/637-1811. www.stelmos.com.* Turn-of-the-century décor and old photographs lend a historic feel to this longstanding and popular spot. Steak menu. Closed major holidays. Dinner. Bar. $$$
D

Plainfield

★ **PLAINFIELD DINER.** *3122 E Main St, Plainfield (46168). Phone 317/839-9464.* Diner menu. Breakfast, lunch, dinner. One of the oldest historic roadside diners in Indiana. $

Richmond

★★ **OLDE RICHMOND INN.** *138 S 5th St, Richmond (47374). Phone 765/962-2247.* Continental menu. Closed Jan 1, Labor Day, Dec 25. Lunch, dinner. Bar. Children's menu. Restored mansion built in 1892. Outdoor seating. $$$
D

★ **TASTE OF THE TOWN.** *1616 E Main St, Richmond (47374). Phone 765/935-5464.* Italian, American menu. Closed Sun; some major holidays. Lunch, dinner. Bar. Children's menu. Casual attire. $$
D

Terre Haute

★★ **FROG'S BISTRO AND WINE SHOPPE.** *810 Wabash Ave, Terre Haute (47802). Phone 812/478-9663.* American menu. Lunch, dinner. This New Orleans-style bistro has an emphasis on wine and jazz. $$

Ohio River Scenic Byway

✳ INDIANA

Part of a multistate Byway; see also IL, OH.

Quick Facts

LENGTH: 303 miles.

TIME TO ALLOW: 2 days.

BEST TIME TO DRIVE: This route is a beautiful drive in any season. High season is July through October.

BYWAY TRAVEL INFORMATION: Chamber of Commerce: 812/838-3639; Historic Southern Indiana: 800/489-4474; Byway local Web site: www.ohioriverscenicroute.org.

SPECIAL CONSIDERATIONS: Some areas along this Byway are prone to flooding during the fall and spring, causing occasional closures. The road can become slippery in the winter or when it rains. The road is also narrow in some spots, and sometimes it is winding and hilly.

BICYCLE/PEDESTRIAN FACILITIES: This Byway has narrow shoulders and winding roads, so only off-road biking and hiking are recommended. The new bicycle trail along the Byway includes Perry and Crawford counties. There are also many hiking and walking trails near the Byway in the Angel Mounds State Historic Site, the Hoosier National Forest, Clifty Falls State Park, and along the Ohio River. Just off the Byway, cyclists and hikers will find the American Discovery Trail, a nationally designated trail that will reach from the east coast to the west coast when it is completed.

This winding, hilly route follows the Ohio River, which has had a tremendous impact on this area's history. The route offers a pleasant escape from suburban concerns as it passes villages, well-kept barns, vineyards, and orchards. Historic architecture along the way retains a charm that is often missing from modern development.

Here, tucked away in the very toe of southwestern Indiana, are swamps full of water lilies and rare birds. The most rugged part of this Byway features rock outcroppings, forested hills, caves, and scenic waterways. The limestone bluffs (dotted with cave entrances) are abundant with wildlife.

THE BYWAY STORY

The Ohio River Scenic Byway tells archaeological, historical, natural, recreational, and scenic stories that make it a unique and treasured Byway.

Archaeological

Angel Mounds State Park is located on the banks of the Ohio River near Evansville. It is one of the best-preserved Native American sites in North America, where an advanced culture lived between AD 900 and 1600. These people were named Mississippian by archaeologists. The town served as an important center for religion, politics, and trade. Noted archaeologist Glenn A. Black directed excavation of the site from 1938 until his death in 1964. At the park is an interpretive center, where artifacts are displayed and explained.

Indiana

✻ *Ohio River Scenic Byway*

Historical

The 981-mile Ohio River goes through six states. In the early days, the river was used by those living along its shores to transport their goods, and the Ohio River was the primary way west for early settlers of the frontier. Later, with the coming of the steamboat, the Ohio River became the center of the transportation and industrial revolution. Prior to the Civil War, the river had great significance as the boundary between slave and free states. Today, the river is still used to transport coal to generating plants, but mostly it is used for recreation.

Newburgh, once a large commercial port between Cincinnati and New Orleans, is now a historic district. This quaint town overlooks the Ohio River and offers a multitude of unique shopping and dining opportunities in its downtown district.

Abraham Lincoln's family built a farmstead along Little Pigeon Creek, not far from the Ohio River. At the Lincoln Boyhood National Memorial, you can see and help with the daily chores the Lincoln family performed on the Indiana frontier. Log farm buildings are staffed during the summer months by costumed interpreters who assist those delving into history to make butter, split wood, or break flax.

Corydon was the place where the Indiana State Constitution was drafted in 1816. The Corydon Capitol State Historic Site preserves the state's first capitol building, constructed of Indiana limestone.

The Culbertson Mansion State Historic Site in New Albany preserves a 22-room French Second Empire home built in 1869. It was built by one of the merchants whose wealth was derived from this location.

The Howard Steamboat Museum in Jeffersonville is housed in the mansion of the founder of the largest inland shipyard in the United States. It depicts the fascinating history of riverboats and their construction.

Madison, a prosperous town in the 19th century, is now a venue for historic homes. The Lanier Mansion State Historic Site found in this town is a Greek Revival home designed by architect Francis Costigan, who also designed other historic homes in Madison. While in Madison, stop by the Early American Trades Museum to view demonstrations on wheel wrighting, carpentry, blacksmithing, and other trades common to the 19th century.

Vevay, originally settled in 1802 by the Swiss, transformed the area into the first commercial vineyards and winery in the United States. Also, Rising Sun holds the oldest continuously operating courthouse in Indiana. A trolley runs through the town, offering a tour of all the historic sites.

The Levi Coffin House in Fountain City was the place where over 2,000 freedom seekers found refuge. Levi Coffin, known to many as the President of the Underground Railroad, opened the doors to his Newport home to offer food, shelter, and clothing to runaway slaves on their journey to freedom.

Natural

As a traveler of this route, you will enjoy agricultural countryside dotted with well-kept barns, vineyards, and orchards. Vistas of rural villages dominated by church spires and historic courthouses span the Byway, and thriving cities with imposing architecture can be seen as well. Tucked away in southwestern Indiana, you can find cypress swamps, water lilies, and rare birds.

One of the natural features that is found along the Ohio River Scenic Byway is the Hovey Lake State Fish and Wildlife Area. This area is a 4,300-acre wetland. Adjoining the lake is the Twin Swamps Nature Preserve, the highest-quality cypress swamp in Indiana.

In Evansville, nature can be found even in the middle of the city at the Wesselman Woods Nature Preserve, a 200-acre stand of virgin timber. Hoosier National Forest offers 80,000 acres of forest, along with four lakes, scenic drives, river overlooks, and Ohio River access sites. You'll find plenty of opportunities for camping, fishing, hiking, swimming, horseback riding, or just enjoying the shade and scenery.

The Harrison-Crawford State Forest includes Wyandotte Woods, with its breathtaking natural escarpments overlooking the Ohio River, and Wyandotte Caves. Here, you can tour the caverns used by past settlers for chert mining.

The Needmore Buffalo Farm is home to a sizeable North American bison herd. Visitors can discover the role of the buffalo in southern Indiana and purchase buffalo meat and craft items.

Recreational

The Hoosier National Forest offers plenty of opportunities for camping, fishing, hiking, swimming, and horseback riding. The Falls of the Ohio in Clarksville (phone 812/280-9970) offers many nature hikes where you can look at fossil beds and various aquatic habitats. There are also picnic areas and a museum.

Scenic

The beauty of the Ohio River Scenic Byway is unmatched, especially in the fall when the forests surrounding the Byway change to red and golden tones. These beautiful colors are reflected in the blue serenity of the Ohio River. You can also take a side trip off the Byway that leads to breathtaking views and opportunities for exciting sightseeing in the national and state forests surrounding the Byway.

Also, you don't want to overlook the charm of the Indiana towns with their historic districts that stand proud with regal Victorian homes.

HIGHLIGHTS

Heading east to west on the Byway, consider using the following itinerary. If you're traveling in the other direction, simply read this list from the bottom and work your way up.

- **Hillforest Mansion:** Begin at Hillforest Mansion in Aurora and sightsee. Travel along Highway 56 and continue on Highway 156 to Madison. This stretch is approximately 60 miles, but it's on very curvy roads and will take longer than you may expect.

- **J. F. D. Lanier State Historic Site:** The J. F. D. Lanier State Historic Site is located in Madison, and Clifty Falls State Park, with its many trails (many of them rugged), is located just outside the town on Highway 56. The town of Madison also has several historic sites open to the public as well as a wonderful Main Street.

- **Howard Steamboat Museum and The Falls of the Ohio State Park:** Follow Highway 56 out of Madison and head south on Highway 62, just past the town of Hanover. Highway 62 will take you into Jeffersonville, Clarksville, and New Albany. These are also known as the Falls Cities and are directly across the Ohio River from Louisville, Kentucky. The Howard Steamboat Museum and The Falls of the Ohio State Park are located here.

- **Corydon Capitol State Historic Site:** Continue on Highway 62 out of New Albany to Corydon. This is where the Corydon Capitol State Historic Site is located— Indiana's first state capitol. Other historic sites and a Civil War battle memorial are also located in Corydon. Corydon is approximately 20 miles from New Albany.

- **Wyandotte Caves and Woods:** About 10 miles west of Corydon on Highway 62, you come to the Wyandotte Caves and Wyandotte Woods.

55

Indiana

✹ *Ohio River Scenic Byway*

A variety of cave tours are available, and Wyandotte Woods offers hiking, camping, and picnic areas. Just past the town of Leavenworth, about 8 miles from the Wyandotte area, take Highway 66. Around this area, you'll enter the Hoosier National Forest, where you will find a variety of activities.

- **Lincoln Boyhood National Memorial:** Stay on Highway 66 through the Ohio River towns of Cannelton, site of the Cannelton Cotton Mill (a National Register site), Tell City, and Grandview. As a side trip, take Highway 231 north just past Grandview to Lincoln Boyhood National Memorial. This national park celebrates the life of Abraham Lincoln. He lived in a cabin at this site from the ages of 7 to 21.
- **Angel Mounds Historic Site:** Back at Grandview, stay on Highway 66 to Newburgh. Angel Mounds State Historic Site is between Newburgh and Evansville.
- **The Reitz Home Museum:** The Reitz Home Museum is located in the historic downtown Riverside District of Evansville. This is just a short distance from the Ohio River. Take Highway 62 west out of Evansville. It's about 25 miles to the Illinois border and the end of the Indiana portion of the Ohio River Scenic Route.

THINGS TO SEE AND DO

Driving along the Ohio River Scenic Byway will certainly keep your senses engaged, but if you yearn to get out of the car and stretch your legs, or if you'd like to make a mini-vacation out of your trip, check out these attractions along the route.

✪ **ANGEL MOUNDS STATE HISTORIC SITE.** *8215 Pollack Ave, Evansville (47715). Phone 812/853-3956. www.angelmounds.org.* Largest and best-preserved group of prehistoric mounds (AD 1100-1450) in Indiana. Approximately 100 acres. Interpretive center has a film, exhibits, and artifacts; reconstructed dwellings on grounds. Open mid-Mar-Dec, Tues-Sat 9 am-5 pm; Sun 1-5 pm; closed holidays. **FREE**

BLUE RIVER CANOE TRIPS. CAVE COUNTRY CANOES. *Main St, Milltown (47145). Phone 812/365-2705. www.cavecountrycanoes.com.* Canoe trips (7 to 58 miles); some include camping (2 to 4 days). Open Apr-Oct. **$$$$**

CARNEGIE CENTER FOR ART AND HISTORY. *201 E Spring St, New Albany (47510). Phone 812/944-7336. www.carnegiecenter.org.* History and heritage of the area; changing exhibits. Hand-carved animated diorama. Art gallery has works by local and regional artists; lectures, demonstrations, workshops. Open Tues-Sat 10 am-5:30 pm; closed holidays. **FREE**

CLIFTY FALLS STATE PARK. *2221 Clifty Dr, Madison (47250). Phone 812/265-1331 or 812/273-8885. www.state.in.us/dnr/parklake/parks/cliftyfalls.html.* From a high, wooded plateau, this 1,360-acre park offers a view of the Ohio River and its traffic, as well as hills on the Kentucky shore. It also contains waterfalls of Clifty Creek and Little Clifty Creek, bedrock exposures, numerous fossil beds, and a deep boulder-strewn canyon reached by the sun at high noon only; variety of wildlife, regional winter vulture roost. Swimming pool (Memorial Day-Labor Day; fee); tennis, picnicking (shelters, fireplaces), playground, concession. The inn in the park has lodgings all year (for reservations, call 812/265-4135). Primitive and developed camping (fee). Naturalist service; nature center. **$$**

CORYDON CAPITOL STATE HISTORIC SITE. *202 E Walnut St, Corydon (47112). Phone 812/738-4890. www.indianahistory.org/heritage/corydon.html.* Corydon was the seat of the Indiana Territorial government (1813-1816) when the first constitutional convention assembled here. Following Indiana's admission to the Union in 1816, this building was the state capitol, housing the first sessions of the state legislature and supreme court, until 1825. Construction of the blue limestone building started in 1814 and was completed in 1816. Nearby is Governor Hendricks' headquarters, home of Indiana's second governor; restored.

Open Tues-Sat 9 am-5 pm, Sun 1-5 pm; closed Jan 1, Thanksgiving, Dec 25. **FREE**

EVANSVILLE MUSEUM OF ARTS, HISTORY, AND SCIENCE. *411 SE Riverside Dr, Evansville (47713). Phone 812/425-2406. www.emuseum.org.* Permanent art, history, and science exhibits; sculpture garden, Koch Planetarium (fee). Rivertown USA, re-creation of a turn-of-the-century village. Tours. Open Tues-Sat 10 am-5 pm, Sun noon-5 pm; closed Mon, holidays. **FREE**

HOLIDAY WORLD & SPLASHIN' SAFARI. *452 E Christmas Blvd, Santa Claus (47579). Phone toll-free 877/GO-FAMILY or 812/937-4401. www.holidayworld.com.* Holiday World theme park includes more than 60 rides, games, shows, exhibits, attractions, and water rides themed around Christmas, July 4, and Halloween. Live music, high-dive shows, Lincoln-era exhibit, wax museum, antique toy and doll museums, craftspersons at work, petting zoo, and Santa himself. Sidewalk and indoor restaurants. Open mid-May-Aug, daily; early May and Sept-early Oct, weekends. **$$$$**

LEVI COFFIN HOUSE STATE HISTORIC SITE. *113 US 27 N, Fountain City (47341). Phone 765/847-2432.* Federal-style brick home of the Quaker abolitionist who helped 2,000 fugitive slaves escape to Canada; period furnishings. Tours. Open June-late Aug, Tues-Sat 1-4 pm; Sept-late Oct, Sat 1-4 pm; closed July 4. **$**

LINCOLN BOYHOOD NATIONAL MEMORIAL & LINCOLN STATE PARK. *Hwy 162, Lincoln City (47579). Phone 812/937-4541. www.nps.gov/libo.* Lincoln spent his boyhood years (1816-1830) in this area, reading books, clerking at James Gentry's store, and helping his father with farm work. When Lincoln was 21, his family moved to Illinois, where his political career began. The 200-acre wooded and landscaped park includes the grave of Nancy Hanks Lincoln, mother of Abraham Lincoln. She was 35 years old and Abraham was 9 when she died on October 5, 1818.

The Memorial Visitor Center has information available on the park, including the Cabin Site Memorial, the park's 2 miles of walking trails, and the gravesite. A film is shown at the visitor center every hour depicting Lincoln's Indiana years. Nearby, on the original Thomas Lincoln tract, is the Lincoln Living Historical Farm, with a furnished log cabin similar to the one the Lincolns lived in, plus log buildings, animals, and crops of a pioneer farm. Costumed pioneers carry out family living and farming activities typical of an early 19th-century farm. Farm (open May-Sept). Park open Dec-Feb, 8 am-4:30 pm; Mar-Nov, 8 am-5 pm; closed Jan 1, Thanksgiving, Dec 25. Per family **$**

The **Lincoln Amphitheatre** has an outdoor musical/theatrical production about the life of Lincoln when he lived in Indiana between the ages of 7 and 21. Drama is in a covered amphitheater near the site of Lincoln's home. Open mid-June-mid-Aug, Tues-Sun. Phone toll-free 800/264-4-ABE. **$$**

The park also includes approximately 1,700 acres with a 58-acre lake. Swimming, bathhouse, fishing, boating (no motors; rentals); hiking trails, picnic areas, concessions, primitive and improved camping, cabins, group camp. Naturalist service (June-Aug).

MESKER PARK ZOO. *2421 Bement Ave, Evansville (47720). Phone 812/435-6143. www.meskerparkzoo.com.* The zoo features more than 700 animals, a bird collection, and a children's petting zoo. Also visit the Discovery Center Education Building. Tour train and paddleboats (Apr-Oct). Open daily 9 am-5 pm. **$$**

WESSELMAN PARK. *551 N Boeke Rd, Evansville (47711). Phone 812/479-0771. www.wesselman.evansville.net.* Approximately 400 acres; picnicking, tennis, handball, softball, basketball, bike trails, jogging trail, playground, 18-hole golf course (fee). Half of the park is devoted to a nature preserve. Hartkey swimming

Indiana

✽ *Ohio River Scenic Byway*

pool and Swonder ice rink (fees) adjacent. Tues-Sun 8 am-4 pm; closed Mon, major holidays. **FREE**

✣ **WYANDOTTE CAVES.** 7315 S Wyandotte Cave Rd, Leavenworth (47137). Phone 812/738-2782. www.wyandottecaves.com. Approximately 7 miles of mapped passages. Features include Garden of Helictites, a large collection of gravity-defying formations; Rothrock's Cathedral, an underground mountain 105 feet high, 140 feet wide, and 360 feet long; and Pillar of the Constitution, a stalagmite approximately 35 feet high and 71 feet in circumference. The cave was used by prehistoric Native Americans for mining aragonite and is known to have been the source of saltpeter and Epsom salts around 1812. Jacket recommended, cave temperature 52 degrees. One-hour guided tours (Memorial Day-Labor Day, daily). Two-hour guided tours (Memorial Day-Labor Day, daily; rest of year, Tues-Sun; closed holidays). Two-, three-, five-, and eight-hour tours (Sat and Sun, by reservations only). Mar-late May 9 am-5 pm; late May-early Sept 9 am-6 pm; early Sept-late Oct 9 am-5 pm. **$$$$**

PLACES TO STAY

If you choose to include an overnight stay in your trip along this Byway, Mobil Travel Guide recommends the following lodgings.

Corydon

★ **BEST WESTERN OLD CAPITOL INN.** I-64 and IN 135, Corydon (47112). Phone 812/738-4192; toll-free 800/780-7234. www.bestwestern.com. 77 rooms, 2 story. Check-out noon. TV. Pool. ¢

★★ **KINTNER HOUSE INN.** 101 S Capitol Ave, Corydon (47112). Phone 812/738-2020. 15 rooms, 3 story. Complimentary full breakfast. Check-out 11 am, check-in 1 pm. TV; cable (premium); VCR available. Brick Victorian house (1873); antique furnishings. Totally nonsmoking. **$**

Evansville

★ **DRURY INN.** 3901 US 41 N, Evansville (47711). Phone 812/423-5818; toll-free 800/DRURY-INN. www.druryinn.com. 151 rooms, 4 story. Pet accepted, some restrictions. Complimentary continental breakfast. Check-out noon. TV. In-room modem link. Laundry services. In-house fitness room. Indoor pool, whirlpool. ¢

★ **FAIRFIELD INN BY MARRIOTT.** 5400 Weston Rd, Evansville (47712). Phone 812/429-0900; toll-free 800/228-2800. www.fairfieldinn.com. 110 rooms, 4 story. Complimentary continental breakfast. Check-out noon. TV; cable (premium). In-room modem link. In-house fitness room. Indoor pool. ¢

★★ **HOLIDAY INN.** 4101 US 41 N, Evansville (47711). Phone 812/424-6400; toll-free 800/465-4329. www.holiday-inn.com. 198 rooms, 1-2 story. Check-in 3 pm, check-out 11 am. TV; cable (premium), VCR available. In-room modem link. Guest laundry. Restaurant, bar. Room service. In-house fitness room, sauna. Game room. Indoor pool, children's pool, whirlpool. Free airport transportation. Business center. ¢

★★★ **MARRIOTT.** 7101 Hwy 41 N, Evansville (47725). Phone 812/867-7999; toll-free 800/228-9290. www.marriott.com. A beautiful tropical, glass-enclosed atrium lobby is the setting for guests as they enter this hotel conveniently located at the airport. 199 rooms, 5 story. Check-in 3 pm, check-out noon. TV; cable (premium). Restaurant, bar. In-house fitness room. Game room. Indoor pool, whirlpool. Free airport transportation. Concierge. **$$**

58

★ **SIGNATURE INN.** *1101 N Green River Rd, Evansville (47715). Phone 812/476-9626; toll-free 800/822-5252. www.signatureinn.com.* 125 rooms, 2 story. Complimentary continental breakfast. Check-out noon. TV; cable (premium). In-room modem link. Pool. ¢
D SC

PLACES TO EAT

A long day of driving is sure to make you hungry. At the end of your journey, take a table at one of the following restaurants.

★ **MAGDALENA'S.** *103 E Chestnut St, Corydon (47112). Phone 812/738-8075.* Closed some major holidays. Lunch, dinner. Casual attire. $
D

★ **THE OLD MILL RESTAURANT.** *5031 New Harmony Rd, Evansville (47720). Phone 812/963-6000.* American menu. Lunch, dinner. A local favorite that hosts the annual Germanfest. $

★ **WHISKY'S.** *334 Front St, Lawrenceburg (47025). Phone 812/537-4239.* Closed Sun; Easter, July 4, Dec 25. Lunch, dinner. Bar. Children's menu. Dining in two restored buildings (circa 1850 and 1835) joined together. $$
D

★★ **KEY WEST SHRIMP HOUSE.** *117 Ferry St (IN 56), Madison (47250). Phone 812/265-2831. www.keywestshrimphouse.com.* Closed Mon. Lunch, dinner. Children's menu. Century-old building; fireplace. $$
D

★ **BEARNO'S NEW ALBANY.** *3002 Charlestown Crossing, New Albany (47510). Phone 812/949-7914.* Italian menu. Lunch, dinner. $$
D

★ **PIGASUS.** *223 W 5th St, New Albany (47150). Phone 812/941-1349.* Barbecue menu. Closed Sun. Lunch, dinner. Bar. Casual attire. Outdoor seating. $
D

The Great River Road
❋ IOWA

Part of a multistate Byway; see also IL, MN, WI.

Quick Facts

LENGTH: 326 miles.

TIME TO ALLOW: 2 days.

BEST TIME TO DRIVE: Any time of the year brings dynamic and beautiful views of the river and the surrounding landscape. Fall is a particularly popular time with its beautiful foliage displays.

BYWAY TRAVEL INFORMATION: Iowa Department of Tourism: 515/242-4705; Byway local Web site: www.mississippi-river.com/mrpc/fiaframe.htm.

RESTRICTIONS: Short portions of the road may be closed once every few years due to winter snowstorms. Such interruptions usually last less than 36 hours.

BICYCLE/PEDESTRIAN FACILITIES: In Iowa, bicyclists enjoy the same rights as motorists and are not prohibited from using any portion of the Iowa Great River Road. The majority of the roadway includes improved shoulders that are available for use by bicyclists. Numerous separated and multi-use recreation trails intersect and parallel the Great River Road as well. Frequent roadside stops provide parking areas, trails, walkways, public rest rooms, and other service amenities.

"The Mississippi is in all ways remarkable," said the famous writer Mark Twain. Join travelers from around the world to discover dramatic vistas of Old Man River during all seasons. View soaring eagles and 100,000 migrating geese and ducks. Experience Midwest hospitality on the main streets of river towns and cities or visit sacred sites and landscape effigies of Native Americans. You can also experience the Mississippi River on steamboats, commercial barges, and recreational crafts. This 326-mile Byway provides a look into America's past, present, and future.

The Byway's story begins with the landscape: abrupt and dramatic limestone bluffs cut by glacial meltwater in the north contrast with broad sandy floodplains in the south. For thousands of years, Native Americans knew the importance of the continent's largest river. Later, its meandering course marked the political boundaries of territories, towns, cities, states, and counties of the advancing society. Today, the Upper Mississippi River and the Great River Road are national repositories of geological wonders, unparalleled scenic beauty, wildlife, native vegetation, and the miracles of hydrology. The river and road are also milestones to the expansion and development of the United States and the Midwest.

THE BYWAY STORY

The Great River Road tells archaeological, historical, natural, recreational, and scenic stories that make it a unique and treasured Byway.

Iowa

The Great River Road

Archaeological

Regionally and nationally significant archaeological resources of the Iowa Great River Road are numerous and continue to be the focus of research and interpretation. A primary site on the Great River Road is the Effigy Mounds National Monument, the site of 200 mounds. Of these 200 mounds, 29 are effigy outlines of mammals, birds, or reptiles. Eastern Woodland Indian culture built these sites between 500 BC and AD 1300. They are preserved and interpreted for the public. Additional property is being added to the monument site to expand its protection of these unique resources. Other important sites are found at the Mines of Spain State Recreation Area and the Toolesboro Indian Mounds National Historic Landmark.

Historical

The history of the Great River Road is found in its buildings and towns. People who live along the river can't think of a better combination than historic buildings set in the land of a great river. Ideas and dreams of a moving, bustling riverboat society surface over and over in restored cathedrals and forts or in riverboat casinos.

Although most of Iowa was settled in the mid-1800s, the Mississippi River made it an accessible territory long before the United States became a nation. Natives and explorers alike saw the raw, unharnessed power of this beautiful passageway. One of the first outposts was Fort Madison, where history is still re-created today. Like much of the United States, the Great River Road was under Spanish and French rule until the time of the Louisiana Purchase in 1802.

Soon after, settlers and industry came to Iowa. Bustling river towns created cultural landmarks like Snake Alley—the curviest road in the United States. Little towns along the road grew and became the cities they are today, full of identity as Mississippi River towns. During the booming days of river trade, writer Samuel Langhorn Clemens (better known as Mark Twain) captured the atmosphere and the time period in his novels.

When the Civil War came to Iowa, most of the people living there fought for the Union. Civil War memorials are now found in several of the towns on the Byway.

Natural

Geology, the hydrologic cycle, and erosion are among the big stories that the Mississippi River and the Great River Road tell in Iowa. The forces of nature can be seen in how the river has cut a deep channel in ancient limestone layers in the northern reaches. The ever-changing channel of the river, the deposition of sediments, and the broad floodplain of the Mississippi River in the southern part of the state speak of a different natural dynamic. The Upper Mississippi River National Wildlife and Fish Refuge is the nation's oldest, longest, and most popular wildlife refuge. Many other state, county, and city parks provide opportunities for spotting and watching wildlife.

Recreational

Nationally and regionally important recreational opportunities abound along the Iowa Great River Road. Abundant water recreation activities include boating, sailing, fishing, waterfowl hunting, and swimming. For decades, the Iowa Great River Road and its side roads have been popular pleasure routes for sightseeing. Although fall is a particularly popular time due to the beautiful foliage displays, any time of the year brings dynamic and beautiful views of the river and its attendant landscape. Numerous multipurpose trails and support facilities are available along the road as well.

Scenic

The magnificent scenery of the Great River Road is centered around the Mississippi River, the landscape it has created, and the cultural expressions that are rooted in the landscape. The river is almost continuously visible from the Great River Road (or is within a few miles of the Byway). Dams along the river create large pools of open water upstream. Along the northern part of the river, steep limestone bluffs descend directly to the bank of the river. Downstream, the floodplain opens to afford long, uninterrupted views of the valley. Collections of stops allow travelers to discover the mix of scenic qualities. Roadside spots, shady parks, and locks and dams of the Mississippi River all offer places for you to stop and take in the scenic beauty of the Great River Road and the Mississippi River.

The four seasons of the upper Midwest provide continually dynamic backgrounds and changes in the vegetation and activity on the water. Culturally, the rural landscape provides a multitude of settings for small farms, protected wetlands, streams and rivers, and intermittent woodlots and forests. The residential and main street architecture of small towns and river cities offers much interest and contrast to the rural images. Many efforts exist to protect the countryside landscape character.

HIGHLIGHTS

Not sure where to begin? Consider taking this Lansing to Guttenberg tour of Iowa's Great River Road.

- The tour begins in **Lansing,** home of Mount Hosmer Park with a panoramic view of the river. Also of interest is the **Fish Farm Mound** (an Indian burial site) and the nearby **Our Lady of the Wayside** shrine.
- The next stop, **Harper's Ferry,** is 15 miles past Lansing. The town is built on a concentrated area of Native American mounds and was an important river town after the introduction of the steamboat. The Mississippi backwaters behind the town still attract hunters, trappers, and commercial fishing.
- Just south of Harper's Ferry lies the **Yellow River Forest State Recreation Area.** This 8,000-acre forest contains some of Iowa's greatest terrain, with high scenic bluffs and cold streams. The Iowa Department of Natural Resources harvests the Yellow River Forest timber for use all over the state. The Paint Rock unit of the forest houses most recreational opportunities, including camping, canoeing, snowmobiling, hunting and fishing, and hiking and equestrian trails.
- **Effigy Mounds National Monument,** the next stop, is just 2 miles south of Yellow River Forest. Prehistoric mounds are common from the plains of the Midwest to the Atlantic seaboard, but only in this area were they constructed in an effigy outline of mammals, birds, or reptiles. The monument contains 1,481 acres with 200 mounds, of which 29 are effigies; the others are conical, linear, and compound. Eastern Woodland Indian cultures built these mounds from about 500 BC to AD 1300. Natural features in the monument include forests, tallgrass prairies, wetlands, and rivers.

see page A5 for color map

- The Effigy Mounds National Monument Visitor Center, located in **Marquette,** includes displays of local Woodland and Mississippian cultures, artifacts, and a herbarium. Riverboat casino gambling is available on the *Miss Marquette* Riverboat Casino.
- **Pike's Peak State Park** is 5 miles south of Marquette. This park boasts one of Iowa's most spectacular views across the Mississippi on the highest bluff along the river. It was named for Zebulon Pike, who was sent in 1805 to scout placement of military posts along the river. A fort was never built on this

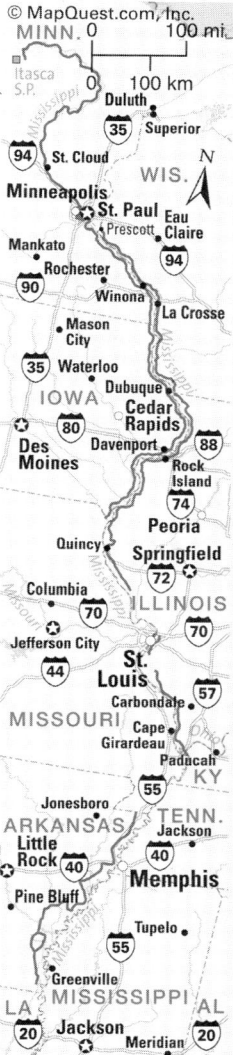

Iowa

The Great River Road

land, and it went into private ownership. Because settlers were not able to build on this property, the peak remains as Pike saw it 200 years ago.

- The tour terminates in **Guttenberg,** 15 miles south of Pike's Peak State Park. Guttenberg boasts two scenic overlooks and a mile-long landscaped park along the river. A copy of the Gutenberg Bible is on display at the local newspaper. The city offers blocks and blocks of historic buildings.

THINGS TO SEE AND DO

Driving along the Great River Road will certainly keep your senses engaged, but if you yearn to get out of the car and stretch your legs, or if you'd like to make a mini-vacation out of your trip, check out these attractions along the route.

THE APPLE TREES HISTORICAL MUSEUM. *1616 Dill St, Burlington (52601). Phone 319/753-2449.* The museum, a remaining wing of railroad magnate Charles E. Perkins's mansion, contains Victorian furnishings; antique tools, costumes, dolls, toys, buttons, glass, and china; Native American artifacts; and changing exhibits. Maintained by the Des Moines County Historical Society. Guided tours (May-Oct, Sat, Sun 1:30-4:30 pm; by appointment Mon-Fri; fee).

BELLEVUE STATE PARK. *24668 Hwy 52, Dubuque (52031). Phone 563/872-3243 or 563/872-4019. www.state.ia.us/dnr/organiza/ppd/bellevue.htm.* Approximately 540 acres on a high bluff; river view. Native American mounds, rugged woodlands. Hiking trails, snowmobiling, picnicking, camping (electricity, dump station). Nature center.

BUFFALO BILL CODY HOMESTEAD. *28050 230th Ave, Princeton (52768). Phone 563/225-2981. www.scottcountyiowa.com/conservation/ buffalobill.html.* Restored boyhood home of Buffalo Bill Cody, built by his father in 1847. Live buffalo are on the grounds. Open Apr-Oct, daily 9 am-5 pm. $

CATHEDRAL SQUARE. *2nd and Bluff sts, Dubuque (52001). Phone 563/582-7646.* Surrounding the square are stylized figures of a lead miner, farmer, farmer's wife, priest, and river hand; opposite the square is the architecturally and historically significant St. Raphael's Cathedral.

CRAPO AND DANKWARDT PARKS. *2700 and 2900 S Main sts, Burlington (52601). Phone 319/753-8110 or -8117.* The parks (approximately 175 acres) are situated along the Mississippi on the site where the American flag first flew over Iowa soil (1805); includes an illuminated fountain, arboretum, and formal flower garden. Swimming. Black Hawk Spring Indian trail; tennis, archery range, ice skating. Picnicking; playground. Open daily. **FREE**

CRYSTAL LAKE CAVE. *7699 Crystal Lake Cave Dr (Hwy 52), Dubuque (52003). Phone 563/556-6451 or 563/872-4111. www.crystallakecave.com.* Network of passageways carved by underground streams; surrounding lake with glittering stalactites and stalagmites. Guided tours. Open Memorial Day-late Oct, daily; May, Sat and Sun. $$$

DAN NAGLE WALNUT GROVE PIONEER VILLAGE. *14910 110th Ave, Davenport (52804). Phone 563/285-9903.* Three-acre walk-through site contains 18 historic buildings moved from various locations in the county. Visitors can explore a blacksmith shop, schoolhouse, and pioneer family home; also St. Anne's Church. Open April-Oct, daily. **FREE**

DAVENPORT AND MISSISSIPPI RIVER ISLANDS HIKE. *Davenport (52801).* This hike explores the old waterfront of Davenport, as well as two park islands in the Mississippi between Illinois and Iowa. Begin on Credit Island Park, the site of a turn-of-the-century amusement park. Scenic trails loop around the island, which is now a community park with a playground area and a municipal golf course. A number of public art pieces are also found here, part of the Quad Cities' Art in the Park project.

Cross over to the Iowa mainland from the east end of Credit Island Park and walk along the Mississippi through two more riverside parks. Centennial Park features riverside walkways past sports fields and stadiums. Atop the bluff on Division Street is Museum Hill, home of the Putnam Museum of Science and Natural History (1717 West 12th Street) and the Davenport Museum of Art (1737 West 12th Street). The Putnam houses two permanent exhibits about the Mississippi River; the art museum's permanent collection indulges works by Midwestern painters, such as Thomas Hart Benton and Grant Wood, an Iowa native famous for his painting *American Gothic*. Just to the east is LeClaire Park, home to summer outdoor events and concerts.

The Davenport Downtown Levee includes a riverboat casino, restaurants, nightclubs, and the renovated Union Station railroad depot, which houses the Quad Cities Convention and Visitors Center. A local farmers market is also held here on Wednesday and Saturday mornings from May through October.

Just downstream from the historic Government Bridge, Dam 15 provides a navigational pool for commercial shipping on the Mississippi. Lock 15 allows boats to transfer between the river's pools. Cross Government Bridge to Arsenal Island, which was acquired by the US government in 1804 under a treaty with the Sauk and Fox Indians. Fort Armstrong was established in 1816 on the tip of the island, where a replica now stands. Manufacturing began on the island in 1840, and in 1869, it became home to the Rock Island Arsenal, a major military manufacturing facility. The island contains a number of historic homes and structures, including the Rock Island Arsenal Museum; the restored Colonel George Davenport Mansion, filled with furnishings from the mid-1800s; and the Mississippi River Visitors Center, with exhibits about the history of navigation on the river. A Confederate Soldiers' Cemetery and National Military Cemetery date back to the 1800s. Hikers and bikers can enjoy a 5-mile trail around the island.

DIAMOND JO CASINO. *3rd Street, Ice Harbor exit, Dubuque (52001). Phone 563/583-7005 or toll-free 800/LUCKY-JO. www.diamondjo.com.* Casino gambling on the river.

★ EFFIGY MOUNDS NATIONAL MONUMENT. *151 Hwy 76, Harpers Ferry (52146). Phone 563/873-3491. www.nps.gov/efmo/.* Preserves traces of indigenous civilization from 2,500 years ago. Mounds were built in the shapes of animals, birds, and other forms. Area divided by Yellow River; Great Bear Mound is the largest known bear effigy in the state, 70 feet across the shoulders, 137 feet long, and 5 feet high. Footpath leads from headquarters to Fire Point Mound Group and to scenic viewpoints overlooking the Mississippi and Yellow rivers. Guided walks (Memorial Day-Labor Day, daily). Visitor center has a museum and a 15-minute film. Open daily; closed Thanksgiving, Dec 25, Jan 1. **$**

DUBUQUE ARBORETUM AND BOTANICAL GARDENS. *3800 Arboretum Dr, Dubuque (52001). Phone 563/556-2100. www.dubuquearboretum.com.* Features annual and perennial gardens; rose, water, and formal gardens; ornamental trees, woodland, and prairie wildflower walk. Open May-Oct, 7 am-dusk; Nov-May, 9 am-5 pm, Sat 9 am-1 pm; closed Sun (Nov-May), Jan 1, Thanksgiving, Dec 25. **FREE**

Iowa

❋ The Great River Road

DUBUQUE COUNTY COURTHOUSE. *720 Central Ave, Dubuque (52001). Phone 563/589-4445.* This gold-domed courthouse is on the National Historic Register of Places. Open daily; closed holidays. **FREE**

DUBUQUE MUSEUM OF ART/OLD COUNTY JAIL. *Eighth St and Central Ave, Dubuque (52001). Phone 563/557-1851.* The museum is housed in a brand-new facility; the gallery is an example of Egyptian Revival architecture. Open Tues-Fri 10 am-5 pm, Sat and Sun 1-5 pm.

EAGLE POINT PARK. *W on Shiras Ave, off Rhomberg Ave, Dubuque (52732). Phone 563/589-4263. www.dbq.com/parks/eagle.html.* Flower gardens; picnicking (shelters), lodge; playground, observation tower; children's nature center; petting zoo. Open May-late Oct, daily 8 am-dusk; Nov-May 9 am-5 pm, Sat 9 am-1 pm (walk only). **FREE**

FAMILY MUSEUM OF ARTS & SCIENCE. *2900 Learning Campus Dr at 18th St, Bettendorf (52722). Phone 563/344-4106. www.familymuseum.org.* Hands-on exhibits; Rhythm Alley, Heartland, The Homestead, Kinder Garten; also a traveling exhibit gallery and a children's program area. Open daily; closed holidays. **$**

HERITAGE HILL NATIONAL HISTORIC DISTRICT. *Washington and High sts, Burlington (52601). www.visit.burlington.ia.us/history.html.* This 29-square-block area contains churches, mansions, and houses in a wide variety of architectural styles, including a full range of Victorian buildings from the 1870s to the turn of the century. Walking tours, auto cassette tours, and brochures available. Contact the Convention & Tourism Bureau (phone toll-free 800/82-RIVER).

KEOKUK DAM. *523 N Water St, Keokuk (52632). Phone 319/524-4091 or 319/524-9660.* Ameren-Union Electric Power Plant with a mile-long dam (1910-1913) across the Mississippi River to Hamilton, IL. Lock 19 is operated by the Army Corps of Engineers. **FREE**

KEOKUK RIVER MUSEUM. *101 Mississippi Dr, Keokuk (52632). Phone 319/524-4765.* Located in sternwheel towboat *George M. Verity;* houses historical items of the upper Mississippi River valley. Open May-Oct, daily. **$**

LADY LUCK CASINO. *1777 Lady Luck Pkwy, Bettendorf (52722). Phone 563/359-7280; toll-free 800/724-5825.* Casino gambling; restaurant, gift shop, lodging.

LILLIAN RUSSELL THEATRE. *311 Riverview Dr, Clinton (52732). Phone 563/242-6760.* Musicals and comedies aboard the paddlewheel showboat *The City of Clinton.* Open June-Aug. **$$$**

MARK TWAIN OVERLOOK. *Lombard and 2nd sts, Muscatine (52761).* Three acres with a panoramic view of the Mississippi River Valley and downtown Muscatine; picnicking. **FREE**

MISSISSIPPI BELLE II. *311 Riverview Dr, Clinton (52732). Phone 563/243-9000; toll-free 800/457-9975.* Offers year-round casino gambling along the Mississippi River. Entertainment. Concession. Sun-Thurs 9-2 am, Fri-Sat 9-4 am.

MUSCATINE ART CENTER. *1314 Mulberry Ave, Muscatine (52761). Phone 563/263-8282. www.muscatineartcenter.org.* Consists of the Laura Musser Museum and the Stanley Gallery. The museum is housed in an Edwardian mansion; changing art exhibits, special events, Estey player pipe organ with 731 pipes, antiques, and historical displays; Oriental carpets, furniture, paintings, drawings, prints, sculpture, and graphics in the permanent collection. Open Tues-Fri 10 am-5 pm, Thurs evenings 7-9 pm, Sat and Sun 1-5 pm; closed Mon, holidays. **FREE**

NATIONAL CEMETERY. *1701 J St, Keokuk (52632). Phone 563/524-1304. www.cem.va.gov/nchp/keokuk.htm.* This is one of the 12 original national cemeteries designated by the US Congress. Unknown Soldier monument, Civil War graves. Open daily dawn-dusk. **FREE**

PEARL BUTTON MUSEUM. *117 W 2nd St, Muscatine (52761). Phone 563/263-1052. www.pearlbuttoncapital.com.* Dedicated to the pearl button industry. Exhibits on making buttons from Mississippi River mussel shells. Open Tues-Sat noon-4 pm (or by appointment). **FREE**

PIKES PEAK STATE PARK. *15316 Great River Rd, McGregor (52157). Phone 563/873-2341. www.state.ia.us/dnr/organiza/ppd/pikepeak.htm.* Park of 970 acres on bluffs overlooking the Mississippi River. Native American mounds; colored sandstone outcroppings; woods and wildflowers. Trail leads across rugged terrain to Bridal Veil Falls. Hiking. Picnicking. Camping (electricity, dump station). Observation point. Boardwalks.

***RHYTHM CITY* CASINO.** *200 E 3rd St, Davenport (52801). Phone toll-free 800/BOAT-711.* This riverboat casino departs from River Drive, between Centennial and Government bridges.

RIVERVIEW PARK. *6th Ave N and Riverview, Clinton (52732). Phone 563/243-1260.* Swimming pool (Memorial Day-Labor Day); marina, boat ramp; lighted tennis courts, fountain; playground, recreational trail, baseball stadium, picnicking; RV parking (fee). Open daily.

SAMUEL F. MILLER HOUSE AND MUSEUM. *318 N 5th St, Keokuk (52632). Phone 319/524-5599. keokuk-ia.com/tourism/samuelmillerhouse.htm.* Restored home of the US Supreme Court Justice appointed by Abraham Lincoln. Open Memorial Day weekend to Labor Day weekend, Fri-Sun 1-4 pm. **$**

SHADY CREEK RECREATION AREA. *3550 Hwy 22, Muscatine (52761). Phone 563/263-7913.* 16 acres. Boat ramp. Picnicking, playground, shelter. Improved camping (from May-Oct; **$$$**). Park. Open daily.

SPOOK CAVE AND CAMPGROUND. *13299 Spook Cave Rd, McGregor (52157). Phone 563/873-2144.* Guided 35-minute tour of an underground cavern via power boat. Campground has a swimming beach, lake fishing, hiking trails, and picnic areas. Open May-Oct, daily. Camping **$$$**.

SCOTT COUNTY PARK. *270th St, Long Grove (52756). Phone 563/285-9656. www.scottcountyiowa.com/conservation/scottco.html.* More than 1,000 acres with a pioneer village and nature center. Swimming (fee), fishing; ball fields, 18-hole golf, skiing, tobogganing, ice skating, picnicking, camping, and trailer sites (fee). Open daily. **FREE**

WILDCAT DEN STATE PARK. *1884 Wildcat Den Rd, Muscatine (52761). Phone 563/263-4337. www.state.ia.us/dnr/organiza/ppd/wildcat.htm.* A 321-acre park with historic mid-19th-century gristmill, one-room schoolhouse, Pine Creek Bridge; scenic overlook. Hiking trails, working mill, picnicking, and primitive camping. Open daily 4 am-10:30 pm.

PLACES TO STAY

If you choose to include an overnight stay in your trip along this Byway, Mobil Travel Guide recommends the following lodgings.

Bettendorf

★★★ **ABBEY HOTEL.** *1401 Central Ave, Bettendorf (52722). Phone 563/355-0291; res 800/438-7535. www.theabbeyhotel.com.* This breathtaking Romanesque hotel, once a monastery, overlooks the Mississippi River and is surrounded by beautiful gardens and landscaping. With luxurious guest suites and top-rated service, guests can indulge in fine elegance and simple serenity. 19 rooms, 3 story. Complimentary full breakfast. Check-out noon, check-in 2 pm. TV; cable (premium), VCR available. In-room modem link. Bar. Room service. In-house fitness room. Free airport transportation. **$**

Iowa

❋ *The Great River Road*

★★ **JUMERS CASTLE LODGE.** *I-74 at Spruce Hills Dr, Bettendorf (52722). Phone 563/359-7141; toll-free 800/285-8637. www.jumers.com.* At this upscale hotel, guests will find rich tapestries, fine antiques, and grand elegance. Guest rooms have four-poster beds. The award-winning restaurant serves both American and German cuisine. 210 rooms. Pet accepted, some restrictions; fee. Check-out noon. TV; cable (premium). Restaurant, bar; entertainment. In-house fitness room, sauna. Health club privileges. Indoor, outdoor pools; whirlpool. Lawn games. Free airport transportation. $

★ **SIGNATURE INN.** *3020 Utica Ridge Rd, Bettendorf (52722). Phone 563/355-7575; toll-free 800/822-5252. www.signatureinn.com.* 119 rooms, 3 story. Complimentary full breakfast. Check-out noon. TV; cable (premium), VCR available (movies). In-room modem link. In-house fitness room. Health club privileges. Pool. Free airport transportation. Business center. ¢

Burlington

★★ **BEST WESTERN PZAZZ MOTOR INN.** *3001 Winegard Dr, Burlington (52601). Phone 319/753-2223; toll-free 800/780-7234. www.bestwestern.com.* 151 rooms, 3 story. Pet accepted. Check-out noon. TV; cable (premium), VCR available. In-room modem link. Restaurant, bar; entertainment. Room service. In-house fitness room, sauna. Game room. Indoor pool, whirlpool. Airport transportation. ¢

★ **COMFORT INN.** *3051 Kirkwood, Burlington (52601). Phone 319/753-0000; toll-free 877/424-6423. www.comfortinn.com.* 52 rooms, 2 story. Complimentary continental breakfast. Check-out 11 am. Pet accepted, some restrictions; fee. TV; cable (premium). In-room modem link. Pool. ¢

Clinton

★★ **BEST WESTERN FRONTIER MOTOR INN.** *2300 Lincoln Way St, Clinton (52732). Phone 563/242-7112; toll-free 800/780-7234. www.bestwestern.com.* 113 rooms, 1-2 story. Pet accepted; fee. Check-out noon. TV; cable (premium). Restaurant, bar. Room service. In-house fitness room. Indoor pool, whirlpool. Cross-country ski 5 miles. ¢

★ **RAMADA INN.** *1522 Lincoln Way, Clinton (52734). Phone 563/243-8841; toll-free 800/272-6232. www.ramada.com.* 115 rooms, 2 story. Pet accepted, some restrictions; fee. Check-out noon. TV; cable (premium), VCR available. Bar. Room service. Game room. Indoor pool. ¢

Davenport

★★ **BEST WESTERN STEEPLEGATE INN.** *100 W 76th St, Davenport (52806). Phone 563/386-6900; toll-free 800/780-7234. www.bestwestern.com.* 121 rooms, 2 story. Pet accepted; fee. Check-out noon. TV; cable (premium). Restaurant, bar; entertainment. Room service. In-house fitness room. Game room. Indoor pool, whirlpool. Free airport transportation. ¢

★★ **PRESIDENT CASINO'S BLACKHAWK HOTEL.** *200 E 3rd St, Davenport (52801). Phone 563/328-8000.* 189 rooms, 11 story. Check-out noon. TV; cable (premium). Restaurant, bar. In-house fitness room. Free airport transportation. ¢

★ **SUPER 8.** *410 E 65th St, Davenport (52807). Phone 563/388-9810; toll-free 800/800-8000. www.super8.com.* 61 rooms, 2 story. Pet accepted, some restrictions; fee. Complimentary continental breakfast. Check-out 11 am. TV; cable (premium). ¢

Dubuque

★ **DAYS INN.** *1111 Dodge St, Dubuque (52003). Phone 563/583-3297; toll-free 800/544-8313. www.daysinn.com.* 154 rooms, 2 story. Pet accepted, some restrictions; fee. Complimentary continental breakfast. Check-out 11 am. TV; cable (premium), VCR available. Restaurant, bar. In-house fitness room. Pool. Free airport transportation. $

★★ **THE REDSTONE INN.** *504 Bluff St, Dubuque (52001). Phone 563/582-1894. www.theredstoneinn.com.* 14 rooms, 3 story. Complimentary continental breakfast. Check-out 11 am, check-in 3 pm. TV. In-room modem link. Victorian mansion (1894) built by a prominent Dubuque industrialist; fireplaces, antiques. Totally nonsmoking. ¢

Guttenberg

★★★ **THE LANDING.** *703 S River Park Dr, Guttenberg (52052). Phone 563/252-1615. www.thelanding615.com.* 5 rooms, 3 story. This German-built limestone structure was erected in the late 1800s and features great views of the Mississippi River. $$

Lansing

★★★ **SUZANNE'S BED & BREAKFAST.** *120 N 3rd St, Lansing (52151). Phone 563/538-3040.* 4 rooms, 3 story. This beautiful Victorian-era bed-and-breakfast was built in 1865. $

★★★ **THORNTON HOUSE.** *371 Diagonal St, Lansing (52151). Phone 563/538-4878. www.thorntonhouse.net.* 4 rooms, 3 story. This brick mansion has over 3,300 square feet of living space with a two-story enclosed deck. $$$$

Keokuk

★ **HOLIDAY INN EXPRESS.** *4th and Main sts, Keokuk (52632). Phone 319/524-8000; toll-free 800/465-4329. www.holiday-inn.com.* 80 rooms, 5 story. Complimentary continental breakfast. Check-out noon. TV; cable (premium). In-room modem link. In-house fitness room, sauna. Game room. Indoor pool, whirlpool. ¢

McGregor

★ **HOLIDAY SHORES.** *Business Hwy 18, McGregor (52157). Phone 563/873-3449. www.holidayshoresmotel.com.* 33 rooms, 2-3 story. No elevator. Check-out 10:30 am. TV. Game room. Indoor pool, whirlpool. Overlooks the Mississippi River. ¢

Muscatine

★ **ECONO LODGE.** *2402 Park Ave, Muscatine (52761). Phone 563/264-3337; toll-free 877/424-6423. www.econolodge.com.* 91 rooms, 2 story. Complimentary full breakfast. Check-in 4 pm, check-out 11 am. TV; cable (premium). Coin laundry. Restaurant. Indoor pool. ¢

★★ **HOLIDAY INN.** *2915 N Hwy 61, Muscatine (52761). Phone 563/264-5550; toll-free 800/465-4329. www.holiday-inn.com.* 112 rooms, 3 story. Pet accepted; fee. Check-out noon. TV; cable (premium). In-room modem link. Laundry services. Restaurant, bar. Room service. In-house fitness room. Health club privileges. Sauna. Indoor pool, children's pool, whirlpool. Business center. ¢

Iowa

✤ *The Great River Road*

PLACES TO EAT

A long day of driving is sure to make you hungry. At the end of your journey, take a table at one of the following restaurants.

★★★ **JUMER'S.** *I-74 at Spruce Hills Dr, Bettendorf (52722). Phone 319/359-1607. www.jumers.com.* Located inside of Jumers Castle Lodge, this restaurant features a German menu, with specialties such as regensburg goulash, chicken von jumer, and rack of lamb. With a German décor throughout, the restaurant is a true dining experience. American, German menu. Breakfast, lunch, dinner, brunch. Bar. Children's menu. $$
D

★★ **STUBBS EDDY RESTAURANT/PUB.** *1716 State St, Bettendorf (52722). Phone 319/355-0073.* Seafood menu. Closed Sun; holidays. Dinner. Bar. Entertainment Tues, Thurs. Children's menu. $$$
D

★ **IOWA MACHINE SHED.** *7250 NW Blvd, Davenport (52806). Phone 563/391-2427. www.machineshed.com/davenport/.* American menu. Closed Jan 1, Thanksgiving, Dec 25. Breakfast, lunch, dinner. Bar. Children's menu. $$
D

★★ **THUNDER BAY GRILLE.** *6511 Brady St, Davenport (52806). Phone 563/386-2722. www.thunderbaygrille.com/davenport/.* American menu. Closed holidays. Lunch, dinner, Sat-Sun brunch. Bar. Children's menu. Bi-level dining. $$
D

★ **MARIO'S.** *1298 Main St, Dubuque (52001). Phone 563/556-9424.* Italian, American menu. Closed holidays. Lunch, dinner. Bar. $$
D

★ **YEN CHING.** *926 Main St, Dubuque (52001). Phone 563/556-2574.* Mandarin, Hunan menu. Closed Sun, holidays. Lunch, dinner. $
D SC

★★ **CAFÉ MISSISSIPPI.** *431 S River Park Dr, Guttenberg, IA (52052). Phone 563/252-4405.* American menu. Lunch, dinner. Located along the Mississippi River Lock and Dam 10. $

★ **HARBOR HOUSE.** *126 N 1st St, Harpers Ferry (52146). Phone 563/586-2586.* American menu. Closed Mon. Lunch, dinner. Regional American cuisine just a short walk from the ferry. $

★ **MILTY'S.** *200 Main St, Lansing (52151). Phone 563/538-4585.* American menu. Closed Sun. Lunch, dinner. This family-style eatery is a local favorite. $

★★★ **THE FAITHFUL PILOT CAFÉ.** *117 N Cody Rd, Le Claire (53753). Phone 563/289-4156. www.faithfulpilotcafe.com.* American menu. Dinner. Featuring progressive American cuisine in an elegant, modern dining room. $$

★ **GRAMMA'S KITCHEN.** *I-80, exit 284, Walcott (52773). Phone 563/284-5055.* American menu specializing in country cuisine. Closed Jan 1, Thanksgiving, Dec 25. Breakfast, lunch, dinner. $
D

Loess Hills Scenic Byway
✹ IOWA

Quick Facts

LENGTH: 220 miles.

TIME TO ALLOW: 7 hours.

BEST TIME TO DRIVE: Spring through fall.

BYWAY TRAVEL INFORMATION: Loess Hills Scenic Byway Council: 712/482-3029; Western Iowa Tourism Region: toll-free 888/623-4232; Byway local Web site: www.loesshillstours.com.

RESTRICTIONS: Occasionally, winter weather may necessitate a slower rate of travel. Also, the gravel roads on some of the excursion loops may also become somewhat degraded during periods of high rainfall.

BICYCLE/PEDESTRIAN FACILITIES: With the addition of the Wabash Trace Bicycle and Pedestrian Trail, the Byway now includes opportunities for biking and walking. There are many smaller trails that bicyclists and pedestrians can travel on throughout the corridor as well, but the Wabash Trace was specifically created for biking and walking.

The Loess Hills Scenic Byway weaves through a landform of windblown silt deposits along the eastern edge of the Missouri River Valley. This unique American treasure possesses natural features that are found in only two places in the world: western Iowa and the Yellow River Valley of China. Travelers are intrigued by the extraordinary landscape of prairies and forest-covered bluffs.

The loess (pronounced LUSS) soil deposits were initially left by glacial melt waters on the floodplain of the Missouri River. These deposits were then blown upward by strong winds. The steep, sharply ridged topography of this area was formed over thousands of years by the deposition and erosion of the wind-blown silt. Today, the rugged landscape and strong local contrasts in weather and soil conditions provide refuge for a number of rare plants and animals, many of which can be found only in the Loess Hills.

As you drive the western edge of Iowa, you pass through dozens of prairie towns. Larger cities like Council Bluffs and Sioux City offer venues of recreation, culture, and history. Learn a little about some of the people who passed through the area on a trek west and discover more about the people who stayed here.

You will want to enjoy the many nature areas along the way as well. The Dorothy Pecaut Nature Center, the Hitchcock Nature Area, and the Loess Hills State Forest are just a few of the many places on the Byway that are dedicated to preserving and restoring the native prairies of western Iowa.

Iowa

❋ Loess Hills Scenic Byway

THE BYWAY STORY

The Loess Hills Scenic Byway tells archaeological, cultural, historical, natural, recreational, and scenic stories that make it a unique and treasured Byway.

Archaeological

As scientists and archaeologists continue to prod the hills for clues to the past, the hills are still revealing artifacts and remnants that point to cultures from many thousands of years ago. Archaeological studies reveal places in the Loess Hills that have been continuously occupied for 12,000 years. Cultures from 12,000 years ago until recent times have been studied and cataloged in order to provide an idea of what human existence has been like in the Loess Hills. Hints of past civilizations are enough to make the Loess Hills an archaeologically significant area.

Evidence of a nomadic culture of hunters and gatherers was found in Turin (which is on the Byway) in the middle of the last century. The site yielded some of the oldest human remains in North America. The bones date back nearly 8,000 years and provided an inside look into life during that time period. In the city of Glenwood, stop at the Mills County Historical Museum to explore a reconstructed earth lodge and artifacts that were discovered in the area. The structure and artifacts are remnants of a culture more recent than that of Turin. You may also want to stop at Blood Run National Historic Landmark, where there was once a center of commerce and society for the Oneota Indians. During the period of AD 1200 to 1700, these people constructed buildings, homes, and effigy mounds in the area. The collection of resources from Loess Hills is informative to both archaeologists and visitors to the Byway wanting to catch a glimpse at life on the Byway thousands of years ago.

Cultural

Predominantly agricultural, the cultures on the Byway have influenced one another over the years and evolved with the rest of the nation. Many of the first European settlers of the Loess Hills of Iowa came from Danish, German, or Swedish cultures that settled in the United States nearly 200 years ago to create a unique heritage. These people learned to live among each other, creating a diverse culture characteristic of the United States.

The Byway culture today is a blend of the old and new living side by side. Urban centers like Sioux City and Council Bluffs give way to hidden corners of agricultural hamlets.

From cities to villages, cultural events occur regularly on the Byway. While some visitors may choose to attend theatrical events and tour museums, others may want to taste some of the local flavor at a farmers market, county fair, or heritage celebration. When travelers come to the Loess Hills Scenic Byway, many of them take part in activities like the rodeo or apple harvest festivals. To learn more about the people of the Loess Hills, visit the Moorhead Cultural Center, which has displays and activities that tell the story of people and culture on the Byway.

Historical

To complement the area's fascinating natural history, the Loess Hills have another history to tell. The history of human settlement in the Loess Hills has left many stories behind in old buildings and sacred places. A home and hunting ground to some of the continent's earliest people, human habitation in the Loess Hills has been developing for many years. The people native to this land had a great respect for the Loess Hills and recognized them for the natural anomaly that they are. The land was greatly honored until the early 1800s; when explorers began to wind their way through this land in the 1700s, it was the end of an old way of life around the Loess Hills.

Significant historical treasures are located within the Loess Hills from the period of European settlement as well. To see pieces of this history, follow the Lewis and Clark Trail north and south along the hills or prepare for a trip along one of the many trails that traveled to the

West. The Oregon Trail, Mormon Trail, and California Trail all traversed the Loess Hills in their route westward. In fact, the Mormon Trail had a stopover point in the hills called the Great Encampment that was used during the winter months. At this point, permanent settlement in the Loess Hills became an option for pioneers crossing the plains.

The places left behind are protected as designated historic sites. You will find National Historic Landmarks, places on the National Register of Historic Places, and National Historic Trails. Museums and information centers along the Byway allow visitors to study the history that surrounds the hills. Buildings like the General Dodge House and the Woodbury County Courthouse display styles of architecture from another time. Many sites in the Loess Hills were also used on the Underground Railroad to transport escaped slaves to the north. And you will find monuments to the first explorers, including Sergeant Floyd of the Lewis and Clark expedition, who was the only explorer to die during the journey. Museums cover everything from Civil War history to prehistoric life in this part of Iowa.

Natural

Along the Loess Hills Scenic Byway, the rare kind of soil known as loess has been formed into hills that allow a different kind of ecosystem to develop. This ecosystem features plants and animals that are rarely found anywhere else. Not only do the hills represent a rare kind of soil, but they are also a slice of the once vast prairie lands of the United States. The hills contain most of Iowa's remaining native prairie, making the Byway a site that preserves natural history.

The area of the Loess Hills has been dubbed a National Natural Landmark in order to further promote its protection. There are also four Iowa State Preserves in the Loess Hills. You may want to stop at Five Ridge Prairie Preserve, Mount Talbot Preserve, Turin Hills Preserve, or Sylvan Runkel Preserve to observe the untouched habitat of the prairie. When you drive through places like Broken Kettle Grassland on the Byway, you may see unique plants like the ten-petal Blazing Star or hear a fat prairie rattlesnake shaking its tail in the distance.

Because many of the creatures along the Byway are threatened or endangered, you will also find wildlife refuges along the Byway. The De Soto National Wildlife Refuge is maintained by the US Fish and Wildlife Service; you may be able to glimpse migratory waterfowl nesting and feeding in the area. About 350,000 snow geese stop here in the fall as they travel south. Many other species of birds and geese stop and stay at the wildlife refuge. The Loess Ridge Nature Center is an excellent place to find out more about the habitat and ecosystem of the Loess Hills. The center provides many engaging exhibits, including live animal displays and a

see page A10 for color map

73

Iowa

✵ Loess Hills Scenic Byway

butterfly garden. Other preserves and parks along the Byway have their own information centers where visitors can find out more about a particular place or part of the Loess Hills. After you have stopped at the centers and preserves, the unique traits of the Loess Hills will be apparent.

Recreational

During a drive on the Loess Hills Scenic Byway, you will want to get out of the car and stretch. And there are several places along the Byway that are perfect for more than just stretching. Opportunities for outdoor recreation are around every corner. Between preserves and state parks, you will have an excellent chance to view unique wildlife. De Soto National Wildlife Refuge is a home and hotel to many waterfowl and migratory birds. Wildlife watching is a popular activity on the Byway's four preserves. At Stone State Park, try camping or tour the trails in your own way. Whether you love hiking, biking, or horseback riding, all these modes of transportation are welcomed. Even snowmobilers ride through these wooded trails onto the white prairie in the wintertime. Stop at an orchard or Small's Fruit Farm to pick your own apples.

If outdoor recreation isn't a priority, try touring historical sites and monuments. Historical museums are located throughout the Byway, along with many unusual buildings and historic districts. Gaming is also a popular activity on the Byway at the casinos in the Byway communities. Visitors will find slot machines, table games, and nightlife. In addition, the best antique shopping in Iowa is rumored to be found in the communities along the Byway. Cross the river from Council Bluffs to Omaha, Nebraska, to explore this busy city full of distractions. Or simply settle down at a quiet restaurant for a bite to eat and time to look at the map.

Scenic

The rolling hills created by the loess soil of the Loess Hills make driving this Byway a pleasant experience from any angle. The hills themselves are in the setting of the Missouri River Flood Plain, where they create a view of a unique landform that is characteristic of the Byway. Visitors who drive the Loess Hills Scenic Byway enjoy the scenic overlooks and the sight of the hills rolling on and on. Viewing the unique formations of the Loess Hills creates a sensation of continuity as you see the prairie as a whole. Because of the unique properties of loess soil, you can enjoy "cat steps" in the hills where the loess has slumped off, creating a unified ledge. In the distance, you may catch a glimpse of the Missouri River as it meanders beside the hills.

Throughout the year, the prairie rolls through the seasons. Fall is one of the favorites of travelers who come to see the hardwood forests and prairie vegetation change to rich hues of red and orange. Pieces of an agricultural lifestyle form a patchwork of fields and historic communities along the Byway. Pioneer cemeteries next to country churches tell the story of earlier settlers who came through this place and the hardships they had to face. Travelers also experience the sights of the cities on the

Byway, like Sioux City and Council Bluffs, that remain great stopping points at any time of year. Parks, museums, and historic buildings offer a taste of the Byway cities.

HIGHLIGHTS

When driving this Byway, consider using this Loess Hills prairie tour as your itinerary.

- The tour starts in the **Broken Kettle Grasslands,** just south of Akron. The preserve is half prairie and constitutes the largest remaining section of the vast prairie that once covered most of Iowa. The preserve contains some flora and fauna not found in any other part of the Loess Hills to the south or the state of Iowa; these include the prairie rattlesnake and the ten-petal Blazing Star.

- **Five Ridge Prairie,** located on the Ridge Road Loop, about 5 miles south of Broken Kettle Grasslands, is a combination of prairie and woodlands. This is one of the best sites of unbroken prairie remnants in Iowa. You'll notice the climate changes between open grasslands, which are warmed by the sun and dry prairie breezes, and the shadowy woods, which remain cooler and more humid. Expect to find some rugged hiking trails at this site.

- The **Dorothy Pecaut Nature Center** is on the Stone Park Loop just south of the Highway 12 entrance to Stone State Park. The center is devoted wholly to Iowa's Loess Hills. The center has live animal displays, hands-on exhibits, a butterfly garden, and a walk-through exhibit showing life under the prairie.

- Your final stop, **Stone State Park,** is located on Sioux City's interpretive northwest side. It has 1,069 acres of prairie-topped ridges and dense woodlands. Dakota Point and Elk Point provide scenic overlooks of Nebraska, South Dakota, and Iowa. The multi-use trails handle hikers, bicyclists, horseback riders, and snow-mobilers, and campsites with showers are available. This park is a site on the Lewis and Clark National Historic Trail.

THINGS TO SEE AND DO

Driving along the Loess Hills Scenic Byway will certainly keep your senses engaged, but if you yearn to get out of the car and stretch your legs, or if you'd like to make a mini-vacation out of your trip, check out these attractions along the route.

BELLE OF SIOUX CITY. *100 Larsen Park Rd, Sioux City (51101). Phone toll-free 800/424-0080.* Tri-level riverboat casino. Open May-Oct, daily.

GOLDEN SPIKE MONUMENT. *S 21st St and 9th Ave, Council Bluffs (51501).* Erected in 1939, this 56-foot golden concrete spike commemorates the junction of the Union Pacific and Central Pacific railroads in Council Bluffs.

HISTORIC GENERAL DODGE HOUSE. *605 3rd St, Council Bluffs (51503). Phone 712/322-2406. www.omaha.org/oma/dodge.htm.* Restored Victorian home (1869) built by Grenville M. Dodge, chief construction engineer for Union Pacific Railroad and a general in the Civil War. Guided tours. Open Tues-Sat 10 am-5 pm, Sun 1-5 pm; closed Mon, holidays; also Jan. **$$**

HISTORIC POTTAWATTAMIE COUNTY JAIL. *226 Pearl St, Council Bluffs (51501). Phone 712/323-2509.* This unique and historic (1885) three-story rotary jail is sometimes referred to as the "human squirrel cage" or "lazy Susan jail." Open Mar-Dec, Mon, Wed-Sat 10 am-5 pm, Sun 1-5 pm; closed holidays. **$**

LAKE MANAWA STATE PARK. *1100 S Shore Dr, Council Bluffs (51501). Phone 712/366-0220. www.state.ia.us/dnr/organiza/ppd/manawa.htm.* More than 1,500 acres with a 660-acre lake. Dream Playground designed by and for children. Swimming, supervised beach, fishing, boating (ramps, rentals); hiking trails, bicycle trails, snowmobiling, picnicking, camping (electricity).

Iowa

❋ *Loess Hills Scenic Byway*

LEWIS AND CLARK MONUMENT. *19962 Monument Rd, Council Bluffs (51501). Phone 712/328-4650.* Shaft of native stone on bluffs depicts Lewis and Clark holding council with Oto and Missouri Indians. Open daily 6 am-10 pm. **FREE**

LEWIS AND CLARK STATE PARK. *21914 Park Loop, Onawa (51040). Phone 712/423-2829. www.state.ia.us/government/dnr/organiza/ppd/ lewisclk.htm.* Park of 176 acres on a 250-acre lake. Swimming; fishing; boating (ramp). Hiking trails. Snowmobiling. Picnicking. Replica of Lewis and Clark keelboat, *Discovery*. Camping (electricity, dump station).

LINCOLN MONUMENT. *323 Lafayette Ave, Council Bluffs (51501).* A granite shaft marks the spot from which Lincoln designated the town as the eastern terminus of the Union Pacific Railroad. Erected in 1911.

MISSISSIPPI BELLE II. *311 Riverview Dr, Clinton (52732). Phone 563/243-9000; toll-free 800/457-9975.* Offers year-round casino gambling along the Mississippi River. Entertainment. Concession. Open Sun-Thurs 9-2 am, Fri-Sat 9-4 am.

MORMON TRAIL MEMORIAL. *530 1st Ave, Council Bluffs (51503).* A huge boulder marks the passage of Mormons out of the city on their trek to Utah.

RAILSWEST RAILROAD MUSEUM. *16th Ave and S Main St, Council Bluffs (51503). Phone 712/323-5182 or 712/322-0612.* Historic Rock Island depot (1899); railroad memorabilia; model trains on display. Open Memorial Day-Labor Day, Mon-Tues, Thurs-Sat 10 am-4 pm; Sun 1-5 pm; rest of year, by appointment; closed major holidays. **$$**

RUTH ANNE DODGE MEMORIAL. *N 2nd and Lafayette aves, Council Bluffs (51503).* Commissioned by the daughters of G. M. Dodge in memory of their mother, this bronze statue of an angel is the work of Daniel Chester French.

SERGEANT FLOYD MONUMENT. *1000 Larsen Park Rd, Sioux City (51103). Phone 712/279-0198. www.cr.nps.gov/nr/travel/lewisandclark/ser.htm.* First registered national historic landmark in the United States. The 100-foot obelisk marks the burial place of Sergeant Charles Floyd, the only casualty of the Lewis and Clark Corps of Discovery. Open daily.

SERGEANT FLOYD WELCOME CENTER AND MUSEUM. *1000 Larsen Park Rd, Sioux City (51103). Phone 712/279-0198.* Former Missouri River inspection ship now houses a museum, information center, and gift shop. Open Oct-Apr 9 am-5 pm, May-Sept 8 am-6 pm. **FREE**

SIOUX CITY PUBLIC MUSEUM. *2901 Jackson St, Sioux City (51104). Phone 712/279-6174. www.sioux-city.org/museum/.* Exhibits show Sioux City history and life in pioneer days; geological, archaeological, and Native American materials. Located in a Romanesque 23-room mansion. Open Tues 9 am-8 pm, Wed-Sat 9 am-5 pm, Sun 1-5 pm; closed major holidays. **FREE**

STONE STATE PARK. *5001 Talbot Rd, Sioux City (51103). Phone 712/255-4698. www.state.ia.us/dnr/organiza/ppd/stone.htm.* On 1,069 acres. The park overlooks the Missouri and Big Sioux river valleys; view of three states from Dakota Point Lookout near Big Sioux River. Fishing, bridle and hiking trails, snowmobiling, picnicking, camping. Ten-acre **Dorothy Pecant Nature Center.** Open Tues-Sun; closed holidays.

TRINITY HEIGHTS. *33rd and Floyd Blvd, Sioux City (51103). Phone 712/239-8670. www.sctrinityheights.org.* A 30-foot stainless steel statue of the Immaculate Heart of Mary. Life-size carving of the Last Supper. Outdoor cathedral with a 33-foot statue of Christ. Open daily. **FREE**

WOODBURY COUNTY COURTHOUSE. *620 Douglas St, Sioux City (51101). Phone 712/279-6624.* Courthouse was the largest structure ever completed in the architectural style of Chicago's Prairie School (1916-1918). Designed by Purcell and Elmslie, longtime associates of Louis Sullivan, the city block-long building is constructed of Roman brick, ornamented with massive pieces of Sullivanesque terra-cotta, stained glass, and relief sculpture by Alfonso Ianelli, who also worked with Frank Lloyd Wright. Both the exterior and the highly detailed interior are in near-pristine condition; courtrooms still contain original architect-designed furniture and lighting fixtures. Open Mon-Fri.

PLACES TO STAY

If you choose to include an overnight stay in your trip along this Byway, Mobil Travel Guide recommends the following lodgings.

Council Bluffs

★★ **AMERISTAR CASINO HOTEL.** *2200 River Rd, Council Bluffs (51501). Phone 712/328-8888; toll-free 877/462-7827.* The only gaming riverboat and hotel in the area. 160 rooms, 5 story. Check-out 11 am. TV; VCR available. Restaurant, bar; entertainment. Room service. In-house fitness room, sauna. Game room. Indoor pool, whirlpool. ¢

★ **FAIRFIELD INN BY MARRIOTT.** *520 30th Ave, Council Bluffs (51501). Phone 712/366-1330; toll-free 800/228-2800. www.fairfieldinn.com.* 62 rooms, 3 story. Complimentary continental breakfast. Check-out noon. TV; cable (premium). VCR available. Indoor pool, whirlpool. ¢

★ **HEARTLAND INN.** *1000 Woodbury Ave, Council Bluffs (51503). Phone 712/322-8400.* 89 rooms, 2 story. Complimentary continental breakfast. Check-out 11 am. TV; cable (premium). Sauna. ¢

★★ **QUALITY INN.** *3537 W Broadway, Council Bluffs (51501). Phone 712/328-3171; toll-free 877/424-6423. www.qualityinn.com.* 89 rooms, 2 story. Pet accepted, some restrictions. Complimentary continental breakfast. Check-out noon. TV; cable (premium). In-room modem link. Restaurant. Indoor pool. Free airport transportation. ¢

Onawa

★ **SUPER 8.** *22868 Filbert Ave, Onawa (51040). Phone 712/423-2101; toll-free 800/800-8000. www.super8.com.* 80 rooms. Pet accepted, some restrictions; fee. Check-out 11 am. TV. ¢

Sioux City

★★ **BEST WESTERN CITY CENTRE.** *130 Nebraska St, Sioux City (51101). Phone 712/277-1550; toll-free 800/528-1234. www.bestwestern.com.* 114 rooms, 2 story. Pet accepted, some restrictions; fee. Check-out noon. TV. Laundry services. Bar. Pool. Free airport transportation. ¢

★ **HOLIDAY INN EXPRESS.** *4230 S Lakeport St, Sioux City (51106). Phone 712/274-1400; toll-free 800/465-4329. www.holiday-inn.com.* 58 rooms, 2 story. Complimentary continental breakfast. Check-out 11 am. TV; VCR available (free movies). In-room modem link. In-house fitness room. ¢

★★ **THE INNS OF ROSE HILL.** *1525 Douglass St, Sioux City (51105). Phone 712/277-1386. www.innsofrosehill.com.* 11 rooms, 3 story. Featuring luxury private rooms and amenities in two restored English mansions in the Rose Hill Historic District. $$

Iowa

❋ *Loess Hills Scenic Byway*

★★ **SIOUX CITY PLAZA HOTEL.** *707 Fourth St, Sioux City (51101). Phone 712/277-4101.* Conveniently located on the river downtown. 193 rooms, 12 story. Pet accepted, some restrictions; fee. Check-out noon. TV; cable (premium). In-room modem link. Restaurant, bar. In-house fitness room, sauna. Indoor pool. Free airport transportation. ¢

★ **SUPER 8.** *4307 Stone Ave, Sioux City (51106). Phone 712/274-1520; toll-free 800/800-8000. www.super8.com.* 60 rooms, 2 story. Pet accepted, some restrictions; fee. Continental breakfast. Check-out 11 am. TV. Restaurant. ¢

PLACES TO EAT

A long day of driving is sure to make you hungry. At the end of your journey, take a table at one of the following restaurants.

Council Bluffs

★ **CHRISTY CRÈME.** *2853 N Broadway, Council Bluffs (51503). Phone 712/322-2778.* American menu. Lunch, dinner. Children's menu. Outdoor seating. $

Crescent

★ **PINK POODLE RESTAURANT.** *633 N Old Lincoln Hwy, Crescent (51526). Phone 712/545-3744.* American menu. Breakfast, lunch, dinner. A quaint small-town favorite known for prime rib. $

Sioux City

★ **GREEN GABLES.** *1800 Pierce St, Sioux City (51105). Phone 712/258-4246.* Steak menu. Closed Dec 25. Lunch, dinner. Children's menu. $$

★ **HUNAN PALACE.** *4280 Sergeant Rd, Sioux City (51106). Phone 712/274-2336.* Chinese menu. Closed Thanksgiving, Dec 25. Lunch, dinner. Bar. $

★★ **MINERVA'S RESTAURANT & BAR.** *2901 Hamilton Blvd, Sioux City (51104). Phone 712/252-0994.* American menu. Lunch, dinner. This comfortable, family-style establishment features steaks and seafood. $

★★★ **THE VICTORIAN OPERA COMPANY.** *1021 4th St, Sioux City (51104). Phone 712/255-4821.* American menu. Closed Sun-Mon. Lunch. Known for interesting sandwiches and desserts. $

Woodward Avenue (M-1)
✺ MICHIGAN

Quick Facts

LENGTH: 27 miles.

TIME TO ALLOW: 5 hours.

BEST TIME TO DRIVE: The busiest months are August because of the Dream Cruise event and November because of America's Thanksgiving Day Parade.

BYWAY TRAVEL INFORMATION: Detroit Metro Convention and Visitors Bureau: 313/202-1800; Byway local Web site: www.woodwardavenue.com.

SPECIAL CONSIDERATIONS: Because this Byway passes through downtown Detroit, avoid driving it during rush hour.

BICYCLE/PEDESTRIAN FACILITIES: Woodward Avenue has six major central business districts, three major cultural centers, and numerous other activity centers. The cultural centers in Detroit and Bloomfield Hills also allow for and encourage pedestrian travel. Many of the communities in the corridor are actively involved in the Walkable Communities program sponsored by the Southeast Michigan Council of Governments.

The Motor City put the world on wheels, so welcome to Detroit's main drag: Woodward Avenue. Stretching out from the base of Detroit at the Detroit River, Woodward follows the pathway of growth from the heart of the city. Lined with history, cultural institutions, and beautiful architecture, Woodward Avenue travels through downtown Detroit, the Boston Edison neighborhood, past Highland Park, the Detroit Zoo, the delightful city of Birmingham, the Cranbrook Educational Community in Bloomfield Hills, and into the city of Pontiac. Nearly every mile of this Byway has historical sites to see that have shaped the industrial life of our nation. Perhaps the most famous are the Ford buildings that stand as monuments to the automobile revolution, although many other equally impressive automobile companies are still found along the Byway.

Woodward Avenue includes both landmarks of the past and monuments to the future. You can find one of the largest public libraries in the nation as well as one of the five largest art museums. From the Freedom Fireworks Display on the Detroit River to the annual Thanksgiving Day Parade to the phenomenal Woodward Dream Cruise, it's as if this street were meant to blend memories with the future.

THE BYWAY STORY

Woodward Avenue tells cultural, historical, recreational, and scenic stories that make it a unique and treasured Byway.

Michigan

✤ Woodward Avenue (M-1)

Cultural

Much of the culture in southeast Michigan stems from the automobile and those who sacrificed much in its development. When visiting museums such as the Detroit Institute of Arts (DIA) and the Detroit Historical Museum, the impact of the automobile is evident. Your first steps into the DIA will take you to huge two-story murals painted by Diego Rivera that depict life on the assembly line. The Woodward Dream Cruise represents more popular culture and is included in the Library of Congress' permanent collection as a local legacy. The Woodward corridor is also one of six areas comprising the Automobile National Heritage Area (under the jurisdiction of the National Park Service).

Woodward Avenue is home to other major cultural institutions as well, including the Cranbrook Educational Community, Orchestra Hall, and the Detroit Public Library, which is among the top ten largest library's in the nation.

The region's culture can also be defined by its faith. Traveling from Detroit to Pontiac, you pass by more than 50 churches, many of which are on national or state historic registers. Prominent among these is the National Shrine of the Little Flower in Royal Oak.

Historical

Woodward Avenue is the signature route for Detroit and southeast Michigan. The avenue gets its name from Judge Augustus B. Woodward, an early leader in the city's history.

Detroit is known internationally as the Motor City because of its role in the development of the automobile. This is the birthplace of the assembly line (Ford's 1913 Model T Plant on Woodward), a major technological innovation that made the automobile affordable to most families. Most people will agree that the assembly line had a major impact on American society and in the development of our urban areas.

The first people who may come to mind when talking about the automobile industry are the Fords, Chryslers, and Durants. But you should also remember the thousands of nameless autoworkers who, with blood and sweat, established the industry and carry on to this day—the United Automobile Workers (UAW). Henry Ford's $5-a-day wage attracted thousands to his factories, but the working conditions were brutal—this was true of all auto plants. In the late 1930s, the UAW was born and, with fair wages and better working conditions, auto workers and their families now enjoy a decent quality of life. Some of the former homes of early auto barons are located in historic neighborhoods, such as the Boston Edison Historic District.

Recreational

The city of Detroit and neighboring communities have worked hard and invested a lot of energy to develop safe, fun activities that the whole family can enjoy. It just so happens that a great majority of these activities are located on Woodward Avenue.

Some of the highlights found along this Byway include watching animals at the world-class Detroit Zoo, attending the Woodward Dream Cruise (where thousands from around the world gather to view hundreds of classic cars from around the country), touring historic homes and mansions, making maple syrup at the festival at Cranbrook, standing on the sidewalk during the famous America's Thanksgiving Day Parade, or watching the largest fireworks display in the nation that explodes over the shimmering Detroit River. This Byway, above most others in the nation, has such an abundance of activities and places to see that you could take part in different activities and visit various sites for months without repeating any of them.

Scenic

A unique scenic quality of Woodward Avenue and the region is the geographic location of Detroit at the Canadian border. Driving southeast on Woodward from Pontiac, you conclude your journey at Hart Plaza, located on the

Detroit River in Detroit's central business district. Approaching the plaza, the Windsor skyline unfolds in front of you. Noticeable at a closer distance are Belle Isle to the east and the Ambassador Bridge to the west. The Detroit/Windsor border is the only geographic location along the entire American-Canadian border (of the contiguous states) where the United States is north of Canada.

If you visit during the International Freedom Festival in July, you will notice hundreds of sail and power boats and get to enjoy the largest fireworks display in the country over the river. This fun-filled family event has a million people celebrating our nation's heritage, watching hundreds of exploding shells reflecting off the glass of the high-rise office buildings and the river.

The Detroit River is a major international waterway for commercial freighter traffic delivering goods to the world. Many of these vessels reach 1,000 feet in length. Detroit has also become an embarkation point for international cruise ships plying the Great Lakes.

HIGHLIGHTS

This Byway begins in Pontiac and ends in Detroit, where you can take the following downtown Detroit historical tour.

- The tour begins just south of Woodward Avenue at the **Detroit River,** a designated American Heritage River. It is the most frequently traveled major international boundary between Canada and the United States. It is also a major national and international waterway for the movement of freight and other commodities from the United States to foreign markets. In 1701, French explorer Antoine DeMothe Cadillac founded Detroit at the foot of the present-day Woodward at the Detroit River. In 2001, Detroit celebrated its 300th birthday.

- Located at the intersection of Michigan, Monroe, Cadillac Square, and Fort Street on Woodward Avenue, **Campus Martius** was part of the historic 1807 plan of Detroit by Judge Woodward. Today, it is being rebuilt as a public park and home to new office developments. This site includes the Michigan Soldiers' and Sailors' Monument of 1872.

- The next stop, the **Lower Woodward Historic District,** is located on Woodward north of State Street and south of Clifford. This district contains numerous former retail and office buildings constructed from 1886 to 1936. Significant loft development activity is underway in these buildings.

- **Woodward East Historic District (Brush Park)** is located east of Woodward, bounded by Watson and Alfred streets within the Brush Park neighborhood. Known for its high Victorian style residences constructed for Detroit's elite in the late 1800s, it is the location of a major urban townhome development.

81

Michigan

❋ Woodward Avenue (M-1)

- The **Peterboro-Charlotte Historic District** is located on Peterboro at Woodward Avenue. The architecture represents a study in late 19th-century middle-class single-family dwellings and early 20th-century apartment buildings.
- Located in Detroit's Cultural Center at 315 E Warren, the **Charles Wright Museum of African American History** offers one of the United States' largest collections of African-American history and culture.
- The final stop, the **Detroit Historical Museum,** is located at 5401 Woodward in Detroit's Cultural Center. The museum specializes in telling the history of the Detroit area from its founding in 1701 to the present, including permanent exhibits, temporary exhibitions, programs, and events. It is especially noted for its Streets of Old Detroit exhibit.

THINGS TO SEE AND DO

Driving along Woodward Avenue will certainly keep your senses engaged, but if you yearn to get out of the car and stretch your legs, or if you'd like to make a mini-vacation out of your trip, check out these attractions along the route.

ALPINE VALLEY SKI RESORT. *6775 E Highland, White Lake (48383). Phone 248/887-2180 or 248/887-4183 (snow conditions). www.skialpinevalley.com.* Ten chairlifts, ten rope tows; patrol, school, rentals, snowmaking; bar, cafeteria. Longest run 1/3 mile; vertical drop 320 feet. Open Nov-Mar, daily; closed Dec 24 afternoon and Dec 25 morning. **$$$$**

★ **BELLE ISLE.** *E Jefferson Ave and E Grand Blvd, Detroit (48207).* Located between the United States and Canada, in sight of downtown Detroit, this 1,000-acre island park offers nine-hole golf, a nature center, guided nature walks, swimming, and fishing (piers, docks). Also picnicking, ball fields, tennis and lighted handball courts. **FREE**

The **Aquarium on Belle Isle** (phone 313/852-4141) is one of the largest and oldest freshwater collections in the country (open daily, $).

Dossin Great Lakes Museum (phone 313/852-4051) on the island includes scale models of Great Lakes ships; restored "Gothic salon" from a Great Lakes liner; marine paintings, reconstructed ship's bridge, and a full-scale racing boat, *Miss Pepsi* (open Wed-Sun, **FREE**).

The **Belle Isle Zoo** (phone 313/852-4083) has animals in natural habitats (open May-Oct, daily, $$).

CHILDREN'S MUSEUM. *67 E Kirby Ave, Detroit (48202). Phone 313/873-8100.* Exhibits include America Discovered, Inuit culture, children's art, folk crafts, birds and mammals of Michigan, holiday themes. Participatory activities relate to exhibits. Special workshops and programs and planetarium demonstrations on Sat and during school vacations. Open Oct-May Mon-Fri 1-4 pm, Sat 9 am-4 pm; hours vary the rest of the year; closed holidays. **FREE**

CIVIC CENTER. *Woodward and Jefferson aves, Detroit (48226).* Dramatic group of buildings in a 95-acre downtown riverfront setting.

COBO HALL-COBO ARENA. *One Washington Blvd, Detroit (48226). Phone 313/877-8777. www.cobocenter.com.* Designed to be the world's finest convention-exposition-recreation building; features an 11,561-seat arena and 720,000 square feet of exhibit area and related facilities. Open daily.

CRANBROOK ACADEMY OF ART AND MUSEUM. *39221 Woodward Ave, Bloomfield Hills (48303). Phone 248/645-3312. www.cranbrook.edu/art/museum/.* Graduate school for design, architecture, and the fine arts. The museum has international arts exhibits and collections. Open Tues-Sun 11 am-5 pm, fourth Fri 11 am-9 pm; closed holidays. $$

CRANBROOK GARDENS. *39221 Woodward Ave, Bloomfield Hills (48303). Phone 248/645-3147. www.cranbrook.edu.* 40 acres of formal and informal gardens; trails, fountains, outdoor Greek theater. Open May-Aug, daily; Sept, afternoons only; Oct, Sat and Sun afternoons. $$

CRANBROOK HOUSE. *380 Lone Pine Rd, Bloomfield Hills (48304). Phone 248/645-3149.* Tudor-style structure designed by Albert Kahn in 1908. Contains exceptional examples of decorative and fine art from the late 19th and early 20th centuries. $$

CRANBROOK INSTITUTE OF SCIENCE. *39221 Woodward Ave, Bloomfield Hills (48303). Phone 248/645-3200; toll-free 877/462-7262. www.cranbrook.edu.* Natural history and science museum with exhibits, observatory, nature center; planetarium and laser demonstrations. Open daily 10 am-5 pm (Fri evenings until 10 pm); closed holidays. $$

DETROIT GRAND PRIX. *1249 Washington Blvd, Detroit (48226).* Indy car race every June on Belle Isle. Friday is Free Prix Day.

DETROIT HISTORICAL MUSEUM. *5401 Woodward Ave, Detroit (48202). Phone 313/833-1805. www.detroithistorical.org.* Presents a walk through history along reconstructed streets of Old Detroit, period alcoves, costumes; changing exhibits portray city life. The museum showcases an automotive exhibition celebrating the 100th anniversary of the automotive industry. Open Tues-Fri 9:30 am-5 pm, Sat 10 am-5 pm, Sun 11 am-5 pm; closed holidays. $

DETROIT INSTITUTE OF ARTS. *5200 Woodward Ave, Detroit (48202). Phone 313/833-7900. www.dia.org.* One of the great art museums of the world, DIA tells the history of humankind through artistic creations. Every significant art-producing culture is represented. Exhibits include The Detroit Industry murals by Diego Rivera, Van Eyck's *St. Jerome,* Bruegel's *Wedding Dance,* and Van Gogh's *Self-Portrait;* African, American, Indian, Dutch, French, Flemish, and Italian collections; medieval arms and armor; and an 18th-century American country house reconstructed with period furnishings. Frequent special exhibitions (fee); lectures, films. Open Wed-Thurs 10 am-4 pm, Fri 10 am-9 pm, Sat-Sun 10 am-5 pm; closed holidays. $

DETROIT LIONS (NFL). *Ford Field, 2000 Brush St, Detroit (48226). Phone 248/335-4131. www.detroitlions.com.* Professional football team.

DETROIT PISTONS (NBA). *The Palace of Auburn Hills, 2 Championship Dr, Auburn Hills (48326). Phone 248/377-0100. www.nba.com/pistons/.* Professional basketball team.

DETROIT PUBLIC LIBRARY. *5201 Woodward Ave, 313/833-1000. www.detroit.lib.mi.us.* Murals by Coppin, Sheets, Melchers, and Blashfield; special collections include National Automotive History, Burton Historical (Old Northwest Territory), Hackley (African Americans in performing arts), Labor, Maps, Rare Books, and US Patents Collection from 1790 to the present. Open Tues-Sat; closed holidays. **FREE**

DETROIT RED WINGS (NHL). *Joe Louis Arena, 600 Civic Center Dr, Detroit (48226). Phone 810/645-6666. www.detroitredwings.com.* Professional hockey team.

Michigan

❋ *Woodward Avenue (M-1)*

DETROIT SHOCK (WNBA). *The Palace of Auburn Hills, 2 Championship Dr, Detroit (48326). Phone 248/377-0100. www.wnba.com/shock/.* Professional basketball team.

DETROIT SYMPHONY ORCHESTRA HALL. *3711 Woodward Ave, Detroit (48201). Phone 313/576-5100. www.detroitsymphony.org.* Restored public concert hall (built in 1919) features classical programs. The Detroit Symphony Orchestra performs here.

DETROIT TIGERS (MLB). *Comerica Park, 2100 Woodward Ave, Detroit (48226). Phone 313/962-4000. tigers.mlb.com.* Professional baseball team.

DETROIT ZOO. *8450 W Ten Mile Rd, Royal Oak (48068). Phone 248/398-0900. www.detroitzoo.org.* One of the world's outstanding zoos, with 40 exhibits of more than 1,200 animals in natural habitats. Outstanding chimpanzee, reptile, bear, penguin, and bird exhibits. Open daily; closed Jan 1, Thanksgiving, Dec 25. **$$**

EASTERN MARKET. *2934 Russell, Detroit (48207). Phone 313/833-1560.* Built on the site of an early hay and wood market in 1892, this and the Chene-Ferry Market are the only two remaining produce/wholesale markets. Today, the Eastern Market encompasses produce and meat-packing houses, fish markets, and storefronts offering items ranging from spices to paper. It is also recognized as the world's largest bedding flower market. Open Mon-Sat; closed holidays. **FREE**

FISHER BUILDING. *3011 W Grand Blvd, Detroit (48202). Phone 313/874-4444.* Designed by architect Albert Kahn, this building was recognized in 1928 as the most beautiful commercial building erected and was given a silver medal by the Architectural League of New York. The building consists of a 28-story central tower and two 11-story wings. Housed here are the Fisher Theater, shops, restaurants, art galleries, and offices. Underground pedestrian walkways and skywalk bridges connect to a parking deck and 11 separate structures, including General Motors World Headquarters and New Center One.

FORD INTERNATIONAL DETROIT JAZZ FESTIVAL. *1 Hart Plaza, Detroit (48226). Phone 313/963-7622. www.detroitjazzfest.com.* Five days of jazz concerts in late August at Hart Plaza. **FREE**

HART PLAZA AND DODGE FOUNTAIN. *1 Hart Plaza, Detroit (48226). Phone 313/877-8077.* A $2 million water display designed by sculptor Isamu Noguchi. **FREE**

HISTORIC TRINITY LUTHERAN CHURCH. *1345 Gratiot Ave, Detroit (48207). Phone 313/567-3100. www.historictrinity.org.* (1931) Third church of congregation founded in 1850, this small neo-Gothic cathedral, built in 1931, contains a 16th-century-style pier-and-clerestory. Luther tower is a copy of the tower at a monastery in Erfurt, Germany. Much statuary and stained glass. Bell tower. Tours by appointment. **FREE**

INTERNATIONAL AUTO SHOW. *1 Washington Blvd, Detroit (48226). Phone 248/643-0250. www.naias.com.* Internationally known and acclaimed auto show at Cobo Hall each January.

INTERNATIONAL INSTITUTE OF METROPOLITAN DETROIT. *111 E Kirby Rd, Detroit (48202). Phone 313/871-8600. www.iimd.org.* Hall of Nations has cultural exhibits from five continents. Cultural programs and ethnic festivals throughout the year. Open Mon-Fri; closed holidays. **FREE**

MARINERS' CHURCH. *170 E Jefferson Ave, Detroit (48226). Phone 313/259-2206. www.marinerschurchofdetroit.org.* Oldest stone church in the city, completed in 1849, was moved 800 feet to its present site as part of the Civic Center plan. Since that time, it has been extensively restored, and a belltower with carillon has been added. Tours (by appointment). **FREE**

MEADOW BROOK ART GALLERY. *2200 N Squirrel Rd, Pontiac (48309). Phone 248/370-3005.* Series of contemporary, primitive, and Asian art exhibitions, including permanent collection of African art; outdoor sculpture garden adjacent to music festival grounds. Open Oct-May, Tues-Sun. **FREE**

MEADOW BROOK HALL. *2200 N Squirrel Rd, Pontiac (48309). Phone 248/370-3140. www.meadowbrookhall.org.* A 100-room English Tudor mansion (1926-1929) with nearly all original furnishings and art objects; antique needlepoint draperies, 24 fireplaces; library has hand-carved paneling; dining room has sculptured ceiling; ballroom has elaborate stone and woodwork. Serves as the cultural and conference center of the university. **Meadow Brook Music Festival** features popular and classical artists (dining and picnicking facilities, phone 248/567-6000, open mid-June-Aug). Open July-Aug, afternoons; rest of year, Sun afternoons. **$$$**

MICHIGAN CONSOLIDATED GAS COMPANY BUILDING. *500 Griswold St, Detroit (48226).* Glass-walled skyscraper designed by Minoru Yamasaki.

MOTOWN MUSEUM. *2648 W Grand Blvd, Detroit (48208). Phone 313/875-2264. www.motownmuseum.org.* This is Hitsville USA: the house where legends like Diana Ross and the Supremes, Stevie Wonder, Marvin Gaye, the Jackson Five, and the Temptations recorded their first hits. Motown's original recording Studio A; artifacts, photographs, gold and platinum records, memorabilia. Guided tours. Open Tues-Sat 10 am-6 pm. **$$**

MUSEUM OF AFRICAN-AMERICAN HISTORY. *315 E Warren Ave, Detroit (48201). Phone 313/494-5800. www.maah-detroit.org.* Exhibits trace the history and achievements of African Americans. Open Wed-Sat 9:30 am-5 pm, Sun 1-5 pm; closed holidays. **$**

RENAISSANCE CENTER. *100-400 E Jefferson Ave, Detroit (48207). Phone 313/568-5600.* Seven-tower complex on the riverfront; includes a 73-story hotel, offices, restaurants, bars, movie theaters, retail shops, and business services. **FREE**

NATIONAL SHRINE OF THE LITTLE FLOWER CATHOLIC CHURCH. *2123 Roseland Ave, Royal Oak (48073). Phone 248/541-4122. www.shrinechurch.com.* Built in 1925 in honor of Saint Thérèse of Lisieux, the church—originally located in a largely Protestant area—was torched by the Ku Klux Klan in 1936. Rebuilt out of copper and stone, a dramatic stone tower displays a cross bearing a 28-foot-high figure of Jesus. On the surrounding wall is a carved portrait of Saint Thérèse, who was also known as the Little Flower. The interior of the octagon-shaped national shrine rises 38 feet; a walnut and granite alter stands in the middle of the church, with seats for parishioners surrounding it. Stone angels guard every exterior doorway. Tours available. **FREE**

VETERANS' MEMORIAL BUILDING. *151 W Jefferson Ave, Detroit (48226). Phone 313/877-8111.* Rises on the site where Cadillac and the first French settlers landed in 1701. This $5.75-million monument to the Detroit area's war dead was the first unit of the $180-million Civic Center to be completed. The massive sculptured-marble eagle on the front of the building is by Marshall Fredericks, who also sculpted "Spirit of Detroit" at the Coleman A. Young Municipal Center (2 Woodward Ave, phone 313/224-5585), a $27-million, 13-story white marble office building and 19-story tower housing more than 36 government departments and courtrooms. Open Mon-Fri; closed holidays. **FREE**

WASHINGTON BOULEVARD TROLLEY CAR. *1301 E Warren Ave, Detroit (48207). Phone 313/933-1300.* Antique electric trolley cars

Michigan

❋ *Woodward Avenue (M-1)*

provide a unique transit service to downtown hotels, the Civic Center, and the Renaissance Center. From late May to Labor Day, Detroit operates the only open-top double-decker trolley car in the world. Open daily. $

WHITCOMB CONSERVATORY. *E Jefferson Ave and E Grand Blvd, Detroit (48213). Phone 313/852-4064.* Exhibits of ferns, cacti, palms, and orchids; special exhibits. Open Wed-Sun. $

WOODWARD DREAM CRUISE. *Phone toll-free 800/338-7648. www.woodwarddreamcruise.com.* Each year, more than 1.5 million people watch or ride in 30,000 classic and vintage cars touring Woodward Avenue; the event is billed as the world's largest one-day celebration of car culture. A parade-like atmosphere means you'll find plenty to eat and drink, with vendors selling T-shirts and other items commemorating the event and street entertainers providing music and other events. Held the third Sat in Aug, 9 am-9 pm, with some events held the preceding Thurs and Fri. FREE

PLACES TO STAY

If you choose to include an overnight stay in your trip along this Byway, Mobil Travel Guide recommends the following lodgings.

Bloomfield Hills

★★★ **KINGSLEY HOTEL AND SUITES.** *39475 Woodward Ave, Bloomfield Hills (48304). Phone 248/644-1400; toll-free 800/544-6835. www.whghotels.com/kingsley/.* This quiet hotel offers the friendly service expected from a small inn, along with the amenities and facilities of a large hotel. Located near shopping, this hotel offers a seafood restaurant, deli, and much more. 160 rooms, 3 story. Crib free. Check-out noon. TV. Some balconies, refrigerators. Restaurant, bar; entertainment Tues-Sat. Room service. In-house fitness room. Indoor pool, whirlpool. Barber, beauty shop. Business services. $$

Birmingham

★ **HOLIDAY INN EXPRESS.** *34952 Woodward Ave, Birmingham (48009). Phone 248/646-7300; toll-free 800/465-4329. www.holiday-inn.com.* 126 rooms, 2-5 story. Complimentary continental breakfast. Check-out noon. TV; cable (premium). In-room modem link. Health club privileges. $

★★★★ **THE TOWNSEND HOTEL.** *100 Townsend St, Birmingham (48009). Phone 248/642-7900; toll-free 800/548-4172. www.townsendhotel.com.* The Townsend Hotel brings the refinement of Europe to the heart of Michigan. Tucked away in the quiet community of Birmingham, where tree-lined streets brim with unique stores and bustling cafés, the hotel is conveniently located less than an hour from Detroit. The guest rooms are handsomely furnished with regal décor. Jewel tones add panache, four-poster beds add charm, and full kitchens in the suites and penthouses ensure that guests never need to leave the comforts of home behind. The cherry-wood paneling and the warm glow of the fireplace make for a particularly inviting space at the Rugby Grille, and its airy gallery filled with the fragrance of fresh flowers is a sunny alternative. The sleek city chic of the Townsend Corner Bar has made it one of the hotspots on the local scene. Its appealing interiors are a perfect match for its Asian-inspired appetizers and creative cocktails. 150 rooms, 4 story. Check-out noon. TV; cable (premium), VCR (movies). In-room modem link. Restaurant, bar. Room service 24 hours. Health club privileges. Cross-country ski 5 miles. Covered parking. Valet parking available. Business center. Concierge. Afternoon tea Tues-Sat by reservation; pianist. Located opposite a park. $$$

Detroit

★★★ ATHENEUM SUITE HOTEL. 1000 Brush St, Detroit (48226). Phone 313/962-2323; 800/772-2323. www.atheneumsuites.com. This hotel is located downtown in Greektown, near many restaurants, shops, and other area attractions. 174 rooms, 10 story. Check-out noon. TV; cable (premium), VCR available. In-room modem link. Bar. Room service 24 hours. In-house fitness room. Health club privileges. Valet parking available. Concierge. Neoclassical structure adjacent to the International Center. $$

★★ COURTYARD BY MARRIOTT. 333 E Jefferson St, Detroit (48226). Phone 313/222-7700; toll-free 800/321-2211. www.courtyard.com. 255 rooms, 21 story. TV. In-room modem link. Restaurant, bar. In-house fitness room, sauna. Indoor pool, whirlpool, poolside service. Outdoor tennis. Racquetball. Valet parking available. Business center. Opposite river. $$

★★★ INN ON FERRY STREET. 84 E Ferry St, Detroit (48202). Phone 313/871-6000; toll-free 800/207-6900. 42 rooms between four restored Victorian mansions and two carriage houses. Complimentary continental breakfast, afternoon social reception. TV. Health club privileges. Laundry services. Concierge. Valet parking available. $

★★ MARRIOTT RENAISSANCE CENTER. Renaissance Center, Detroit (48243). Phone 313/568-8000; toll-free 800/352-0831. www.marriott.com. 1,298 rooms, 72 story. Check-out 1 pm. TV; VCR available (movies). In-room modem link. Restaurant, bar. Room service 24 hours. In-house fitness room, sauna. Indoor pool. Business center. Luxury level. $$

★★★ OMNI DETROIT RIVER PLACE. 1000 River Place Dr, Detroit (48207). Phone 313/259-9500; toll-free 800/843-6664. www.omnihotels.com. This elegant hotel is located downtown on the historic waterfront. Guest rooms boast views of the river and the Canadian border. The hotel has a championship croquet court, the only USCA-sanctioned croquet court in Michigan. 108 rooms, 5 story. Pet accepted, some restrictions; fee. Check-out noon. TV; cable (premium), VCR available. In-room modem link. Restaurant, bar. In-house fitness room, massage, sauna. Indoor pool, whirlpool. Valet parking available. $

★ THE SHORECREST MOTOR INN. 1316 E Jefferson at Rivard, Detroit (48207). Phone 313/568-3000; toll-free 800/992-9616. www.shorecrestmi.com. 54 rooms, 2 story. Check-out noon. TV; cable (premium). Internet access. Restaurant. Room service. ¢

Pontiac

★★★ MARRIOTT. 3600 Centerpoint Pkwy, Pontiac (48341). Phone 248/253-9800; toll-free 800/228-9290. www.marriott.com. 290 rooms, 11 story. Check-out noon. TV; cable (premium), VCR available. Restaurant. In-house fitness room. Indoor pool. Business center. $$

PLACES TO EAT

A long day of driving is sure to make you hungry. At the end of your journey, take a table at one of the following restaurants.

Birmingham

★★★ FORTÉ. 201 S Old Woodward Ave, Birmingham (48009). Phone 248/594-7300. www.forterestaurant.com. Contemporary American menu. Closed Sun; holidays. Lunch, dinner. Entertainment. Valet parking available. Business casual attire. $$$

Michigan

✻ *Woodward Avenue (M-1)*

★★★ **RUGBY GRILLE.** *100 Townsend St, Birmingham (48009). Phone 248/642-5999. www.townsendhotel.com.* Located in the European-style Townsend Hotel, this internationally inspired restaurant's dinner menu features fine steak and chops, fresh seafood, and homemade pastas. Breakfast and lunch, with slightly more standard menu options, are also available in the polished-wood, burgundy-toned space. Continental menu. Breakfast, lunch, dinner, brunch. Bar. Valet parking available. $$$
D

Detroit

★★★ **BARON'S STEAKHOUSE.** *1000 River Place Dr, Detroit (48207). Phone 313/259-4855. www.omnihotels.com.* The snug dining room and patio in the River Palace Hotel overlook the Detroit River. Seafood menu. Breakfast, lunch, dinner. Bar. Children's menu. Outdoor seating. $$$

★★★ **CAUCUS CLUB.** *150 W Congress St, Detroit (48226). Phone 313/965-4970.* One of the city's culinary legends, this English-style dining room serves American cuisine with European accents. The jumbo Dover sole in lemon butter is a signature dish, and the cozy, dimly lit bar a popular, after-work hangout. Continental menu. Closed Sun; major holidays. Lunch, dinner. Entertainment Fri-Sat. $$
D

★★ **FISHBONE'S RHYTHM KITCHEN CAFÉ.** *400 Monroe St, Detroit (48226). Phone 313/965-4600.* Cajun/Creole menu. Closed Dec 25. Lunch, dinner, Sun brunch. Bar. Valet parking available. $$

★ **FLUX.** *630 Woodward Ave, Detroit (48226). Phone 313/962-1200.* American menu. Closed Sun. Lunch, dinner. Bar. Casual attire. $$
D

★★★ **INTERMEZZO.** *1435 Randolph St, Detroit (48226). Phone 313/961-0707. www.intermezzodetroit.com.* Located in the heart of the Harmonie Park theater district, this contemporary restaurant serves Italian cuisine in a lively atmosphere. Closed Sun, Mon; major holidays. Lunch, dinner. Bar. Entertainment Fri-Sat. Outdoor seating. $$$
D

★★★ **OPUS ONE.** *565 E Larned St, Detroit (48226). Phone 313/961-7766. www.opus-one.com.* Partners Jim Kokas and Ed Mandziara have overseen this dressy dining room for more than ten years. Executive chef Tim Giznsky creates inventive American-continental cuisine that can be enjoyed for dinner, for weekday power lunches, or paired with local theater tickets. Closed Sun; major holidays. Lunch, dinner. Bar. Pianist Tues-Sat. Valet parking available. $$$
D

★★ **PEGASUS TAVERNA.** *558 Monroe St, Detroit (48226). Phone 313/964-6800.* Greek menu. Lunch, dinner. Bar. Children's menu. Lattice-worked ceiling; hanging grape vines. $$
D

★ **TRAFFIC JAM & SNUG.** *511 W Canfield St, Detroit (48201). Phone 313/831-9470. www.traffic-jam.com.* Closed Sun; most major holidays. Dinner. Microbrewery and dairy on the premises. Children's menu. $$
D

Ferndale

★ **ANGEL'S CAFÉ.** *214 W Nine Mile Rd, Ferndale (48220). Phone 248/541-0888.* International menu. Closed Mon. Lunch, dinner. A vegetarian-friendly café with an art gallery inside. $

★ **THE BLUE NILE.** 545 W Nine Mile Rd, Ferndale (48220). Phone 248/547-6699. Ethiopian menu. Closed Sun. Lunch, dinner. **$**

★★ **NAMI SUSHI BAR.** 201 W Nine Mile Rd, Ferndale (48220). Phone 248/542-6458. Japanese menu. Lunch, dinner. **$$**

Royal Oak

★★**ANDIAMO OSTERIA OF ROYAL OAK.** 129 S Main St, Royal Oak (48067). Phone 248/582-9300. www.andiamoitalia.com. Italian menu. Closed major holidays. Lunch, dinner. Bar. Children's menu. Casual attire. **$$**
[D]

★ **ATHENS CONEY ISLAND.** 32657 Woodward Ave, Royal Oak (48073). Phone 248/549-1488. A Woodward tradition since the 1950s, Athens Coney Island is a retro-style diner known for the freshest all-beef coney dogs and the quickest service in town. On many summer nights, the parking lots is packed with classic cars. American menu. Lunch, dinner. No credit cards accepted. **$**

Edge of the Wilderness
✳ MINNESOTA

Quick Facts

LENGTH: 47 miles.

TIME TO ALLOW: 3 hours.

BEST TIME TO DRIVE: Fall foliage season (end of September, beginning of October). High season includes July and August.

BYWAY TRAVEL INFORMATION: Byway local Web site: www.scenicbyway.com.

SPECIAL CONSIDERATIONS: Highway 38 courses up and down and curves often—that's part of its charm. Locals named it Highway Loop-de-Loop in the early days. As you drive the Byway, be aware of the lower speed limits, other traffic on the highway, and weather conditions. This is a working roadway, with trucks carrying logs and other local products.

BICYCLE/PEDESTRIAN FACILITIES: Pedestrian and bicyclist facilities are not provided along the route. However, numerous hiking and mountain biking trails spur off the Byway. Facilities are available at the Scenic State Park near Bigfork, as well as a bike trail that runs from Gunn Park to Grand Rapids.

Celebrate northern hospitality, hometown pride, and the treasures of our natural heritage. Minnesota, midway between America's east and west coasts, is the home to 12,000 lakes. It is filled with beautiful country and all the pleasures of the four seasons. The Edge of the Wilderness is the rustic slice of this great state, with more than 1,000 lakes and one mighty river, the Mississippi, all in landscapes of remarkable natural beauty. There are still more trees than people here, offering classic North Woods seclusion.

The Edge of the Wilderness begins in Grand Rapids with meadows and lakes and winds through mixed hardwoods and stands of conifers and aspens in the Chippewa National Forest. Rounding bends and cresting hills, you will find breathtaking views that, during the fall, are ablaze with the brilliant red of sugar maples, the glowing gold of aspen and birch, and the deep bronze of oak.

The Edge of the Wilderness offers some of Minnesota's most popular sporting and resorting opportunities in its unique environment of clear lakes, vast shorelines, and hills blanketed in hardwood forests and northern pines. Recreation seekers will find hiking, camping, fishing, cross-country skiing, and snowmobiling within the Byway corridor. You will find that you are really living on the Edge.

THE BYWAY STORY

Edge of the Wilderness tells historical, natural, recreational, and scenic stories that make it a unique and treasured Byway.

Minnesota

✻ *Edge of the Wilderness*

Historical

At the height of the Great Depression, President Franklin D. Roosevelt formed the Civilian Conservation Corps (CCC) to provide jobs and restore the environment. Nationally, the CCC program cost $3 billion. It removed 15 million families from the welfare rolls and employed 3 million young men. In Minnesota, $85 million was spent and 84,000 enrollees participated. Nationwide, CCC camp crews often were called the Tree Army. They were responsible for planting over 2 billion trees across the United States in nine years. Other tasks included road construction; site preparation; surveys of lakes, wildlife, and streams; and even rodent control.

The Day Lake CCC Camp was one of 20 camps established in Minnesota during the Great Depression. Its first enrollees were African Americans from Kansas and Missouri. Day Lake was the only camp in the forest to host African Americans in the segregated CCC program, where the men came in 1933 to work in Company 786. Day Lake CCC Camp was one of only six that lasted past the CCC era; it became one of four camps in the Chippewa National Forest that housed German prisoners of war during World War II (1943-1945). As the Allied forces gained in the battles of North Africa, they shipped prisoners to the United States. The prisoners worked in local sawmills and for private logging firms. Their wages went to the US Army guards who were present. Security was minimal, and few escapes were reported.

Today, on the west side of Highway 38 are found the remains of a concrete shower. East of the highway and up the hill can be seen the outside stone stairway and a chimney that are remnants of the camp mess hall. Many old camp foundations and sites are also visible. Ironically, after its closing, much of the Day Lake Camp was replanted with red pine, hiding many of the signs of this historical story.

Natural

Aspen, birch, pines, balsam fir, and maples blanket the rolling uplands of the forest along the Edge of the Wilderness Byway. In between these trees, water is abundant, with over 700 lakes, 920 miles of rivers and streams, and 150,000 acres of wetlands. The forest landscape is a reminder of the glaciers that covered northern Minnesota some 10,000 years ago. From the silent flutter of butterflies to the noisy squeal of wood ducks, and from the graceful turn of deer to the busy work of raccoons and beavers, this place of peace is bustling with activity. Many travelers try to identify the laugh of the loon, the honk of the goose, and the chorus of sparrows, red-winged blackbirds, goldfinches, and crickets.

Look skyward to glimpse an eagle, osprey, or turkey vulture. To distinguish among these three mighty birds, watch their manner of flight. The bald eagle has flat wings, the turkey vulture has upswept wings, and the osprey has a crook in its wings. There are more bald eagles on the Edge of the Wilderness than any other part of the lower 48 states. An eagle nest may measure up to 10 feet in diameter and weigh 4,000 pounds. Ospreys live in high nests in treetops, wintering as far away as South America but returning to the same nest year after year.

Minnesota has the greatest number of timber wolves in the lower 48 states as well. They are considered a threatened but not an endangered species in Minnesota. Less often seen but still present on the Edge of the Wilderness are coyotes, bears, and moose. Whitetail deer, ruffed grouse, and waterfowl offer good hunting and wildlife viewing opportunities as well.

During the autumn and spring, you see woody white strands of birch lacing the forest floor along the Byway. If you are a longtime resident or resort vacationer, you may no longer notice this phenomenon. But to newcomers, the question arises, "What happened to make all these trees fall to the ground?"

Paper birch trees, which live only 40 to 80 years, are found throughout Minnesota's northern woodlands. As late as the 1900s, Chippewa tribes in the north built birch bark canoes. They constructed canoes in the early summer when they could easily remove the lightweight bark. They also rolled up birch bark and lighted it, creating a torch. Early settlers prepared birch for railroad ties for the trains that edged northward. In today's economy, birch is used as lumber and firewood and for veneers. Birch also contributes nutrients to the forest floor and has served as food for various insects. Stands of birch often begin to grow after a fire, windstorm, or timber harvest. Another reason so many birch trees lay on the ground is that birch loses out to the taller-growing aspen as both sun-loving species compete. Along the Highway 38 corridor, particularly at Pughole Lake, you can see how birch trees topple due to competition from other trees, disease, and insects, as well as from northern Minnesota's light soils. A look across the lake shows how the pines and hardwoods are gradually taking over a notable birch stand.

Recreational

When the railroad was built, it brought tourists who were interested in fishing and hunting on the Edge of the Wilderness. It also brought the tools and supplies needed to begin the construction of Minnesota Highway 38 in the 1920s. The story of Camp Idlewild on North Star Lake is typical of Itasca County's early development. Two Indiana brothers and a friend, Walter and Lloyd Stickler and Phil Ernest, took the railroad as far as it went in 1907, then walked to what is now known as Camp Idlewild. They build a few guest cabins, and a magazine article was written about the camp. Soon a full-fledged resort was born, thanks to the magazine promotion.

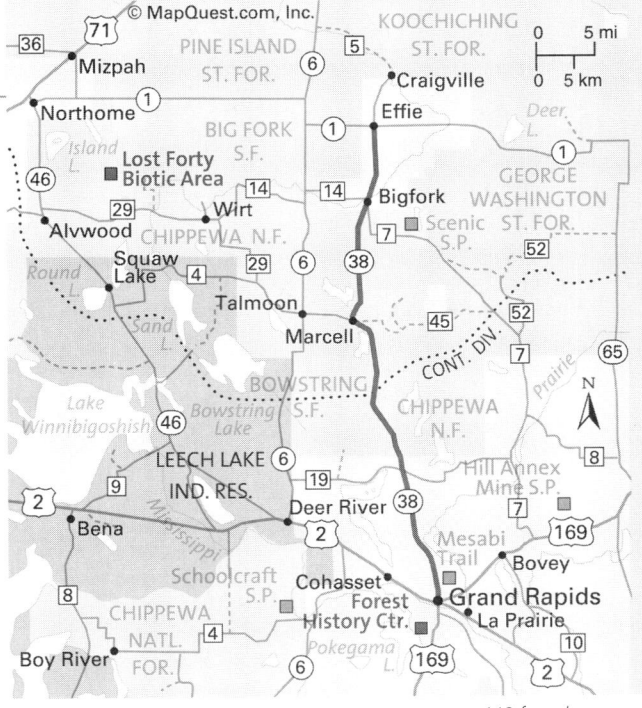

see page A12 for color map

Following World War II, northern Minnesota's tourist and resort industry grew rapidly. Itasca County had a peak of about 300 resorts in the 1940s and 1950s. Today, there are approximately 100 resorts and vacation sites in the area, many vacationers returning year after year. North Star Lake—at more than 3 miles long and about 1/2 mile wide—is considered one of the best fishing and recreational lakes in the area. It is representative of many of the lakes on the Edge of the Wilderness, with its scenic beauty, islands and bays, clear waters, and recreational offerings. The still-visible remnants of the railroad trestle provide good habitat for the lake's many fish. The lake is 90 feet deep and is managed for muskie. Visitors also catch walleye (the state fish of Minnesota), northern pike, largemouth bass, smallmouth bass, bluegills, and crappies.

With a wingspan of over 6 feet, keen vision, and white head and tail feathers, the bald eagle is truly a magnificent bird. People often make a special trip to the Chippewa National Forest just to observe bald eagles. Spending time along the shorelines of the forest's larger lakes is the best way to treat yourself to the sight of an eagle in flight. The forest surrounding the Edge of the Wilderness supports the highest breeding

Minnesota

Edge of the Wilderness

density of eagles in the United States outside of Alaska. Large, fertile lakes; towering red and white pines; and remote areas provide ideal nesting and feeding habitat. Nesting birds return in late February and early March, although a few birds spend the entire winter in the forest. Eggs are laid in early April, and the young leave their lofty nests in mid-July. Eagles occupy their breeding areas until the lakes freeze over in November or December.

The best opportunity for viewing bald eagles is from a boat; in fact, one of the best opportunities to see eagles is to canoe down the Mississippi River between Cass Lake and Lake Winnie. You can search the lakeshore with binoculars to spot eagles that are eating fish on the beaches. These birds of prey often perch in trees found around the larger lakes, such as Winnie, Cass, Leech, and Bowstring. The Bigfork and Leech rivers are also favorite eagle areas. However, if you do not have a boat, you can simply find an area along the beach with a good panoramic view of the lake. Campgrounds, picnic areas, and boat landings are good places to visit. Eagles also frequent Leech Lake Dam and Winnie Dam, particularly in the spring before the lake ice goes out.

Scenic

What makes the Edge of the Wilderness truly unique is its rich and wide variety of upper Minnesota terrain, vegetation, wildlife, and history. While some elements are fairly common in other areas, no other route exposes travelers to so much variety in such a short distance along such a beautiful and accessible corridor. The Edge of the Wilderness is definitely a road to take slowly in order to enjoy the scenery of forests and meadows.

On the outskirts of Grand Rapids, the corridor begins to hint at the landscape to come. At first, the route is flat and flanked by mixed lowland meadows, swamps, and lakes. Very quickly, however, the corridor leaves most signs of the city and begins its rolling journey through mixed hardwoods and stands of conifers with aspen. With so many curves and hills, the corridor hides from view many memorable scenes until the traveler is upon them. Seemingly innocent turns in the road yield eye-popping surprises. The terrain continues in this way for half its length until the town of Marcell, where the terrain flattens slightly and offers more conifers. Between Bigfork and Effie, the corridor's terminus, the landscape introduces lowland wetlands, a flatter landscape that served as the bed of glacial Lake Agassiz thousands of years ago. Surrounding forests continue to contain aspen and lowland conifers such as jack pine and spruce.

A visitor to the Edge of the Wilderness could simply travel the route without stopping to take advantage of its recreational and interpretive opportunities, yet still leave with many vivid memories of the corridor. The Byway hugs the terrain, rising above lakes and then sloping down to meet their shores before rising up again through the trees and down into wetlands. Throughout the southern half of the route, maples, paper birch, and quaking aspen branches provide a canopy that envelops travelers in the lush forest. During the fall color season, the corridor displays bright red sugar maples, warm gold birch and aspens, and maroon red oaks. After the leaves have fallen and the ground is covered with snow, the forest opens up and offers new opportunities to see the terrain and spy on wildlife.

HIGHLIGHTS

For your convenience, the Edge of the Wilderness has milepost markers that guide you on your tour. The Byway officially begins at Grand Rapids, Minnesota. Designated as milepost 0, the Byway proceeds north with sites of interest marked consecutively as follows:

- **Mile 0:** Grand Rapids, a historic logging and paper-making region
- **Mile 3.4:** Lind-Greenway Mine, a historic iron mine
- **Mile 8.5:** Black Spruce/Tamarack Bog Habitat, one of the largest and most mature bogs in the area
- **Mile 12.6:** Trout Lake and Joyce Estate, an impressive estate created in the 1920s
- **Mile 13.4:** Birch Stand at Pughole Lake, an excellent place for viewing wildlife
- **Mile 19.0:** Day Lake CCC Camp, which has a long and varied history of use as both a Depression-era work camp and a German POW camp during World War II
- **Mile 20.9:** Laurentian Divide
- **Mile 23.5:** White Cedar Stand, an important habitat area for indigenous wildlife
- **Mile 24.5:** Scenic overlook at North Star Lake
- **Mile 27.7:** Marcell, a historic logging and rail town
- **Mile 28.7:** Chippewa National Forest Ranger Station, which offers information and displays
- **Mile 30.5:** Gut and Liver Line, the historic railway that once hauled logs and supplies for the area's inhabitants
- **Mile 40.0:** Bigfork, a small logging community
- **Mile 47.0:** Effie, a small agricultural community

THINGS TO SEE AND DO

Driving along the Edge of the Wilderness will certainly keep your senses engaged, but if you yearn to get out of the car and stretch your legs, or if you'd like to make a mini-vacation out of your trip, check out these attractions along the route.

CENTRAL SCHOOL. *10 NW 5th St, Grand Rapids (55744). Phone 218/326-6431.* Heritage center housing a historical museum, a Judy Garland display, antiques, shops, and a restaurant. Open Mon-Fri 9:30 am-5 pm, Sat 10 am-4 pm, Sun (summer) noon-4 pm.

CHIPPEWA NATIONAL FOREST. *200 Ash Ave NW, Cass Lake (56633). Phone 218/335-8600. www.fs.fed.us/recreation/map/state_list.shtml.* Has 661,400 acres of timbered land; 1,321 lakes, with 699 larger than 10 acres; swimming, boating, canoeing, hiking, hunting, fishing, picnicking, camping (fee); winter sports. Bald eagle viewing; several historic sites.

FOREST HISTORY CENTER. *2609 County Rd 76, Grand Rapids (55744). Phone 218/327-4482. www.mnhs.org/places/sites/fhc/.* The center includes a museum building, re-created 1900 logging camp, and log drive wanigan (a floating cook shack used when the logs and men traveled downstream to the mills) maintained by the Minnesota Historical Society as part of an interpretive program. Early Forest Service cabin and fire tower, modern pine plantation, living-history exhibits; nature trails. Open June 1-Labor Day, Mon-Sat 10 am-5 pm, Sun noon-5 pm. $$

JUDY GARLAND BIRTHPLACE AND CHILDREN'S MUSEUM. *2727 US 169 S, Grand Rapids (55744). Phone 218/327-9276; toll-free 800/664-JUDY. www.judygarlandmuseum.com.* Childhood home of the actress. Open daily 10 am-5 pm; closed major holidays (except Memorial Day, July 4, and Labor Day). $

Minnesota

❋ *Edge of the Wilderness*

POKEGAMA DAM. *34385 S US Hwy 2, Grand Rapids (55744). Phone 218/326-6128.* Camping on 21 trailer sites (hookups, dump station; 14-day maximum); picnicking, fishing. $$$

QUADNA MOUNTAIN RESORT AREA CONVENTION CENTER. *100 Quadna Rd, Hill City (55748). Phone 218/697-8444.* Quad chairlift, two T-bars, rope tow; patrol, school, rentals, snowmaking; motel, lodge, and restaurant. Cross-country trails. 16 runs, longest run 26,430 feet, vertical drop 350 feet. Open Thanksgiving-mid-Mar, Fri-Tues. Also golf, outdoor tennis, horseback riding, and lake activities in summer. $$$$

SCENIC STATE PARK. *56956 Scenic Hwy #7, Bigfork (56628). Phone 218/743-3362.* Primitive area of 3,360 acres with seven lakes. Swimming, fishing, boating (ramp, rentals), hiking; cross-country skiing, snowmobiling; picnicking; camping (electrical hookups), lodging; interpretive programs.

PLACES TO STAY

If you choose to include an overnight stay in your trip along this Byway, Mobil Travel Guide recommends the following lodgings.

Bigfork

★★ **ARCADIA LODGE.** *52001 County Rd 284, Bigfork (56628). Phone 218/832-3852; toll-free 888/832-3852. www.arcadialodge.com.* 17 rooms, 2 story. Unique and secluded cabins nestled in the forests with plenty of fishing, swimming, and other outdoor sports available. $$$

★★ **THE WILDERNESS LODGE.** *51760 Becker Rd, Bigfork (56628). Phone 218/743-3458; toll-free 877/865-6343. www.thewildernesslodge.com.* 4 rooms, 2 story. Hidden away in the pine forests, these cabins offer plenty of outdoor activities. $$$

Grand Rapids

★ **AMERICINN.** *1812 S Pokegama Ave, Grand Rapids (55744). Phone 218/326-8999; toll-free 800/634-3444. www.americinn.com.* 43 rooms, 2 story. Complimentary continental breakfast. Check-out 11 am. TV; cable (premium). Sauna. Indoor pool, whirlpool. ¢

★ **COUNTRY INN BY CARLSON.** *2601 S US 169, Grand Rapids (55744). Phone 218/327-4960; toll-free 800/456-4000. www.countryinns.com.* 59 rooms, 2 story. Pet accepted, some restrictions. Complimentary continental breakfast. Check-out noon. TV; cable (premium). Indoor pool, whirlpool. Downhill ski 10 miles, cross-country ski 2 miles. ¢

★★ **SAWMILL INN.** *2301 S Pokegama Ave, Grand Rapids (55744). Phone 218/326-8501; toll-free 800/235-6455. www.sawmillinn.com.* 124 rooms, 2 story. Pet accepted, some restrictions. Check-out noon. TV. Restaurant, bar. Room service. Sauna. Game room. Indoor pool, whirlpool, poolside service. Free airport transportation. ¢

★ **SUPER 8.** *1702 S Pokegama Ave, Grand Rapids (55744). Phone 218/327-1108; toll-free 800/800-8000. www.super8.com.* 58 rooms, 2 story. Complimentary continental breakfast. Check-out 11 am. TV; cable (premium). In-room modem link. Guest laundry. Cross-country ski 1 mile. ¢

Marcell

★★ **TIMBERWOLF INN.** *PO Box 8, Marcell (56657). Phone 218/832-3990.* 12 rooms, 2 story. This log-sided inn features a large corner fireplace and is ideal for skiing, fishing, and snowmobiling. $$

★★ **NORTH STAR LAKE RESORT.** *PO Box 128, Marcell (56657). Phone 218/832-3131; toll-free 800/356-6784. www.northstarlakeresort.com.* 19 rooms, 2 story. Excellent lakeside resort with plenty of activities for children. **$$**

PLACES TO EAT

★★ **THE TIMBERWOLF INN.** *PO Box 8, Marcell (56657). Phone 218/832-3990.* American menu. Lunch, dinner. Featuring American regional cuisine in an attractive wilderness setting. **$**

★★ **SAWMILL INN.** *2301 Pokegama Ave, Grand Rapids (55744). Phone 218/326-8501. www.sawmillinn.com.* American menu. Lunch, dinner. Known for choice prime rib. **$$**

The Grand Rounds Scenic Byway
✹ MINNESOTA

Quick Facts

LENGTH: 52 miles.

TIME TO ALLOW: 3 to 4 hours.

BEST TIME TO DRIVE: Spring through fall. High season is summer.

BYWAY TRAVEL INFORMATION: Minneapolis Park & Recreation Board: 612/661-4800; Byway local Web site: www.mnmississippiriver.com.

SPECIAL CONSIDERATIONS: No significant or unusual seasonal accessibility limitations apply to the Grand Rounds Scenic Byway. With only a few exceptions, the route is generally above local flood plains. During the winter, in addition to the roadway surfaces, a single paved trail is plowed throughout the system for shared use by pedestrians and cyclists.

RESTRICTIONS: Truck traffic is prohibited from using all but several short segments of the Byway.

BICYCLE/PEDESTRIAN FACILITIES: Throughout the entire Grand Rounds Scenic Byway System, paved pedestrian trails (45 miles) and bicycle trails (43 miles) are separated from the adjacent parkway road surfaces. The entire route is designed and maintained for passenger vehicles, bicycles, and pedestrians. The 8-foot-wide paths are designed for two-directional travel, with one exception: the bike path around the Chain of Lakes. When on the paths, please obey the 10 mph speed limit, be considerate of others, keep to the right side of all paths, sound off when passing, and observe all current bike regulations.

Offering a unique Byway experience, the Grand Rounds Scenic Byway is a continuous course of paved pathways. It encompasses more than 50 miles of parks, parkways, bike paths, and pedestrian paths that encircle the Byway's host city, Minneapolis. The system includes the Chain of Lakes, Lake Nokomis, Lake Hiawatha, and the Mississippi River, as well as Minnehaha, Shingle, and Bassetts Creek. Currently, the Byway is divided into seven main sections: the Downtown Riverfront, the Mississippi River, Minnehaha Park, Chain of Lakes, Theodore Wirth, Victory Memorial District, and the Northeast District.

Parks, recreation centers, historic districts, and lakes are all defining characteristics of the Grand Rounds Scenic Byway. Depending on the season, you interact with a landscape encompassing combinations of natural areas, historical features, and cultural amenities. The Grand Rounds Scenic Byway has been the crème de la crème of urban Byways for more than a century, being the longest continuous system of public urban parkways and providing motorists with access to an unprecedented combination of intrinsic qualities of the highest caliber. So spend a day in the park and enjoy the best of Minneapolis.

THE BYWAY STORY

The Grand Rounds Scenic Byway tells archaeological, cultural, historical, natural, recreational, and scenic stories that make it a unique and treasured Byway.

Minnesota

❊ The Grand Rounds Scenic Byway

Archaeological

Recent archaeological digs occurring in the old milling district of Minneapolis have revealed artifacts from the time when Minneapolis was a milling capital. Excavations of this area have allowed archaeologists to uncover pieces of the past that indicate the way ordinary people in the milling districts lived and worked. Visitors can explore the way life used to be in Minneapolis at the newly constructed Mill Ruins Park. This park is a restoration of original walls, canals, and buildings of the West Side Milling District. Another of the archaeological features along the Grand Rounds Scenic Byway in Minnesota is the Winchell Trail. Located on the Mississippi River, this trail preserves an original Native American trail that linked two celebration sites.

Cultural

An urbanized culture originating in the early 1800s is what visitors to the Grand Rounds Scenic Byway will experience. However, amid this urban culture is a prominent appreciation for nature and the arts. This can be seen in the beautiful parkways, lakes, and parks on the Grand Rounds. These places that combine nature with art and architecture are the clearest picture of Minneapolis culture today. They represent the earliest residents of this area— Native American cultures that are responsible for many of the names and legends visitors find all along the Byway. Hiawatha Avenue and Minnehaha Park are examples of remnants of a culture that held a great respect for the waters of Minnesota.

Architectural styles ranging from fine to functional exist all along the Byway. Unique bridges hold views to the old milling district of the city, where settlers made a living working in the sawmills. Modern art of a more recent period can be found in the uptown area of the Byway as well as in the lakeside areas. In addition to the present city, this art represents the contribution of a culture that has evolved from all the previous inhabitants of the Minneapolis area.

Historical

The Grand Rounds Scenic Byway is one of the nation's longest and oldest connected parkway systems owned entirely by an independent park board. This road has many historical qualities. For example, the Stone Arch Bridge found along the Byway is a National Civil Engineering Historical Landmark Site. Minnehaha Falls, also found along the Byway, is immortalized in Longfellow's epic poem "The Song of Hiawatha." The fine craftsmanship is a significant element of the beauty of the Grand Rounds Scenic Byway, while other forms of architecture are being brought to the surface again. The St. Anthony Heritage Trail offers a look at Minneapolis from a perspective that explores the many highlights in the city's history.

Natural

With the Mississippi running through it and lakes scattered all around it, it's no wonder the Grand Rounds is a scenic attraction for nature lovers. In very few other urban places in the world is nature so artfully integrated with the city. The Grand Rounds is centered around several parks and parkways that bring the best of nature to the fingertips of urban life. St. Anthony Falls is the only natural waterfall and gorge along the entire 2,350-mile course of the Mississippi River; Minnehaha Park contains one of the most enchanting waterfalls in the nation; and the Chain of Lakes is a place on the Byway to enjoy the lake shore or sail a boat onto the water. All in all, the Grand Rounds Scenic Byway connects nine lakes, three streams, two waterfalls, the Mississippi River, and surrounding nature to travelers.

At the Eloise Butler Wildflower Garden, you will find hundreds of plants and flowers native to Minnesota. Eloise Butler feared that Minnesota's natural characteristics would be lost with urban development, so she did some development of her own. The garden she created displayed plants and flowers from many different habitats. The garden has expanded into a preserve that is also home to many birds. Nearly all the parks on the Grand Rounds offer a garden of some type. A rose garden is located on Lake Harriet Parkway, while at Minnehaha Park, the Pergola Garden displays native wildflowers. Theodore Wirth Park includes a section known as Quaking Bog. This 5-acre bog is covered in moss and tamarack trees. Exploring these parks and gardens will give you a taste of the best of nature.

Recreational

With such widely varied attractions in one city, you'll find many places to see and different activities to do at each of those places. Outdoor recreation is not out of reach, while touring historical buildings and parks is a possibility as well. Sailboats and canoes are seen every day on the Chain of Lakes, while bikers and hikers have miles of roads to enjoy. The Grand Rounds is one of the nation's first park systems to create separated walking and biking paths. The Byway is also host to many national running events; for example, the Twin Cities Marathon on the Grand Rounds was named "the Most Beautiful Urban Marathon in the Country." Four golf courses can be found on the Grand Rounds Scenic Byway, each one with its own beautiful and challenging terrain.

If recreation needs to be more thought-provoking, be sure to tour some of the more unusual places on the Byway. Sculpture gardens in uptown Minneapolis offer displays of creativity that will capture your imagination. The St. Anthony Heritage Trail offers a place for touring the historic districts of Minneapolis. When you've read all the informational signs, stop at a café or restaurant for a bite to eat and a bit of ambience. Rumor has it that the Nicolette Island Inn has the best desserts in town!

see page A16 for color map

Scenic

The scenes of the Grand Rounds vary greatly, but they are the city of Minneapolis at its finest. The architecturally diverse Minneapolis skyline is visible from nearly every portion of the Grand Rounds. Watch the buildings sparkle in the daylight or see them silhouetted by a Minnesota sunset. The heart of Minneapolis keeps the entire city healthy with growing prosperity and scenic style. The familiar view of the Stone Arch Bridge leads travelers to the historic district of Minneapolis with brick streets and flour mills. The sight of Minnehaha Falls is also a view linked directly to the ideals that embody Minneapolis. The falls have been admired for years, and their beauty is continually preserved. Landmarks like these characterize Minneapolis as a unique city in the United States.

Within a 50-mile drive, motorists witness diverse landscape settings, including natural, urban, historical, and ethnic settings, all from a unified Byway. The Grand Rounds Scenic Byway also includes outdoor sculptures, fountains, and public gardens that have been created by world-renowned artists. Where else can you find such a creative rendition of a giant cherry captured by an equally giant spoon? Fountains

Minnesota

The Grand Rounds Scenic Byway

and sculptures of Minneapolis are the jewels of the city that set it apart from other urban sections of the country.

Often, the landmarks and artwork of Minneapolis lead visitors to the scenic trails of the St. Anthony Falls or Minnehaha Park. If you want to get a closer look at the Mississippi, follow the River Gorge hiking trail, which offers a natural setting right in the middle of the city. In the fall, these tree-lined trails are painted all the colors of autumn. Deep oranges and yellows are splattered about, creating a natural work of art that visitors can walk through. The lakes and rivers of Minneapolis are treasured, and there is something exhilarating about driving along the shores of sparkling blue water. Perhaps you will catch a glimpse of a sailboat gliding serenely along the water or a canoe passing by an inlet near the shore. And when the water is calm, the city skyline is mirrored in the water to create a perfect picture of symmetry. Trees and buildings line the lakes of Minneapolis, making this Byway the perfect combination of man and nature.

HIGHLIGHTS

Drive the Grand Rounds and you will see virtually all of the following sites.

- Beginning at the northeastern terminus of the Grand Rounds on Highway 35 W, travel west and follow the Byway as it winds around to **Theodore Wirth Park.** This park is well worth the visit and includes the **Eloise Butler Wildflower Garden,** with its spectacular display of wildflowers and other flowering plants.

- Returning to the Byway, follow it around **Cedar Lake** to **Lake Calhoun.** Here, turn northeast and travel around **Lake of the Isles** to **Loring Park.** Loring Park is home to the well-known **Sculpture Garden.** With its impressive display of sculpture and fanciful artwork, the Sculpture Garden is one of the highlights of the Grand Rounds.

- Return in the direction you came, only continue to wind around Lake Calhoun to **Lake Harriet** to visit the **Lake Harriet Refectory,** another popular site along the Grand Rounds.

- Continue following the Byway as it turns and travels east to **Minnehaha Park,** where you may view the falls, and **Lake Hiawatha,** immortalized by Longfellow's poem, "The Song of Hiawatha."

- From here, follow the Byway north to its terminus at **Nicollet Island** and **Boom Island.** Both are interesting places to visit in downtown Minneapolis if you have the time. Otherwise, you are pretty much right back where you started!

THINGS TO SEE AND DO

Driving along the Grand Rounds Scenic Byway will certainly keep your senses engaged, but if you yearn to get out of the car and stretch your legs, or if you'd like to make a mini-vacation out of your trip, check out these attractions along the route.

AMERICAN SWEDISH INSTITUTE. *2600 Park Ave, Minneapolis (55407). Phone 612/871-4907. www.americanswedishinst.org.* This museum is housed in a turn-of-the-century, 33-room mansion and features hand-carved woodwork, porcelainized tile stoves, and sculpted ceilings, plus Swedish fine art and artifacts. Open Tues, Thurs-Sat noon-4 pm; Wed noon-8 pm; Sun 1-5 pm; closed Mon, holidays. **$**

BASILICA OF ST. MARY. *88 N 17th St, Minneapolis (55403). Phone 612/333-1381. www.mary.org.* Renaissance architecture patterned after the Basilica of St. John Lateran in Rome. Open daily.

BELL MUSEUM OF NATURAL HISTORY. *10 Church St SE, Minneapolis (55455). Phone 612/624-7083. www.bellmuseum.org.* Dioramas show Minnesota birds and mammals in natural settings; special traveling exhibits; exhibits on art, photography, and natural history research change frequently. The Touch and See Room encourages hands-on exploration and comparison of natural objects. Free admission Sun. Open Tues-Fri 9 am-5 pm, Sat 10 am-5 pm, Sun noon-5 pm; closed Mon, some holidays. **$**

ELOISE BUTLER WILDFLOWER GARDEN. *Theodore Wirth Pkwy and Glenwood Ave, Minneapolis (55422). Phone 612/370-4903. www.minneapolisparks.org.* Horseshoe-shaped glen contains a natural bog, swamp, and habitat for prairie and woodland flowers and birds. Guided tours. Open Apr 1-Oct 15, daily 7:30 am-30 minutes before sunset. **FREE**

FREDERICK R. WEISMAN ART MUSEUM. *333 E River Rd, Minneapolis (55455). Phone 612/625-9494. hudson.acad.umn.edu.* Striking stainless-steel exterior is oddly shaped and was designed by Frank Gehry. Inside are collections of early 20th-century and contemporary American art, Asian ceramics, and Native American Mimbres pottery. Special exhibits. Group tours (by appointment, phone 612/625-9656 three weeks in advance). Open Tues, Wed, Fri 10 am-5 pm; Thurs 10 am-8 pm; Sat-Sun 11 am-5 pm; closed Mon, holidays. **FREE**

GUTHRIE THEATER. *725 Vineland Pl, Minneapolis (55403). Phone 612/377-2224. www.guthrietheater.org.* Produces classic plays in repertory, as well as new works. Open nightly Tues-Sun; matinees Wed, Sat, Sun.

HENNEPIN HISTORY MUSEUM. *2303 Third Ave S, Minneapolis (55404). Phone 612/870-1329. www.hhmuseum.org.* Permanent and temporary exhibits on the history of Minneapolis and Hennepin County, including a collection of textiles, costumes, toys, and material unique to central Minnesota. Research library and archive. Open Sun, Wed, Fri-Sat 10 am-5 pm; Tues 10 am-2 pm; Thurs 1-8 pm; closed Mon, holidays. **$**

HUBERT H. HUMPHREY METRODOME. *900 S Fifth St, Minneapolis (55415). Phone 612/332-0386. www.ballparks.com/baseball/american/metrod.htm.* Home of the Minnesota Twins (baseball), the Minnesota Vikings (football), and the University of Minnesota football. Seats up to 63,000. Open Mon-Fri and for special events. **$$**

IDS TOWER. *80 S 8th St, Minneapolis (55402).* 775 feet, 57 stories; one of the tallest buildings between Chicago and the West Coast.

LYNDALE PARK GARDENS. *3900 Bryant Ave S, Minneapolis (55409). Phone 612/661-4800. www.minneapolisparks.org.* Four distinctive gardens: one for roses, two for perennials, and

Minnesota

❋ *The Grand Rounds Scenic Byway*

the Peace (rock) Garden. Displays of roses, bulbs, other annuals and perennials; exotic and native trees; two decorative fountains; adjacent to bird sanctuary. Apr-Sept is the best time to visit. Open daily 7:30 am-10 pm. **FREE**

MINNEAPOLIS CITY HALL. *350 S 5th St, Minneapolis (55415). Phone 612/673-2491.* Father of Waters statue in the rotunda is carved of the largest single block of marble produced from the quarries of Carrara, Italy. Self-guided tours of this 1891 building. Guided tours the first Wed of the month. Open Mon-Fri; closed holidays. **FREE**

MINNEAPOLIS GRAIN EXCHANGE. *400 S 4th St, Minneapolis (55415). Phone 612/321-7101. www.mgex.com.* Visit the cash grain market and futures market. Tours (given Tues-Thurs). Visitors' balcony; reservations required. Open mornings; closed holidays. **FREE**

MINNEAPOLIS INSTITUTE OF ARTS. *2400 Third Ave S, Minneapolis (55404). Phone 612/870-3131 or 612/870-3200. www.artsmia.org.* Masterpieces from every age and culture. Collection of more than 80,000 objects covers American and European painting, sculpture, decorative arts; period rooms, prints and drawings, textiles, photography; American, African, Oceanic, Asian, and ancient Asian objects. Lectures, classes, films (fee), special events (fee); restaurants. Open Tues, Wed, Fri-Sat 10 am-5 pm; Thurs 10 am-9 pm; Sun 11 am-5 pm; closed July 4, Thanksgiving, Dec 25. **FREE**

MINNEAPOLIS SCULPTURE GARDEN. *726 Vineland Pl, Minneapolis (55403). Phone 612/370-3996. www.minneapolisparks.org.* Ten-acre urban garden features more than 40 sculptures by leading American and international artists, plus a glass conservatory. Open daily 6 am-midnight. **FREE**

MINNEHAHA PARK. *4801 Minnehaha Ave S, Minneapolis (55417). Phone 612/230-6400.* Minnehaha Falls, immortalized as the "laughing water" of Longfellow's epic poem "Song of Hiawatha;" statue of Hiawatha and Minnehaha; picnicking; Stevens House, the first frame house built west of the Mississippi. Park open daily 6 am-10 pm. **FREE**

MINNESOTA LYNX (WNBA). *Target Center, 600 First Ave N, Minneapolis (55403). Phone 612/673-8400. www.wnba.com/lynx.* Professional basketball team.

MINNESOTA TIMBERWOLVES (NBA). *Target Center, 600 First Ave N, Minneapolis (55403). Phone 612/337-DUNK. www.nba.com/timberwolves.* Professional basketball team.

MINNESOTA TWINS (MLB). *Metrodome, 34 Kirby Puckett Pl, Minneapolis (55415). Phone 612/375-1366. www.twins.mlb.com.* Professional baseball team.

MINNESOTA VIKINGS (NFL). *Metrodome, 900 S Fifth Ave, Minneapolis (55415). Phone 612/338-4537. www.vikings.com.* Professional football team.

★ **NICOLLET MALL.** *700 Nicollet Mall, Minneapolis (55403). Phone 612/332-3101.* A world-famous shopping promenade with a variety of shops, restaurants, museums, and art galleries, as well as entertainment ranging from an art show to symphony orchestra performances. No traffic is allowed on this avenue except for buses and cabs. Beautifully designed with spacious walkways, fountains, shade trees, flowers, and a skyway system, this mall is certainly worth a visit.

RIVER CITY TROLLEY. *1301 2nd Ave S, Minneapolis (55403). Phone 612/204-0000. www.rivercitytrolley.com.* A 40-minute loop traverses the core of downtown, passing through the Mississippi Mile, St. Anthony Falls, and the Warehouse District. A Chain of Lakes tour is also available. On-board narration; tours run approximately every 20 minutes. Operates May-Oct, Tues-Fri 10 am-4 pm, Sat-Sun, holidays 10 am-4 pm. **$$**

ST. ANTHONY FALLS. *125 Main St SE, Minneapolis (55414). Phone 612/332-3600. www.nps.gov/miss/maps/model/stanthony.html.* Head of the navigable Mississippi, site of the village of the St. Anthony. A public vantage point at the upper locks and dam provides a view of the falls and of the operation of the locks. Also includes a renovated warehouse with shops and restaurants. Open May-Oct, Wed-Sun noon-5 pm.

WALKER ART CENTER. *Vineland Pl and Lyndale Ave S, Minneapolis (55403). Phone 612/375-7622 or -7577 (recording). www.walkerart.org.* Permanent collection of 20th-century paintings, sculptures, prints, and photographs; also changing exhibits, performances, concerts, films, and lectures. Gallery open Tues, Wed, Fri-Sat 10 am-5 pm; Thurs 10 am-9 pm; Sun 11 am-5 pm; closed Mon; free admission on the first Thurs and Sat of each month. $$

PLACES TO STAY

If you choose to include an overnight stay in your trip along this Byway, Mobil Travel Guide recommends the following lodgings.

★ **BEST WESTERN DOWNTOWN.** *405 S 8th St, Minneapolis (55404). Phone 612/370-1400; toll-free 800/780-7234. www.bestwestern.com.* 159 rooms, 4 story. Complimentary continental breakfast. Check-out noon, check-in 3 pm. TV; cable (premium). In-room modem link. In-house fitness room, sauna. Indoor pool, whirlpool, poolside service. $

★★ **DOUBLETREE HOTEL PARK PLACE.** *1500 Park Pl Blvd, Minneapolis (55416). Phone 612/542-8600; toll-free 800/222-8733. www.doubletree.com.* This hotel offers a quiet suburban location with the convenience of being near downtown. Visitors can enjoy a Vikings, Twins, or Timberwolves game at the nearby Metrodome. 297 rooms, 15 story. Check-out noon, check-in 3 pm. TV; cable (premium). In-room modem link. Restaurant, bar. Room service 24 hours. In-house fitness room, sauna. Game room. Indoor pool, whirlpool. Cross-country ski 2 miles. $

★★ **ELMWOOD HOUSE.** *1 E Elmwood Pl, Minneapolis (55419). Phone 612/822-4558; toll-free 888/822-4558. www.elmwoodhouse.us.* 4 rooms, 3 story. This 1887 Norman chateau is listed on the National Register of Historic Homes. $

★★ **EMBASSY SUITES HOTEL DOWNTOWN.** *425 S 7th St, Minneapolis (55415). Phone 612/333-3111; toll-free 800/EMBASSY. www.embassysuites.com.* A six-story atrium filled with tropical plants greets visitors as they enter this hotel, located in the business and financial district. The property is near the Target Center, Orpheum Theatre, and fine shops. 216 suites, 6 story. Check-out noon, check-in 3 pm. TV; cable (premium). In-room modem link. Restaurant, bar. In-house fitness room, sauna, steam room. Indoor pool, whirlpool. Cross-country ski 1 mile. $

★★★ **THE GRAND HOTEL.** *615 Second Ave S, Minneapolis (55402). Phone 612/288-8888; toll-free 866/843-4726. www.grandhotelminneapolis.com.* 140 rooms, 12 story. Check-out noon, check-in 3 pm. TV; cable (premium), VCR available. Restaurant, bar. Babysitting services available. In-house fitness room, spa. Indoor pool. Business center. $$

★★ **HOLIDAY INN METRODOME.** *1500 Washington Ave S, Minneapolis (55454). Phone 612/333-4646; toll-free 800/448-3663. www.holiday-inn.com.* 265 rooms, 14 story. Pet accepted; fee. Check-out noon, check-in 3 pm. TV; cable (premium). In-room modem link. Restaurant, bar. In-house fitness room. Indoor pool, whirlpool. Airport transportation. $

Minnesota

✤ *The Grand Rounds Scenic Byway*

★★★ **HYATT REGENCY.** *1300 Nicollet Mall, Minneapolis (55403). Phone 612/370-1234; toll-free 800/233-1234. www.hyatt.com.* Located in the heart of the downtown business and financial district, this hotel offers access to the Minneapolis Convention Center via the city's skywalk. It is near the Guthrie Theatre, Walker Art Center, and much more. 533 rooms, 24 story. Check-out noon, check-in 3 pm. TV; VCR available (movies). In-room modem link. Restaurant, bar. In-house fitness room. Health club privileges. Indoor pool, whirlpool. Cross-country ski 1 mile. Airport transportation. Concierge. Luxury level. **$$**

★★★ **THE MARQUETTE.** *710 Marquette Ave, Minneapolis (55402). Phone 612/333-4545; toll-free 800/328-4782. www.marquettehotel.com.* Located in the downtown area, this hotel is connected to shops, restaurants, and entertainment by the city's skywalk system. 277 rooms, 19 story. Pet accepted, some restrictions; fee. Check-out noon, check-in 3 pm. TV; cable (premium), VCR available. In-room modem link. Restaurant, bar. In-house fitness room. Cross-country ski 1 mile. Business center. Luxury level. **$$$**

★★★ **MARRIOTT CITY CENTER.** *30 S 7th St, Minneapolis (55402). Phone 612/349-4000; toll-free 800/228-9290. www.marriott.com.* Linked by the enclosed skywalk to many of the city's offices and shopping complexes in the downtown area, this hotel is very convenient. Golf courses and tennis facilities are nearby. 583 rooms, 31 story. Pet accepted; fee. Check-out noon, check-in 3 pm. TV; cable (premium), VCR available (movies). In-room modem link. Restaurant, bar. In-house fitness room. Health club privileges. Massage, sauna. Cross-country ski 1 mile. Valet parking available. Business center. Luxury level. **$$**

★★ **NICOLLET ISLAND INN.** *95 Merriam St, Minneapolis (55401). Phone 612/331-1800; toll-free 800/331-6528. www.nicolletislandinn.com.* 24 rooms, 2 story. Check-out noon, check-in 3 pm. TV; cable (premium). In-room modem link. Restaurant, bar. Room service. **$**

★★★ **THE WHITNEY HOTEL.** *150 Portland Ave, Minneapolis (55401). Phone 612/375-1234. www.thewhitneyhotel.com.* Located on the historic riverfront, this hotel offers European style and views of the Mississippi River and St. Anthony Falls. It is near many golf courses, attractions, and a local health club with fitness facilities. 96 rooms, 8 story. Check-out noon, check-in 3 pm. TV; cable (premium), VCR available. In-room modem link. Restaurant, bar. Room service 24 hours. Valet parking available. Airport transportation. Concierge. **$$**

PLACES TO EAT

A long day of driving is sure to make you hungry. At the end of your journey, take a table at one of the following restaurants.

★★ **CAFE BRENDA.** *300 1st Ave N, Minneapolis (55401). Phone 612/342-9230. www.cafebrenda.com.* Vegetarian, seafood menu. Closed Sun; most major holidays. Lunch, dinner. Bar. Totally nonsmoking. **$**

★★★ **CHIANG MAI THAI.** *3001 Hennepin Ave S, Minneapolis (55408). Phone 612/827-1606. www.chiangmaithai.com.* Thai menu. Lunch, dinner. Authentic Thai cuisine in the heart of uptown Minneapolis. **$**

★★★ **D'AMICO CUCINA.** *100 N 6th St, Minneapolis (55403). Phone 612/338-2401. www.damico.com.* This elegant restaurant in a restored warehouse offers pasta dishes and an exceptional

Italian wine list, including many fine reserve selections. Italian menu. Closed Sun; major holidays. Dinner. Bar. Piano Fri-Sat. $$$
D

★★ **FIGLIO.** *3001 Hennepin Ave S, Minneapolis (55408). Phone 612/822-1688.* Italian menu. Lunch, dinner, late night. Bar. Casual attire. Outdoor seating. $$
D

★★ **GARDENS-SALONICA.** *19 5th St NE, Minneapolis (55413). Phone 612/378-0611.* Greek menu. Closed Sun; Thanksgiving, Dec 25. Lunch, dinner. Bar. Casual attire. $$
D

★★★ **GOODFELLOW'S.** *40 S 7th St, Minneapolis (55402). Phone 612/332-4800. www.goodfellows restaurant.com.* This former Forum-Cafeteria space feels like a trip back to the 1930s, with a décor of polished wood, Art Deco fixtures, and jade accents. The chef's regional American offerings highlight local, seasonal ingredients and include dishes like a lamb trio of grilled chop, sweet-onion strudel, and seared leg with marjoram sauce. Closed Sun; holidays. Lunch, dinner. Bar. $$$
D

★ **J. D. HOYT'S.** *301 Washington Ave N, Minneapolis (55401). Phone 612/338-3499. www.jdhoyts.com.* Cajun/Creole menu. Closed some major holidays. Breakfast, lunch, dinner, Sun brunch. Bar. Casual attire. Valet parking available. Outdoor seating. $
D

★★ **JACOB'S 101.** *101 NE Broadway, Minneapolis (55413). Phone 612/379-2508.* Lebanese menu. Breakfast, lunch, dinner. This northeast-side restaurant is a local institution with a great wine list. $

★ **KOYI SUSHI.** *122 N 4th St, Minneapolis (55401). Phone 612/375-9811. www.sawatdee .com/sushi.htm.* Sushi menu. Dinner, late night. Bar. Casual attire. $$$
D

★★★ **MORTON'S OF CHICAGO.** *555 Nicollet Mall, Minneapolis (55402). Phone 612/673-9700. www.mortons.com.* Consistent with expectations, this outlet serves famed steaks and seafood. A knowledgeable staff explains the menu in a fun tableside presentation. A warm, club-like atmosphere welcomes a martini-sipping crowd. Steak menu. Closed Jan 1, Dec 25. Lunch, dinner. Bar. Business casual attire. $$
D

★★ **ORIGAMI.** *30 N 1st St, Minneapolis (55401). Phone 612/333-8430. www.origamirestaurant .com.* Japanese menu. Closed Jan 1, Dec 25. Dinner. Outdoor seating. $$
D

★★ **PARK CAFÉ.** *800 Marquette Ave, Minneapolis (55402). Phone 612/349-5710. www.parkcafemn.com.* French bistro menu. Closed Sat-Sun. Breakfast, lunch. Casual attire. $

★★★ **RUTH'S CHRIS STEAK HOUSE.** *920 2nd Ave S, Minneapolis (55402). Phone 612/672-9000. www.ruthschris.com.* This classic steakhouse serves a fast-paced business crowd at its central downtown location. Closed Thanksgiving, Dec 25. Dinner. Bar. Complimentary valet parking. $$$
D

★ **SHUANG CHENG.** *1320 SE 4th St, Minneapolis (55414). Phone 612/378-0208.* Chinese menu. Lunch, dinner. $$
D

★★ **SOPHIA.** *65 SE Main St, Minneapolis (55414). Phone 612/379-1111. www.sophia-mpls.com.* American menu. Closed Jan 1, Dec 25. Lunch, dinner, late night. Bar. Casual attire. Outdoor seating. $$
D

The Great River Road
✹ MINNESOTA
Part of a multistate Byway; see also IL, IA, WI.

Quick Facts

LENGTH: 575 miles.

TIME TO ALLOW: 2 days or longer.

BEST TIME TO DRIVE: Anytime during the summer; fall leaf color is spectacular.

BYWAY TRAVEL INFORMATION: Mississippi River Parkway Commission of Minnesota: 763/212-2560; Byway local Web site: www.mnmississippiriver.com.

SPECIAL CONSIDERATIONS: Several portions of the northern part of the road have a gravel surface. The speed limit on the Byway varies, but generally it is 55 mph in the country and 30 mph through towns. During the winter, you may find some icy patches in the rural parts of the roadway.

BICYCLE/PEDESTRIAN FACILITIES: Beautiful hiking, skiing, and snowmobiling trails exist all along the Byway. Catering to cyclists, the Great River Road provides easy access to a slew of biking trails along the way.

The Great River Road offers the best of Minnesota. The route encompasses the banks of the Mississippi River, 10,000 lakes, beautiful bluff lands, and a variety of outdoor recreation and wildlife. Brilliant wildflowers, evergreen forests, colored autumn leaves, rainbows, snowflakes, migrating birds, and waving fields of grain make this Byway a photographer's paradise.

Recreational spots have taken over the land of the lumberjack, where Paul Bunyan and his blue ox, Babe, used to roam freely. State parks and lakeside resorts are all part of the fun that visitors will find on the Great River Road today. And then there is the river itself. Minnesota is where the mighty Mississippi begins. Its meanderings make up a trail of cultural, historical, natural, recreational, and scenic sites. Whether camping in the forested areas of the north or relaxing in a Minneapolis hotel, you will find accommodations in the perfect setting.

The Great River Road is an adventure along a myriad of quaint towns and urban cities. The Twin Cities metropolitan area offers the hustle and bustle of a city that is rich in history, culture, and recreational opportunities. Deep wilderness surrounds the river towns of Minnesota and offers relaxation and privacy.

THE BYWAY STORY

The Great River Road tells cultural, historical, natural, recreational, and scenic stories that make it a unique and treasured Byway.

Cultural
The Great River Road in Minnesota takes you along the southeast end of the state, providing a

Minnesota

❋ The Great River Road

look at the culture that began and continued on the banks of the Mississippi. This is where you find the source of the Great River and the source of the development of civilization in Minnesota. It began with the Indian tribes of the Sioux and Chippewa, who lived in the area for many years before they began to interact with the Europeans in fur trade. When wildlife grew scarce, new settlers began a new industry of logging, which forced native tribes out of the area so that logging could proceed.

The European settlers along the Great River Road brought a culture of farming and logging and their own style of folklore. The many lakes in Minnesota have been attributed to the legend of Paul Bunyan and his blue ox, Babe. According to legend, this lumberman of grand proportions was a hero in the area, and the hoof prints of his ox created the lakes that are scattered all over the state.

When tourism became a viable industry, it was time to nurse the forests back to health. Although many people initially came to the area for jobs in the logging industry, the tourism industry was blossoming as people traveled to see the beautiful Minnesota forests. In the 1930s, people began to restore the once-great forests of giant red and white pines. An appreciation of nature is still part of the Byway culture today.

Cultures of the past have left traces of the past. Native American languages can still be found in names like Lake Winnibigoshish and Ah-Gwah-Ching (Leach Lake), while the heritage of the European settlers resounds in the names of communities all along the road. You will also find more recent pieces of American culture in Lindbergh State Park and the Lindbergh home, a memorial and a glimpse at the past of the famous pilot Charles Lindbergh.

The civilizations of Minnesota society burst in full glory on the Great River Road in the cities of Minneapolis and St. Paul. Parks and parkways, bridges and buildings are all part of the scenery that you experience throughout this Byway. But even with the great developments and changes of the present time, visitors and native dwellers all along the Great River Road can be seen returning to a culture of an earlier time—a culture of canoeing on the river and enjoying the wildlife and natural wealth of the area.

The Great River Road communities in Minnesota offer many festivals that celebrate the legends, products, immigrant culture, and art found along the Byway. For example, as you travel from Itasca to Bemidji, you can experience the Annual Ozawindib Walk, the Annual International Snowsnake Games, and the Lake Itasca Region Pioneer Farmers Reunion and Show. You can also enjoy the Annual Winter Bird Count, Art in the Park in July, and the People's Art Festival in November. Finally, you can visit the Paul Bunyan and Babe the Blue Ox statues that were erected in 1937 for a winter carnival.

As you travel the route from Little Falls to Brainerd, be sure to visit the Minnesota Fishing Museum in the Northern Pacific Depot, the Arts and Crafts Fair that takes place downtown the weekend after Labor Day, and the Antique Auto Show that takes place the weekend after Labor Day. Also, the Great River Arts Association sponsors exhibits and performances throughout the year, including Sunday afternoon concerts in Maple Island Park. The Heartland Symphony Orchestra performs in both Little Falls and Brainerd, and the Stroia Ballet Company offers performances throughout the year, including *The Nutcracker* each Christmas. Other things to see include the Morrison County Fair, held just east of Little Falls off of Highway 27, and the Little Falls House Concerts that feature folk musicians performing at a variety of venues. Also, Camp Ripley (Highways 371 and 115) contains the Minnesota Military Museum (open to the public) and the Historic Fort Ripley site (by appointment only). Finally, the Little Elk Heritage Preserve is located 2 miles north of Little Falls. This archaeological preserve has old fur-trading sites.

In the Minneapolis to St. Paul area of the Byway, you can experience 1840s food, dining, and preparation. Special events include Children's Day, Historic Mendota Days, the Mill City Blues Festival, the Capital City Celebration, and a New Year's Eve party. Also, visit the Taste of Minnesota, the Minneapolis Aquatennial, the St. Paul Winter Carnival, and La Fete de Saint Jean-Baptiste. From Red Wing to Winona, stop and visit the Red Wing Shoe Museum, located at 315 Main Street. Also found along the route is the Goodhue County Historical Society, located at 1166 Oak Street. Some of the events that occur along this part of the Byway include a Music Festival, a powwow, River City Days, and an Antique and Classic Boat Rendezvous.

Historical

When Henry Rowe Schoolcraft identified the true source of the Mississippi as the crystal-clear waters that flow from Lake Itasca, the final boundaries of the Louisiana Purchase were set. The Mississippi River, however, was important to Native Americans centuries before the Europeans arrived. In fact, the name Mississippi was an Algonquin name that, when applied to rivers, meant Great River (hence the name Great River Road). During the 1820s and 1830s, Fort Snelling and Grand Portage (on Lake Superior) were the focal points of Euro-American activity in the region. Strategically located at the confluence of the Mississippi and Minnesota rivers, Fort Snelling served as the first US military outpost in the area.

The Great River Road traces the river through the Chippewa National Forest, created by Congress in 1902. Although originally only small areas of pines were preserved, now more than 600,000 acres of land are managed by the National Forest Service. In this forest is historic Sugar Point on Leech Lake, another of the Mississippi River reservoirs. You can also visit Battleground State Forest, the site of the last recorded Indian battle with the US government, and Federal Dam, one of six dams constructed in the area between 1884 and 1912 to stabilize water levels on the Mississippi downstream. The legacies of the great pine forests along the Mississippi live in the majestic trees that remain. In Bemidji, the most recognizable landmark is the 1937 colossal statues of Paul Bunyan and Babe the Blue Ox, located on the shore of Lake Bemidji. The legend of the giant lumberjack illustrates the importance of the lumber industry for many northern Minnesota towns.

Due to its closeness to the river, St. Paul became a transportation hub for opening up the upper Midwest. It also became Minnesota's state capital, and Minneapolis became the nation's main flour-milling district. St. Anthony Falls and the locks and dams provide testimony to man's success at harnessing the power of the river to create a thriving urban center. Flour, beer, textiles, and lumber were produced and successfully transported to the nation and the world through the lock-and-dam system that begins here. In the early 1900s through the 1930s, the Cuyuna Iron Range produced over 100 million tons of ore that was used to build the US military machines of the great world wars. In Crosby, travelers visit the Croft Mine Historical Park—this mine operated between 1916 and 1934 and produced the richest ore found on the Cuyna Range, with a composition of 55 percent iron. Portsmouth Lake in Crosby, once the largest open pit mine on the Cyuna Range, is now one of the deepest lakes in Minnesota at 480 feet.

see page A5 for color map

Minnesota

✻ *The Great River Road*

Camp Ripley, a Minnesota National Guard facility, tells two stories: the history of Minnesota soldiers from the Civil War to the present and the development of the weapons of war.

In Little Falls, the Great River Road follows Lindbergh Drive to the boyhood home of Charles Lindbergh, pilot of the first solo flight across the Atlantic Ocean. The evolution of the United States can be traced along the Great River Road on the Ox Cart trail, still visible though Schoolcraft and Crow Wing state parks. Across the river from the Great River Road, Elk River's most noted resident, Oliver H. Kelley, founded the Patrons of Husbandry (popularly known as the Grange), a group that evolved into the Democratic Farmer Laborer, or DFL, party. Today, the Oliver H. Kelley Farm is a 189-acre living-history farm operated by the Minnesota Historical Society. This historic site explores the life and lifestyle of the Kelleys, who lived here in the late 1800s. Byway travelers also learn through excellent interpretation about the historic importance of the Great River Road and the river as transportation routes and vital links among Mississippi River communities.

Natural

One of the most naturally diverse sections of the Great River Road can be found in Minnesota. Travelers are beckoned to the Great River Road to enjoy a multitude of lakes, native wildlife, and rugged river bluffs. If you seek outdoor beauty, you will find what remains of the once-abundant pine forests that have been drawing travelers to Minnesota for over 100 years. Visitors never tire of seeing the lakes along the Great River Road in Minnesota, and there are plenty to see. Nearly every town has its own lake along the road, and when the lakes end, the vastness of the Mississippi River is just beginning. Natural wildlife abounds in the marshes and prairies along the road. Visitors will pass a major roosting site for bald eagles and wildlife habitats for deer, waterfowl, turkeys, and pheasants.

Between the many lakes, streams and rivers flow, urging travelers on. The landscape along the route is the handiwork of a crowd of glaciers that meandered through the area, leaving pockets and creases for rivers and lakes. Large hills were smoothed into the rolling landforms that visitors can see along the road. The great Mississippi River bluffs are present on the Byway, evidence of a notable geological past. This past combined with a present of hardwood forests, lakes, and wildlife creates a beautiful natural setting that will tempt you to leave your car throughout your journey on the Great River Road.

Recreational

From the headwaters to the Iowa border, Minnesota's portion of the Great River Road allows visitors to partake in a vast variety of recreational opportunities along its route. No matter what the itinerary or expectation, there is something to suit every taste on the Great River Road. And the fun and adventure continues throughout all four of Minnesota's very distinct seasons.

The water itself creates a large amount of the recreational draw to the Great River Road. A taste of the recreational possibilities that are available on the water include swimming, sunning, fishing, boating, jet skiing, sailing, canoeing, kayaking, and water-skiing—just to name a few. Strung along the edge of the river are plenty of things to do as well—particularly for outdoor enthusiasts. One of six state parks may be the perfect stop after a drive on the Byway. At parks like Itasca State Park, visitors can enjoy hiking or biking the trails, picnicking, or even bird-watching. The northern part of the Byway provides a haven for hunters and fishermen. The wilderness in this area is home to a great deal of wildlife.

Winter weather is no excuse to stay indoors in Minnesota. Several places along the Byway offer miles of cross-country skiing and snowmobiling. Visitors also find places for downhill skiing. But if moving through the snow doesn't sound appealing, perhaps drilling through the ice does. Ice fishing is a popular pastime for visitors to the Great River Road; ice-fishing resorts, services, and house rentals can be found from one end of the Great River Road to the other. And large catches from icy waters are not uncommon and are a well-known secret among residents of the state.

Scenic

As you travel along the Great River Road, you can look upon pristine lakes, virgin pine forests, quaint river towns, and a vibrant metropolis. You can see eagles, loons, and deer, as well as blazing hardwood forests and hillsides fluttering with apple blossoms. Experience the awesome power of locks and dams, imagine a longshoreman's life as barges glide past, recall Mark Twain's time on paddlewheelers, feel the wind that pushes the sailboats, and watch water-skiers relive the sport's Mississippi birth.

Travelers along the Great River Road are witnesses to a river that is constantly changing. It first evolves from a clear, shallow stream into a meandering, serpentine watercourse. It then changes into vast marshes and later becomes a canoe route. After this, the waterway becomes a rolling river that powers dams and mills as it squeezes its way past the only gorge on the river. It then passes over large waterfalls, through the first locks, past tall sandstone bluffs, and finally into a mile-wide river that is surrounded by a vast and fruitful valley.

The literary legend Mark Twain described the beautiful scene when he said, "The majestic bluffs that overlook the river, along through this region, charm one with the grace and variety of their forms, and the soft beauty of their adornment. The steep verdant slope, whose base is at the water's edge, is topped by a lofty rampart of broken, turreted rocks, which are exquisitely rich and mellow in color—mainly dark browns and dull greens with other tints. And then you have the shining river, winding here and there and yonder, its sweep inter-rupted at intervals by clusters of wooded islands threaded by silver channels; and you have glimpses of distant villages, asleep upon cares; and of stealthy rafts slipping along in the shade of the forest walls; and of white steamers vanishing around remote points."

HIGHLIGHTS

The following must-see tour of the Great River Road's northern section gives you a sample itinerary to follow, if you so choose.

- The Great River Road begins at **Itasca State Park.** Covering about 32,000 acres, the park embraces the headwaters of the Mississippi River and 157 lakes, the foremost of which is Lake Itasca, the source of the great river. Site of the University Biological Station, the park has stands of virgin Norway pine and specimens of nearly every kind of wild animal, tree, and plant native to the state. Camping and hiking, as well as historic sites, are abundant here. Itasca Indian Cemetery and Wegmann's Cabin are important landmarks in the area.

- Leaving Itasca State Park, the Great River Road heads northeast toward the **Chippewa National Forest** and **Lake Bemidji State Park.** Just 31 miles north of Itasca State Park along Highway 71, the Bemidji area is rich in diverse activities. Stopping at Lake Bemidji State Park provides a lot of fun, including the 2-mile **Bog Walk,** a self-guided nature trail. A $4 fee per car (per day) applies when visiting the park.

Minnesota

✻ *The Great River Road*

- Outside the park, along the shores of Lake Bemidji, lie several historic sites, including the famous statues of **Paul Bunyan and Babe the Blue Ox** and of **Chief Bemidji**. The **Carnegie Library Building,** on the National Register of Historic Places, is also located here. You'll find many campgrounds and resorts in the area for staying the night.
- **Cass Lake** is the next stop along the tour of the Great River Road. It's a popular stop for fishing and camping. This lake is unique, though: **Star Island,** in Cass Lake, is an attractive recreation area because it contains an entire lake within itself. Other fabulous lakes in the area are worth visiting as well.
- Enjoy the scenery of northern Minnesota as you follow the GRR signs into the city of **Grand Rapids.** The city offers many activities, including the **Forest History Center,** a logging camp that highlights the logging culture of Minnesota.
- Heading south out of **Grand Rapids,** follow the signs toward **Brainerd.** On the way to Brainerd, be sure to stop at **Savanna Portage State Park,** with its 15,818 acres of hills, lakes, and bogs. The **Continental Divide** marks the great division of water—where water to the west flows into the Mississippi River and water to the east runs into Lake Superior. Be sure to walk along the **Savanna Portage Trail,** too, a historic trail traveled by fur traders, Dakota and Chippewa Indians, and explorers more than 200 years ago. A $4 fee per car (per day) applies when visiting the park.
- The town of Brainerd is a great place to stop for lunch. Many lake resorts nearby offer camping, hiking, and fishing. After Brainerd and south of **Little Falls,** just off of Highway 371, is **Charles A. Lindbergh State Park.** Look for bald eagles here when visiting in the spring or fall. During your visit, stop in at the historic home of Charles A. Lindbergh, Sr., father of the famous aviator. The home is operated by the Minnesota Historical Society and is adjacent to the park. A $4 fee per car (per day) applies when visiting.
- Once in the city of **St. Cloud,** the **Munsinger and Clemens Gardens** offer a relaxing end to this portion of the Great River Road. The nationally known gardens are located near Riverside Drive and Michigan Avenue, right in town. One of the treats of the gardens are the antique horse troughs filled with unique flowers. The gardens are popular but spacious, so they're hardly ever noticeably crowded.

The southern tour continues from this point all the way through Minneapolis and St. Paul and along the Mississippi River and Wisconsin border, down to the border of Minnesota and Iowa.

THINGS TO SEE AND DO

Driving along the Great River Road will certainly keep your senses engaged, but if you yearn to get out of the car and stretch your legs, or if you'd like to make a mini-vacation out of your trip, check out these attractions along the route.

6TH STREET CENTER SKYWAY. *56 E 6th St, St. Paul (55101).* Created out of the second level of 6th Street Center's five-story parking garage. The center includes shops and restaurants. Open daily; closed major holidays.

ALEXANDER RAMSEY HOUSE. *265 S Exchange St, St. Paul (55102). Phone 651/296-0100 or -8760. www.mnhs.org/places/sites/arh/.* Historic home (1872) of Minnesota's first territorial governor; original furnishings. Guided tours; reservations suggested. Open Fri-Sat 10 am-3 pm; closed Dec 25. $$

BUNNELL HOUSE. *710 Johnson St, Winona (55987). Phone 507/452-7575 or -8760.* Unusual mid-19th-century Steamboat Gothic architecture; period furnishings. Open Memorial Day-Labor Day, Wed-Sun 10 am-5 pm, Sun 1-5 pm; Labor Day-second weekend in Oct, Sat 10 am-5 pm, Sun 1-5 pm; rest of year, by appointment. Also here is **Carriage House Museum Shop** (same schedule). $$

CAPITAL CITY TROLLEY. *807 E 7th St, St. Paul (55106). Phone 651/223-5600. www.capitalcitytrolley.com.* Historical tours of downtown St. Paul. $$$$

CATHEDRAL OF ST. PAUL. *226 Summit Ave, St. Paul (55102). Phone 651/228-1766.* This Roman Catholic cathedral was built in 1915. Dome is 175 feet high; central rose window. Open daily.

CENTRAL SCHOOL. *10 NW 5th St, Grand Rapids (55744). Phone 218/326-6431.* Heritage center housing a historical museum, a Judy Garland display, antiques, shops, and a restaurant. Open Mon-Fri 9:30 am-5 pm, Sat 10 am-4 pm, Sun (summer) noon-4 pm.

CITY HALL AND COURTHOUSE. *15 W Kellogg Blvd, St. Paul (55102). Phone 651/266-8023.* Prominent example of the Art Deco style (built in 1932), with Carl Milles' 60-ton, 36-foot-tall onyx *Vision of Peace* statue in the lobby.

CHARLES A. LINDBERGH HOUSE AND HISTORY CENTER. *1200 Lindbergh Dr S, Little Falls (56345). Phone 320/632-3154. www.mnhs.org/places/sites/lh/index.html.* Home of C. A. Lindbergh, former US congressman, and Charles A. Lindbergh, famous aviator. Homestead has been restored to its 1906-1920 appearance with much original furniture; visitor center has exhibits, an audiovisual program, and a gift shop. Open May-Labor Day, Mon-Sat 10 am-5 pm, Sun noon-5 pm; Labor Day-Oct, Sat 10 am-4 pm, Sun noon-5 pm. $$

CHARLES A. LINDBERGH STATE PARK. *1201 Lindbergh Dr S, Little Falls (56345). Phone 320/616-2525. www.mnhs.org/places/national register/stateparks/Lindbergh.html.* 436 acres. Hiking; cross-country skiing; picnicking; camping (hookups, dump station).

CHIPPEWA NATIONAL FOREST. *200 Ash Ave NW, Cass Lake (56633). Phone 218/335-8600. www.fs.fed.us/recreation/map/state_list.shtml.* Has 661,400 acres of timbered land; 1,321 lakes, with 699 larger than 10 acres; swimming, boating, canoeing, hiking, hunting, fishing, picnicking, camping (fee); winter sports. Bald eagle viewing; several historic sites.

COMO PARK. *Midway and Lexington pkwys, St. Paul (55108). Phone 651/266-6400 or 487-8200 (zoo and conservatory). www.comopark.com.* A 448-acre park with a 70-acre lake. The conservatory features an authentic Japanese garden and tea house; Gates Ajar floral display, zoo. Amusement area (open Memorial Day-Labor Day, daily) with children's rides (fee). 18-hole golf course (fee). Open daily.

FOREST HISTORY CENTER. *2609 County Rd 76, Grand Rapids (55744). Phone 218/327-4482. www.mnhs.org/places/sites/fhc/.* Center includes a museum building, re-created 1900 logging camp, and log drive wanigan (a floating cook shack used when the logs and men traveled downstream to the mills) maintained by the Minnesota Historical Society as part of an interpretive program. Early Forest Service cabin and fire tower, modern pine plantation, living-history exhibits; nature trails. Open June 1-Labor Day, Mon-Sat 10 am-5 pm, Sun noon-5 pm. $$

FORT SNELLING STATE PARK. *Hwy 5 and Post Rd, St. Paul (55111). Phone 612/725-2389 or 612/725-2724. www.nps.gov/miss/maps/model/fsnell.html.* A 4,000-acre park at the confluence of the Minnesota and Mississippi rivers. Swimming, fishing, boating; hiking, biking, cross-country skiing; picnicking; visitor center. The park is open daily 8 am-10 pm; the visitor center is open daily 10 am-4 pm. $

GARVIN HEIGHTS. *Huff St and Garvin Heights Rd, Winona (55987).* Park with a 575-foot bluff, offering majestic views of the Mississippi River Valley. Picnic area. Open dawn-dusk.

GREAT AMERICAN HISTORY THEATRE. *30 E 10th St, St. Paul (55101). Phone 651/292-4323 (box office). www.historytheatre.com.* Original works with American and Midwestern themes. Open Sept-May, Thurs-Sun.

Minnesota

❋ The Great River Road

HISTORIC FORT SNELLING. *Hwys 5 and 55, St. Paul (55111). Phone 612/726-1171. www.mnhs.org/places/sites/hfs/.* Stone frontier fortress restored to its appearance of the 1820s; daily drills and cannon firings; craft demonstrations. Visitor center with films, exhibits. Open June-Aug, Mon-Tues 10 am-5 pm (guided tours), Wed-Sat 10 am-5 pm, Sun noon-5 pm (living history); May and Sept-Oct, Sat 10 am-5 pm, Sun noon-5 pm. **$$**

JUDY GARLAND BIRTHPLACE AND CHILDREN'S MUSEUM. *2727 US 169 S, Grand Rapids (55744). Phone 218/327-9276; toll-free 800/664-JUDY. www.judygarlandmuseum.com.* Childhood home of the actress. Open daily 10 am-5 pm; closed major holidays (except Memorial Day, July 4, and Labor Day). **$**

LANDMARK CENTER. *75 W 5th St, St. Paul (55102). Phone 651/292-3228 or -3230 (tours). www.landmarkcenter.org.* Restored federal courts building constructed in 1902; currently the center for cultural programs and gangster history tours. Houses four courtrooms and a four-story indoor courtyard (the Cortile). Includes a restaurant, an archive gallery, an auditorium, the Schubert Club Keyboard Instrument Collection (open Mon-Fri), and the Minnesota Museum of American Art. 45-minute tours (Thurs and Sun; also by appointment); self-guided tours (daily). Open daily; closed holidays. **FREE**

LUTHER SEMINARY. *2481 Como Ave, St. Paul (55108). Phone 651/641-3456. www.luthersem.edu.* On the campus (started in 1869) is the Old Muskego Church (1844), the first church built by Norse immigrants in America; moved to the present site in 1904. Tours.

★ **MALL OF AMERICA.** *60 E Broadway, Bloomington (55425). Phone 952/883-8800. www.mallofamerica.com.* A retail/family entertainment complex with more than 500 stores and restaurants. The fact that Minnesota charges no sales tax on clothing makes shopping especially appealing here. The mall also features plenty of non-shopping activities for kids, such as Knott's Camp Snoopy, a 7-acre indoor theme park with rides and entertainment; the LEGO Imagination Center, with giant LEGO models and play areas; Golf Mountain miniature golf course; a 14-screen movie complex; and Underwater World, a walk-through aquarium (daily). Fees for activities. Open Mon-Fri 10 am-9:30 pm, Sat 9:30 am-9:30 pm, Sun 11 am-7 pm; closed Thanksgiving, Dec 25.

MILLE LACS LAKE. *124 N 6th St, Brainerd (56401). Phone 320/532-5626; toll-free 888/350-2692. www.millelacs.com.* This is Minnesota's second largest lake and one of the loveliest. It spans more than 200 square miles. Near the lakeshore are nearly 9,000 Native American mounds. Fishing, boating, camping.

MINNESOTA AMATEUR BASEBALL HALL OF FAME. *St. Cloud Civic Center, 10 S 4th Ave, St. Cloud (56301). Phone 320/255-7272. www.mnamateurbaseballhof.com.* Features great moments from amateur and professional baseball. Open Mon-Fri. **FREE**

MINNESOTA CHILDREN'S MUSEUM. *7th and Wabasha sts, St. Paul (55102). Phone 651/225-6000. www.mcm.org.* Hands-on learning exhibits for children up to 10 years old; museum store stocked with unique puzzles, maps, toys, games, and books. Open Tues-Thurs, Sat-Sun 9 am-5 pm; Fri 9 am-8 pm; Memorial Day-Labor Day, also Mon 9 am-5 pm; closed major holidays. **$$**

MINNESOTA HISTORY CENTER. *345 Kellogg Blvd W, St. Paul (55102). Phone 651/296-6126; toll-free 800/657-3773 (except in the Twin Cities). www.mnhs.org/places/historycenter/.* Home to the Historical Society, the center houses a museum with interactive exhibits and an extensive genealogical collection; special events, gift shop, restaurant. Open Tues 10 am-8 pm, Wed-Fri 10 am-3 pm, Sat 10 am-5 pm, Sun noon-5 pm. **FREE**

MINNESOTA TRANSPORTATION MUSEUM. *193 Pennsylvania Ave E, St. Paul (55101). Phone 651/228-0263. www.mtmuseum.org.* Display museum and restoration shop; 2-mile rides in early 1900s electric streetcars along the reconstructed Como-Harriet line. Open Memorial Day weekend-Labor Day, daily; after Labor Day-Oct, weekends. **$**

MINNESOTA WILD (NHL). *Xcel Energy Center, 317 Washington St, St. Paul (55102). Phone 651/222-WILD. www.wild.com.* Professional hockey team.

MOUNDS PARK. *Mounds Blvd and Burns Ave, St. Paul (55106). Phone 651/266-6400.* More than 25 acres of park containing prehistoric Native American burial mounds. Eighteen mounds existed on this site in 1856; six remain. Picnic facilities, ball field, view of the Mississippi River.

MUNSINGER AND CLEMENS GARDENS. *13th St and Kilian Blvd, St. Cloud (56301). Phone 800/264-2940.* Clemens Gardens features, among others, the White Garden, based on Kent, England's White Garden at Sissinghurst Garden. Munsinger Gardens is surrounded by pine and hemlock trees. Open Memorial Day-Labor Day. **FREE**

PAUL BUNYAN STATE TRAIL. *State Hwy 371 and Excelsior Rd, Brainerd (56425). www.paulbunyantrail.com.* This trail is a 100-mile recreational trail for joggers, walkers, bikers, hikers, and snowmobilers (rentals available). The trail passes by six communities, nine rivers, and 21 lakes.

POKEGAMA DAM. *34385 S US Hwy 2, Grand Rapids (55744). Phone 218/326-6128.* Camping on 21 trailer sites (hookups, dump station; 14-day maximum); picnicking, fishing. **$$$**

QUADNA MOUNTAIN RESORT AREA CONVENTION CENTER. *100 Quadna Rd, Hill City (55748). Phone 218/697-8444. www.quadna-resort.com.* Quad chairlift, two T-bars, rope tow; patrol, school, rentals, snowmaking; motel, lodge, and restaurant. Cross-country trails. 16 runs, longest run 26,430 feet, vertical drop 350 feet. Thanksgiving to mid-Mar, Fri-Tues. Also golf, outdoor tennis, horseback riding, and lake activities in summer. **$$$$**

ST. BENEDICT'S CONVENT. *104 Chapel Ln, St. Joseph (56374). Phone 320/363-7100. www.sbm.osb.org.* Community of more than 400 Benedictine (Roman Catholic) women that began in 1857. Tours of historic Sacred Heart Chapel (1913) and archives. Gift and crafts shop; Monastic Gardens.

SCIENCE MUSEUM OF MINNESOTA. *120 W Kellogg, St. Paul (55102). Phone 651/221-9444. www.smm.org.* Technology, anthropology, paleontology, geography, and biology exhibits; 3-D laser show; William L. McKnight 3M Omnitheater. Open daily. **$$$**

★ **SIGHTSEEING CRUISES.** *St. Paul (55102). Phone 651/227-1100.* Authentic Mississippi River sternwheelers *Harriet Bishop, Betsy Northrop,* and *Jonathan Padelford* make 1 3/4-hour narrated trips to Historic Fort Snelling. Sidewheeler *Anson Northrup* makes a trip through the lock at St. Anthony Falls. Dinner, brunch cruises, and showboat tours also available. Showboat tours available. Open Memorial Day-Labor Day, daily; May and Sept, weekends.

STATE CAPITOL. *75 Constitution Ave, St. Paul (55155). Phone 651/296-2881. www.mnhs.org/places/sites/msc/.* Designed in the Italian Renaissance style by Cass Gilbert and decorated with murals, sculpture, stencils, and marble,

Minnesota

✤ *The Great River Road*

the Capitol opened in 1905. Guided tours (45 minutes) leave on the hour; the last tour leaves one hour before closing (group reservations required). Open Mon-Fri 9 am-4 pm, Sat 10 am-3 pm, Sun 1-4 pm; closed major holidays, except President's Day. **FREE**

STEARNS HISTORY MUSEUM. *235 S 33rd Ave, St. Cloud (56301). Phone 320/253-8424.* Located in a 100-acre park, the center showcases cultural and historical aspects of past and present life in central Minnesota; contains a replica of a working granite quarry; agricultural and automobile displays; research center and archives. Open daily; closed holidays. **$$**

UPPER MISSISSIPPI RIVER NATIONAL WILDLIFE AND FISH REFUGE. *51 E 4th St, Winona (55987). Phone 507/452-4232. www.umesc.usgs.gov/umr_refuge.html.* From Wabasha, Minnesota, extending 261 miles to Rock Island, Illinois, the refuge encompasses 200,000 acres of wooded islands, marshes, sloughs, and backwaters. It abounds with fish, wildlife, and plants. Twenty percent of the refuge is closed for hunting and trapping until after duck-hunting season. Boat required for access to most parts of the refuge. Open daily.

UNIVERSITY OF MINNESOTA. *100 Church St SE, Minneapolis (55455). Phone 612/624-6888. www.umn.edu.* Founded in 1851. More than 46,000 students populate one of the largest single campuses in the United States. Several art galleries and museums on campus. Tours.

WELCH VILLAGE. *26685 County 7 Blvd, Red Wing (55089). Phone 651/258-4567. www.welchvillage.com.* Three quad, five double, triple chairlifts; 50 runs; Mitey-mite; patrol, rentals; snowmaking; cafeteria. Longest run 4,000 feet; vertical drop 350 feet. (Nov-Mar, daily; closed Dec 25). **$$$$**

PLACES TO STAY

If you choose to include an overnight stay in your trip along this Byway, Mobil Travel Guide recommends the following lodgings.

Baxter

★ **COUNTRY INN & SUITES BY CARLSON.** *1220 Dellwood Dr N, Baxter (56401). Phone 218/828-2161; toll-free 800/456-4000. www.countryinns.com.* 68 rooms, 2 story. Pet accepted, some restrictions. Complimentary continental breakfast. Check-out noon. TV; cable (premium), VCR available (movies). In-room modem link. Sauna. Indoor pool, whirlpool. Cross-country ski 1 1/2 miles. ¢

★ **PAUL BUNYAN INN.** *1800 Fairview Rd N, Baxter (56425). Phone 218/829-3571; toll-free 877/728-6926. www.paulbunyancenter.com.* 34 rooms. Complimentary continental breakfast. Check-out noon. TV; cable (premium). Sauna. Indoor pool, whirlpool. Downhill ski 15 miles, cross-country ski 1 mile. ¢

Brainerd

★★ **CRAGUN'S PINE BEACH LODGE AND CONFERENCE CENTER.** *11000 Cragun's Rd, Brainerd (56401). Phone 218/829-3591; toll-free 800/272-4867.* 285 rooms, 1-2 story. Service charge 15%. Check-out noon, check-in 5 pm. TV. Fireplaces. Restaurant, dining room, bar; entertainment. Barbecues. Free supervised children's activities (mid-June-mid-Sept), ages 4-12. In-house fitness room, saunas. Indoor, outdoor pools; whirlpools. Six lighted, two indoor tennis courts. Downhill ski 9 miles, cross-country ski on site. Private beaches. Lake excursions. Motorboats, canoes, sailboats, pontoon boats; snowmobile trails. Airport transportation. $

★★ **MADDEN'S ON GULL LAKE.** *11266 Pine Beach Peninsula, Brainerd (56401). Phone 218/829-2811; toll-free 800/642-5363. www.maddens.com.* Minnesota's largest resort, encompassing over 1,000 acres of land. 41 rooms, 3 story. Closed mid-Oct-mid-Apr. Check-out 1 pm, check-in 4:30 pm. TV; VCR available. In-room modem link. Many fireplaces. Dining room, bar; entertainment. Pizzeria. Supervised children's activities (July-mid-Aug), ages 4-12. In-house fitness room, saunas. Game room. Three indoor, two outdoor pools; whirlpools; poolside service. Golf on premise, greens fee $23. Outdoor tennis. Lawn games. Bicycles. Boats: motors, sailboats, kayaks, water bikes, pontoon boats, rowboats, speedboats. Airport transportation. Private beaches. On Gull Lake. Movies. $

★★ **RAMADA INN.** *2115 S 6th St, Brainerd (56401). Phone 218/829-1441; toll-free 800/272-6232. www.ramada.com.* 150 rooms, 2 story. Pet accepted; fee. Check-out noon. TV. In-room modem link. Restaurant, bar. Room service. Sauna. Indoor pool, whirlpool, poolside service. Outdoor tennis. Downhill ski 7 miles, cross-country ski 3 miles. Free airport transportation. ¢

Brooklyn Center

★ **BAYMONT INN.** *6415 James Cir N, Brooklyn Center (55430). Phone 763/561-8400; toll-free 800/301-0200. www.baymontinn.com.* 99 rooms, 3 story. Complimentary continental breakfast. Check-out noon. TV; cable (premium). In-room modem link. Cross-country ski 1/2 mile. ¢

★ **SUPER 8.** *6445 James Cir N, Brooklyn Center (55430). Phone 763/566-9810; toll-free 800/800-8000. www.super8.com.* 102 rooms, 2 story. Complimentary continental breakfast. Check-out 11 am. TV; cable (premium). ¢

Eagan

★★ **CROWNE PLAZA HOTEL.** *2700 Pilot Knob Rd, Eagan (55121). Phone 651/454-3434; toll-free 800/2-CROWNE. www.crowneplaza.com.* Located just 5 miles from the airport and the Mall of America, this full-service hotel offers transportation to and from both locations and free parking. 187 rooms, 6 story. Check-out noon. TV; cable (premium). In-room modem link. Restaurant, bar. In-house fitness room. Pool. Free airport transportation. $

Fridley

★ **BEST WESTERN KELLY INN.** *5201 NE Central Ave, Fridley (55441). Phone 763/571-9440; toll-free 800/780-7234. www.bestwestern.com.* 95 rooms, 2 story. Pet accepted, some restrictions. Complimentary continental breakfast. Check-out 11 am, check-in 3 pm. TV; cable (premium). Room service. Sauna. Game room. Indoor pool, whirlpool. ¢

Grand Rapids

★ **AMERICINN.** *1812 S Pokegama Ave, Grand Rapids (55744). Phone 218/326-8999; toll-free 800/634-3444. www.americinn.com.* 43 rooms, 2 story. Complimentary continental breakfast. Check-out 11 am. TV; cable (premium). Sauna. Indoor pool, whirlpool. ¢

★ **COUNTRY INN BY CARLSON.** *2601 S US 169, Grand Rapids (55744). Phone 218/327-4960; toll-free 800/456-4000. www.countryinns.com.* 59 rooms, 2 story. Pet accepted, some restrictions. Complimentary continental breakfast. Check-out noon. TV; cable (premium). Indoor pool, whirlpool. Downhill ski 10 miles, cross-country ski 2 miles. ¢

Minnesota

❋ *The Great River Road*

★★ **SAWMILL INN.** 2301 S Pokegama Ave, Grand Rapids (55744). Phone 218/326-8501; toll-free 800/235-6455. www.sawmill.com. 124 rooms, 2 story. Pet accepted, some restrictions. Check-out noon. TV. Restaurant, bar. Room service. Sauna. Game room. Indoor pool, whirlpool, poolside service. Free airport transportation. $

Little Falls

★ **SUPER 8.** 300 12th St NE, Little Falls (56345). Phone 320/632-2351; toll-free 800/800-8000. www.super8.com. 51 rooms, 2 story. Check-out 11 am. TV. ¢

Minneapolis

★ **DAYS INN.** 2407 University Ave SE, Minneapolis (55414). Phone 612/623-3999; toll-free 800/544-8313. www.daysinn.com. 131 rooms, 6 story. Complimentary continental breakfast. Check-out 11 am, check-in 3 pm. TV. In-room modem link. Downhill ski 10 miles, cross-country ski 1 mile. Near the university. ¢

★★ **HILTON NORTH.** 2200 Freeway Blvd, Minneapolis (55430). Phone 763/566-8000; toll-free 800/774-1500. www.hilton.com. 176 rooms, 10 story. Pet accepted, some restrictions; fee. Check-out noon, check-in 3 pm. TV; cable (premium), VCR available. Restaurant, bar. In-house fitness room. Pool. Business center. $$

★★★ **THE WHITNEY HOTEL.** 150 Portland Ave, Minneapolis (55401). Phone 612/375-1234. www.thewhitneyhotel.com. Located on the historic riverfront, this hotel offers European style along with views of the Mississippi River and the St. Anthony Falls. It is near many golf courses, attractions, and a local health club with fitness facilities. 96 rooms, 8 story. Check-out noon, check-in 3 pm. TV; cable (premium), VCR available. In-room modem link. Restaurant, bar. Room service 24 hours. Valet parking available. Airport transportation. Concierge. $$

Nisswa

★★ **GRAND VIEW LODGE.** 23521 Nokomis Ave, Nisswa (56468). Phone 218/963-2234; 800/432-3788. www.grandviewlodge.com. 12 rooms, 65 cottages, 2 story. Service charge 15%, no tipping. Check-out 12:30 pm, check-in 4:30 pm. TV; VCR available. Some fireplaces. Dining room, bar. Free supervised children's activities (Memorial Day-Labor Day, daily except Sun), ages 3-12. Game room. Heated pool, whirlpool, poolside service. Golf on premise. Outdoor tennis. Cross-country ski on site. Lawn games, bicycles. Boats, motors, canoes, paddleboats, kayaks, pontoon boat. Water-skiing. Private beach. Overlooks Gull Lake. Airport transportation. $$

Red Wing

★ **BEST WESTERN QUIET HOUSE & SUITES.** 752 Withers Harbor Dr, Red Wing (55066). Phone 651/388-1577; toll-free 800/780-7234. www.bestwestern.com. 51 rooms, 2 story. Pet accepted, some restrictions; fee. Check-out 11 am. TV. In-room modem link. In-house fitness room. Indoor, outdoor pools; whirlpool. ¢

★ **DAYS INN.** 955 E 7th St, Red Wing (55066). Phone 651/388-3568; toll-free 800/544-8313. www.daysinn.com. 48 rooms. Pet accepted; fee. Complimentary continental breakfast. Check-out 11 am. TV. Indoor pool, whirlpool. Downhill ski 7 miles, cross-country ski 1 mile. Municipal park, marinas opposite. ¢

★★ **GOLDEN LANTERN INN.** 721 East Ave, Red Wing (55066). Phone 651/388-3315; toll-free 888/288-3315. www.goldenlantern.com. 5 rooms, 2 story. No room phones. Children accepted by arrangement. Complimentary full breakfast. Check-out 11 am, check-in 4-5 pm. Downhill ski 8 miles, cross-country ski 1 mile. Tudor brick house built in 1932. Totally nonsmoking. $

★★ **ST. JAMES HOTEL.** *406 Main St, Red Wing (55066). Phone 651/388-2846; toll-free 800/252-1875. www.st-james-hotel.com.* Built in 1875, this hotel offers 19th-century Victorian style along with modern amenities. It is located on the bank of the Mississippi River in the heart of Red Wing, near many local attractions, restaurants, and shops. 60 rooms, 2-5 story. TV; cable (premium), VCR available. In-room modem link. Restaurant, bar; entertainment Fri-Sat. Downhill, cross-country ski 10 miles. Free covered parking. Airport transportation. $

St. Cloud

★★ **BEST WESTERN AMERICANNA INN AND CONFERENCE CENTER.** *520 S US 10, St. Cloud (56304). Phone 320/252-8700; toll-free 800/780-7234. www.bestwestern.com.* 63 rooms, 2 story. Pet accepted; fee. Check-out 11 am. TV; cable (premium). In-room modem link. Restaurant, bar; entertainment. Room service. Sauna. Game room. Indoor pool, whirlpool. ¢

★ **COMFORT INN.** *4040 2nd St S, St. Cloud (56301). Phone 320/251-1500; 877/424-6423. www.comfortinn.com.* 63 rooms, 2 story. Complimentary continental breakfast. Check-out 11 am. TV; cable (premium). Coin laundry. In-house fitness room, sauna. ¢

★ **FAIRFIELD INN BY MARRIOTT.** *4120 2nd St S, St. Cloud (56301). Phone 320/654-1881; toll-free 800/228-2800. www.fairfieldinn.com.* 57 rooms, 3 story. Complimentary continental breakfast. Check-out 11 am. TV; cable (premium). In-room modem link. Game room. Indoor pool, whirlpool. Cross-country ski 1 mile. ¢

★★ **HOLIDAY INN.** *75 37th Ave S, St. Cloud (56301). Phone 320/253-9000; toll-free 800/465-4329. www.holiday-inn.com.* 257 rooms, 3 story. Check-out 11 am. TV; cable (premium). Restaurant, bar. In-house fitness room, sauna. Five indoor pools, children's pool, whirlpool, poolside service. ¢

St. Paul

★★ **EMBASSY SUITES.** *175 E 10th St, St. Paul (55101). Phone 651/224-5400; toll-free 800/EMBASSY. www.embassysuites.com.* This all-suite hotel is located near the Science Museum of Minneapolis, Galtier Plaza, the Mall of America, and much more. 210 rooms, 8 story. Check-out noon, check-in 3 pm. TV; cable (premium). In-room modem link. Restaurant, bar. In-house fitness room, sauna, steam room. Indoor pool, whirlpool. Free airport transportation. $$

★ **EXEL INN.** *1739 Old Hudson Rd, St. Paul (55106). Phone 651/771-5566; toll-free 800/367-3935. www.exelinns.com.* 100 rooms, 3 story. Pet accepted, some restrictions. Complimentary continental breakfast. Check-out noon, check-in 3 pm. TV; cable (premium). In-room modem link. Laundry services. Game room. Downhill ski 15 miles, cross-country ski 2 miles. ¢

★★ **FOUR POINTS BY SHERATON.** *400 Hamline Ave N, St. Paul (55104). Phone 651/642-1234; toll-free 800/535-2339. www.starwood.com/fourpoints/.* 197 rooms, 4 story. Check-out noon, check-in 3 pm. TV; cable (premium). In-room modem link. Restaurant, bar. Room service. In-house fitness room. Health club privileges. Beach. Indoor pool, whirlpool. Downhill ski 15 miles, cross-country ski 7 miles. $

Minnesota

❋ *The Great River Road*

★★ **HOLIDAY INN.** *2201 Burns Ave, St. Paul (55119). Phone 651/731-2220; toll-free 800/465-4329. www.holiday-inn.com.* 195 rooms, 8 story. Complimentary continental breakfast. Check-out noon, check-in 3 pm. TV; cable (premium). In-room modem link. Restaurant, bar. Room service. In-house fitness room, sauna. Game room. Indoor pool, whirlpool. Downhill ski 10 miles, cross-country ski 1/2 mile. Luxury level. $

★★★ **THE SAINT PAUL HOTEL.** *350 Market St, St. Paul (55102). Phone 651/292-9292; toll-free 800/292-9292. www.stpaulhotel.com.* A Historic Hotel of America, this beautifully restored property was founded in 1910 by wealthy businessman Lucius Ordway and still maintains an old-style European charm. The hotel has hosted many famous individuals while maintaining a commitment to business and leisure visitors to this capital city. Most rooms have splendid downtown, Rice Park, or St. Paul Cathedral views. 254 rooms, 12 story. Check-out noon, check-in 3 pm. TV. In-room modem link. Restaurant, bar. In-house fitness room. Health club privileges. Business center. Concierge. Connected to downtown skyway system. $$

Waite Park

★ **MOTEL 6.** *815 S 1st St, Waite Park (56387). Phone 320/253-7070; toll-free 800/466-8356. www.motel6.com.* 93 rooms, 2 story. Check-out noon. TV; cable (premium). In-room modem link. Downhill ski 10 miles. ¢

Winona

★ **DAYS INN.** *420 Cottonwood Dr, Winona (55987). Phone 507/454-6930; toll-free 800/544-8313. www.daysinn.com.* 58 rooms, 2 story. Complimentary continental breakfast. Check-out 11 am. TV. Cross-country ski 1 mile. ¢

★★ **BEST WESTERN RIVERPORT INN & SUITES.** *900 Bruski Dr, Winona (55987). Phone 507/452-0606; toll-free 800/780-7234. www.bestwestern.com.* 106 rooms, 3 story. Pet accepted, some restrictions; fee. Complimentary continental breakfast. Check-out 11 am. TV; cable (premium), VCR available. Restaurant, bar. Room service. Game room. Indoor pool, whirlpool. Downhill ski 8 miles, cross-country ski 1 mile. $

★★ **QUALITY INN.** *956 Mankato Ave, Winona (55987). Phone 507/454-4390; toll-free 800/228-5151. www.qualityinn.com.* 112 rooms, 2 story. Check-out 11 am. TV. Restaurant 24 hours, bar. Room service. Indoor pool, whirlpool. Cross-country ski 1 mile. ¢

★ **SUPER 8.** *1025 Sugar Loaf Rd, Winona (55987). Phone 507/454-6066; toll-free 800/800-8000. www.super8.com.* 61 rooms, 3 story. No elevator. Complimentary continental breakfast. Check-out 11 am. TV. Cross-country ski 1 mile. ¢

PLACES TO EAT

A long day of driving is sure to make you hungry. At the end of your journey, take a table at one of the following restaurants.

Brainerd

★ **BAR HARBOR SUPPER CLUB.** *8164 Interlaken Rd, Brainerd (56468). Phone 218/963-2568.* Closed Dec 24-25. Lunch, dinner. Bar; entertainment Wed-Sun. Children's menu. Outdoor seating. $$

★ **IVEN'S ON THE BAY.** *5195 N Hwy 371, Brainerd (56401). Phone 218/829-9872.* Closed some major holidays. Dinner, Sun brunch. Bar. Children's menu. $$

Minneapolis

★★ **510 RESTAURANT.** *510 Groveland Ave, Minneapolis (55403). Phone 612/874-6440. www.510restaurant.com.* Originally a glittery 1920s-era hotel, this restaurant's grand opulence is a step back in time. The food is wonderfully inventive. American, French menu. Closed Sun. Dinner. **$$$**
D

★ **BLACK FOREST INN.** *1 E 26th St, Minneapolis (55404). Phone 612/872-0812. www.blackforestinnmpls.com.* German menu. Lunch, dinner, late night. Bar. Casual attire. Outdoor seating. **$$**
D

★★ **CAMPIELLO.** *1320 W Lake St, Minneapolis (55408). Phone 612/825-2222. www.damico.com.* Italian menu. Closed Dec 25. Dinner, Sun brunch. Bar. Valet parking available. Outdoor seating. **$$**
D

★ **EMILY'S LEBANESE DELI.** *641 University Ave NE, Minneapolis (55413). Phone 612/379-4069.* Lebanese menu. Closed Tues; Easter, Thanksgiving, Dec 25. Lunch, dinner. Casual attire. **$**

★★ **FIRST COURSE.** *5607 Chicago Ave S, Minneapolis (55417). Phone 612/825-6900.* American, Mediterranean menu. Closed Sun; major holidays. Lunch, dinner. Children's menu. Casual attire. Outdoor seating. **$$**
D

★★ **KIKUGAWA.** *43 SE Main St, Minneapolis (55414). Phone 612/378-3006. www.kikugawa-sushi.com.* Japanese menu. Closed Jan 1, Thanksgiving, Dec 25. Lunch, dinner. Bar. Casual attire. **$$$**
D

★★ **NYE'S POLONAISE.** *112 E Hennepin Ave, Minneapolis (55414). Phone 612/379-2021. www.nyespolonaise.com.* American menu. Closed Dec 25. Lunch, dinner, late night. Bar. Casual attire. Polka Thurs-Sat. **$$$**
D

★ **SAWATDEE.** *607 Washington Ave S, Minneapolis (55415). Phone 612/338-6451. www.sawatdee.com.* Thai menu. Lunch, dinner. Bar. Casual attire. **$$**
D

Red Wing

★ **LIBERTY'S.** *303 W 3rd, Red Wing (55066). Phone 651/388-8877.* Continental menu. Closed most major holidays. Breakfast, lunch, dinner, Sun brunch. Bar. Entertainment Fri-Sat. **$$**
D SC

St. Cloud

★ **D. B. SEARLE'S.** *18 5th Ave S, St. Cloud (56301). Phone 320/253-0655.* Closed some holidays. Lunch, dinner. Built in 1886. **$$**
D

St. Paul

★ **CECIL'S DELI.** *651 S Cleveland, St. Paul (55116). Phone 651/698-0334. www.cecilsdeli.com.* Breakfast, lunch, dinner. **$**

★ **CIATTI'S.** *850 Grand Ave, St. Paul (55105). Phone 651/292-9942.* Italian menu. Closed Dec 24-25. Lunch, dinner, Sun brunch. Bar. Children's menu. Casual attire. **$**
D

★★ **DAKOTA BAR AND GRILL.** *1021 E Bandana Blvd, St. Paul (55108). Phone 651/642-1442. www.dakotacooks.com.* Jazz fans will enjoy this modern restaurant in historic Bandana Square. Regional ingredients are used in such eclectic dishes as pheasant fritters. Along with the extensive wine list, guests can enjoy dining on

Minnesota

✽ *The Great River Road*

the outdoor patio in the summer. Closed Sun. Dinner. Bar. Live music. Located in a restored railroad building. Casual attire. Outdoor seating. $$
D

★ **DIXIE'S.** *695 Grand Ave, St. Paul (55105). Phone 651/222-7345. www.dixiesrestaurants.com.* Southern, Cajun menu. Closed Thanksgiving, Dec 25. Lunch, dinner, Sun brunch. Bar. Children's menu. Casual attire. Outdoor seating. $
D

★★★ **FOREPAUGH'S.** *276 S Exchange St, St. Paul (55102). Phone 651/224-5606. www.forepaughs.com.* Some say the ghost of the former owner haunts this romantic three-story home in St. Paul, with its nine dining rooms complete with lace curtains. The French menu has English subtitles, and the restaurant offers shuttle service to nearby theaters. Continental menu. Closed some holidays. Lunch, dinner, Sun brunch. Bar. Children's menu. Restored mansion (1870). Casual attire. Valet parking available. Outdoor seating. $$
D

★ **GALLIVAN'S.** *354 Wabasha St, St. Paul (55102). Phone 651/227-6688. www.minnesotamenus.com/gallivans/.* American menu. Closed Sun; major holidays. Lunch, dinner. Bar. Entertainment Fri-Sat. Casual attire. $
D

★ **LEEANN CHIN.** *214 E 4th St, St. Paul (55101). Phone 651/224-8814. www.leeannchin.com.* Chinese menu. Closed major holidays. Lunch, dinner. Bar. Children's menu. Casual attire. Totally nonsmoking. $$
D

★★ **LEXINGTON.** *1096 Grand Ave, St. Paul (55105). Phone 651/222-5878. www.the-lexington.com.* The comfortable, clubby dining room is chandelier-lit and boasts a curving mahogany bar. Since 1935, this restaurant has prepared such traditional favorites as chateaubriand and prime rib in a simple and straightforward manner. American menu. Closed Dec 25. Lunch, dinner, Sun brunch. Bar. Children's menu. Casual attire. $$$
D

★ **MANCINI'S CHAR HOUSE.** *531 W 7th St, St. Paul (55102). Phone 651/224-7345. www.mancinis.com.* Steak menu. Closed major holidays. Dinner. Bar. Entertainment Wed-Sat. Children's menu. Casual attire. $$$
D

★★ **MUFFULETTA IN THE PARK.** *2260 Como Ave, St. Paul (55108). Phone 651/644-9116.* Eclectic/International menu. Closed some major holidays. Lunch, dinner, Sun brunch. Casual attire. Outdoor seating. Totally nonsmoking. $$
D

★★ **RISTORANTE LUCI.** *470 Cleveland Ave, St. Paul (55116). Phone 651/699-8258. www.ristoranteluci.com.* No surprises, just the kind of food everyone wants to eat most of the time: simple and satisfying. Italian menu. Closed Sun, Mon; most major holidays. Dinner. Children's menu. Casual attire. Totally nonsmoking. $$
D

★★ **SAKURA.** *350 St. Peter St, St. Paul (55102). Phone 651/224-0185. www.sakurastpaul.com.* Japanese menu. Closed Jan 1, Thanksgiving, Dec 25. Lunch, dinner. Bar. Casual attire. $$
D

★★ **THE ST. PAUL GRILL.** *350 Market St, St. Paul (55102). Phone 651/224-7455. www.stpaulhotel.com.* Lunch, dinner, Sun brunch. Bar. Casual attire. Valet parking available. $
D

★★ **TOBY'S ON THE LAKE.** *249 Geneva Ave N, St. Paul (55128). Phone 651/739-1600. www.tobysonthelake.com.* American menu. Closed Dec 25. Lunch, dinner, Sun brunch. Bar. Children's menu. Casual attire. Outdoor seating. $$
D

★★★ **W. A. FROST AND COMPANY.** *374 Selby Ave, St. Paul (55102). Phone 651/224-5715. www.wafrost.com.* Located in the historic Dacotah Building, the three dining rooms of this restaurant are decorated with Victorian-style wallpaper, furnishings, and oil paintings. Closed some major holidays. Dinner, Sun brunch. Bar. Children's menu. Renovated pharmacy (1887). Valet parking available Fri-Sat. Outdoor seating. Totally nonsmoking. **$$$**
D

Historic Bluff Country Scenic Byway

❋ MINNESOTA

Quick Facts

LENGTH: 88 miles.

TIME TO ALLOW: 3 hours.

BEST TIME TO DRIVE: Summer and fall.

BYWAY TRAVEL INFORMATION: Byway local Web site: www.bluffcountry.com.

SPECIAL CONSIDERATIONS: An extremely scenic route through the Yucatan Valley, the segment between Houston and Caledonia to the south curves frequently and climbs many steep grades in response to the area's topography. When traveling the Bluff Country, be careful of the wildlife inhabiting the area. Rattlesnakes are occasionally found on rock outcrops and on river bottoms.

BICYCLE/PEDESTRIAN FACILITIES: Shoulders are very narrow (2 feet) and do not support bicyclist or pedestrian use. The Historic Bluff Country Scenic Byway is unique, however, in that a parallel bicycle and pedestrian facility, the Root River Trail, is provided between Fountain and Houston. Many visitor facilities and interpretive opportunities are located at trailheads for this facility, providing similar experiences for bicyclists and motorists alike. You can ski, hike, and bike at the Root River State Trail, but other trails are available off the Historic Bluff Country Scenic Byway. You can find hiking and cross-country trails at Historic Forestville State Park. Between Houston and Rushford, you'll find the Wetbark Trail. And in towns like Rushford, Preston, and Fountain, trails are also available.

Follow the panoramic Root River Valley to the Mississippi River. The scenery along the western end of the Byway showcases Minnesota's rich and rolling farmland, while the eastern part of the route winds toward the Great River Road along a beautiful trout stream and canoe route through spectacular tree-covered bluffs featuring limestone palisades and the rich hardwoods found in this area.

This valley was untouched by the glaciers and has weathered gradually over time to create a magnificent pastoral setting dotted with small towns, quaint and historic lodgings, and a recreational bike and hiking trail. Whether you come to Historic Bluff Country for the recreational opportunities or for the culture you'll find all along the Byway, you won't be disappointed. Day trips along and adjacent to the Historic Bluff Country Scenic Byway take you back to days when the horse and buggy dominated travel, while sights of places known only in history books are before your very eyes.

THE BYWAY STORY

The Historic Bluff Country Scenic Byway tells archaeological, cultural, historical, natural, recreational, and scenic stories that make it a unique and treasured Byway.

Archaeological

The Huta Wakpa and the Cahheomonah people who were native to the Driftless Area along the Byway realized its virtues as they found good hunting ground and resources for making tools. As a result, many Native American settlements, structures, and sites were once located all along the Byway. Now, many of these sites

Minnesota

Historic Bluff Country Scenic Byway

have been lost by time and nature, but they once represented the complex seasons and lifestyles of the people who first lived there.

Still in existence today is the Grand Meadow Quarry Site. This natural bed of chert provided materials for Native American arrowheads and tools for 10,000 years. Archaeological interpretation for visitors is limited to stories of the past and the occasional piece of chert that might be discovered in the forest.

Cultural

Among the small, historic Minnesota communities of the Historic Bluff area, a trail to the past defines today's cultures. Some of these cultures, like the Old Order Amish population, have changed little since first settling here. Others have evolved with the rest of America and yet keep a firm hold on their heritage. The descendents of Norwegians and Germans still celebrate many of the same events that their ancestors did 150 years ago.

Little is left, however, of the Native American culture that once dominated the lands along the Byway. At one time, the Winnebago and the Dakota developed celebrations and rituals around the Root River. The few tribes that still survive appreciate cultural sites and sacred places in Bluff Country, and archaeological sites provide a look at places that were important to these cultures. Above all, harmony with the land was a part of life for these people. Europeans who eventually took over the river and bluffs have their own appreciation for the lands they settled. Names like Spring Grove and Grand Meadow reflect the natural beauty of the past that can still be visited today on the Byway.

The settlers who built the towns in Bluff Country were industrious. Towns flourished, and their downtown sections contained splendid architecture that can still be observed today. The attitude of growth amid small-town life is still present in Byway towns today. Meanwhile, the past is still untouched in Amish communities, where people live without electricity, automobiles, and many of the cultural ideals that most Americans hold. Amish buggies can be seen near Harmony, Preston, and Lanesboro; the occupants may be traveling to a local store to sell homemade baked goods and crafts. Furniture created by Amish craftsmen can be found at shops in Byway towns.

Historical

Bluff Country is a unique place where modern ways of life have not entirely overshadowed memories of the past. Historic sites and districts are still evident throughout this historic Byway. Towns along the route were built on a very successful agricultural past. The agricultural society that developed in southeastern Minnesota in the 1800s flourished for many years and then evolved with the coming and going of the railroad. Barns, mills, and silos stand as evidence of the past in many different places along the Byway. Historic centers in Houston, Lanesboro, and Fillmore County display other parts of the past along the Byway.

When southeastern Minnesota was settled, little towns sprang up as the centers of agricultural trade and social activity. Farmers congregated to sell their crops and buy needed supplies, while grain was stored in silos or grain elevators like the one in Preston that once belonged to the Milwaukee Elevator Company. This grain elevator is now a rest stop for travelers along the Root River Trail. The towns profited from agricultural success, and now many of them display historic districts and buildings with elaborate architecture. Many of the buildings, including churches, libraries, and banks, are available to travelers as they explore history on the Byway.

As a result of a successful farming industry, mills began to appear in Bluff Country. In Lanesboro, a dam on the river provided power for three flour mills. The commercial district thrived there from the 1860s through the 1920s, and visitors today can tour these brick buildings and the shops and restaurants inside

see page A13 for color map

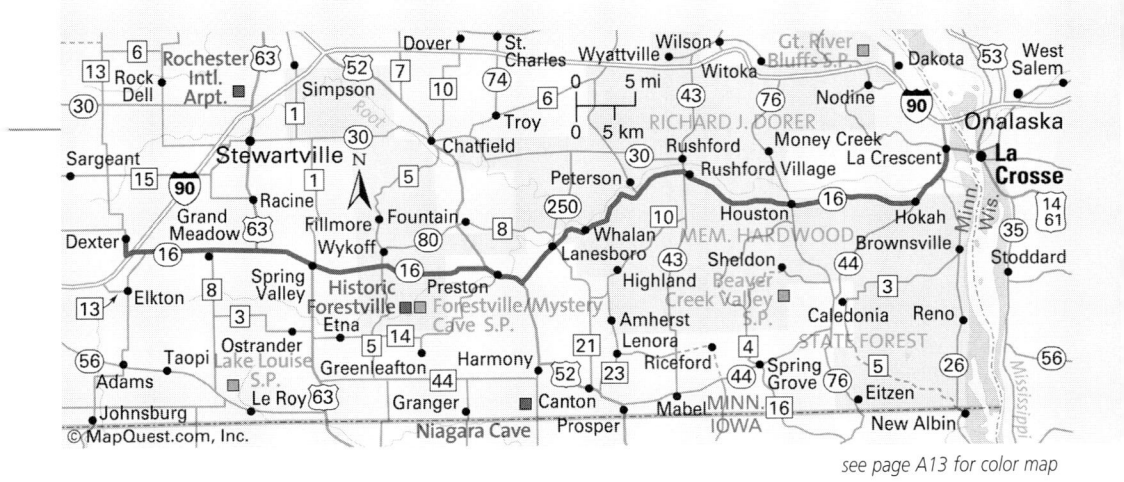

them. Towns along the Byway also benefited from the coming of the railroad. The entire town of Wykoff was platted and settled when the railroad passed through. The old Southern Minnesota Depot in Rushford now serves as a trail center. When the railroad left, the tracks lay in disrepair for many years until they were converted to the Root River Trail.

Natural

Readily accessible at every turn, the nature of Historic Bluff Country is available at overlooks, parks, riverbanks, and vistas. Deep river valleys, sinkholes, caves, and bluffs are all natural features of the Byway that travelers will want to explore. Bluff Country is part of the Driftless Area left untouched by the glaciers that once covered most of Minnesota and Wisconsin. The Root River flows along the Byway, calling to explorers from a tree-lined bank. The Byway celebrates its natural qualities with trails, parks, and places to stop and learn more about Byway surroundings.

Because of the unique situation of Bluff Country's natural history, unique habitats are found throughout the area. Labeled as Scientific and Natural Areas, these places often support unusual plant and animal species, because natural features of the land remained unchanged during the time that the rest of the state was changed by glaciers.

The caves and underground streams along the Byway are a result of karst terrain, which is created when rainwater is absorbed into the ground to dissolve the calcite limestone beneath the ground. As this thick layer of limestone weakens, sinkholes form. Streams that continue to flow through the dissolved limestone create caves and caverns. The Mystery Cave and Niagara Cave are two places where you can see the results of a few streams and a little rainwater. You may also want a good look at a sinkhole. The nearby city of Fountain is known as the Sinkhole Capital of the United States. Here, you find more evidence of the unique geological features along the Byway.

If you like to get out of the caves and enjoy nature above ground, the Byway offers forests, rivers, and bluffs. The Root River and the Richard J. Dorer Memorial Forest provide places to enjoy the greenery of trees and the sunlight as it sparkles on moving water. Maple, oak, and birch create a setting for bird-watching, and in the fall, their colors draw audiences from miles around. As the Root River flows along the Byway and separates into two branches, canoe access points allow you to get a closer look. And of course, the natural feature for which the Byway was named is prevalent along the way. Travelers drive over and next to bluffs on the Byway. In the city of Hokah, a bluff known as Mount Tom stands over Como Park, where you can see the enchanting Como Falls.

Recreational

The forests, the trails, and the river are the central places for recreation on the Byway. And because most of the Byway goes through these places, recreation is never far away. State parks and natural areas provide places for hiking, camping, and exploring. With two caves within the Byway vicinity, spelunking is a skill that every traveler will want to develop. Driving through friendly towns, beautiful river valleys,

Minnesota

✳ Historic Bluff Country Scenic Byway

and the bluff-covered countryside, you will certainly find a place to stop and have some fun. Classic outdoor recreational opportunities are around every corner.

Most travelers like to explore places with unique names and histories. The Forestville/Mystery Cave State Park has both. Not only do visitors have the opportunity to go spelunking in the Mystery Cave, but they also are able to explore the town site of one of Minnesota's oldest communities, Forestville. For visitors who want to continue spelunking, one of the largest caves in the Midwest, Niagara Cave, is located south of Harmony. You can camp among the forests and bluffs at Beaver Creek Valley State Park.

Two state trails on the Byway are the perfect place for bikers, hikers, and, in wintertime, cross-country skiers. The Root River State Trail and the Preston-Harmony Valley State Trail connect to provide miles of scenic trail for visitors to enjoy. Along the way, signs provide information about the surrounding area. When you visit the Richard J. Dorer Memorial Harwood Forest, you will be pleased to find trails for hiking, horseback riding, and off-road vehicles.

Perhaps the most inviting trail on the Byway is one made of water. The Root River is perfect for an afternoon or a day in a canoe. Several branches and tributaries give you the chance to get to know Bluff Country a little better. The Root River is not just a waterway, but also home to a population of trout. Anglers enjoy the wooded atmosphere of the Root River as they try to tempt a prize-winning fish. On the banks of the river, picnickers enjoy the shade of a maple tree and the peaceful sound of flowing river water. The Root River is central to many of the activities along the Byway, but even if you stray from the riverside, you'll be sure to find something to do.

Scenic

Bluff Country is a land of rolling hills, pastoral fields, and scenic rivers. Several places along the way highlight the beauty of a river valley or a hardwood forest. The colors of the landscape change with each season to provide rich greens, beautiful golds, and a winter white. Each season has its own spin on the scenery, but you'll find the same scenic place whenever you drive the Historic Bluff Country Scenic Byway. And when you aren't enjoying the scenery from your car, you can be out on the Byway experiencing it.

From the car, you'll find dramatic views of steep river bluffs, or you can watch the Root River as it flows alongside the Byway. As the Byway parallels the river, enjoy the tree-lined banks of the river and the occasional tuft of wildflowers growing nearby. Of course, there are places to stop along the way and get a closer look at the river. The trail that parallels the river provides visitors with a scenic hike along the Byway.

Historic buildings, bridges, and walls along the Byway only enhance the scenery, making the Byway one of the most picturesque corners of the country. As civilization has developed in this natural prairie, industrious cultures and families left churches, homes, and parks for future generations to enjoy. Their efforts create an added sense of pastoral perfection along the Byway as the fields and forests are accented with classic buildings that continue to stand the test of time.

In addition to historic buildings, the Byway features a number of historic waysides and parks. These parks feature inviting picnic areas with beautiful overlooks. At Magelssen Park, you will find yourself at the top of a bluff overlooking Rushford. The fifth largest tree in the state, a Burr Oak tree, is one of the sights you will want to look for in the park. At many of the parks and waysides, you can enjoy an afternoon walk along one of the quaint stone walls.

THINGS TO SEE AND DO

Driving along the Historic Bluff Country Scenic Byway will certainly keep your senses engaged, but if you yearn to get out of the car and stretch your legs, or if you'd like to make a mini-vacation out of your trip, check out these attractions along the route.

FORESTVILLE/MYSTERY CAVE STATE PARK. *Hwy 118, Spring Valley (55965). Phone 507/352-5111 or 507/937-3251 (Mystery Cave). www.dnr.state.mn.us/state_parks/forestville_mystery_cave/.* A 3,170-acre park in the Root River valley with a historic town site. Fishing; hiking; bridle trails. Cross-country skiing, snowmobiling; picnicking; camping. Open Mon-Thurs 8 am-4 pm, Fri 8 am-10 pm, Sat 9 am-9 pm, Sun 9 am-6 pm. $$

METHODIST CHURCH. *221 W Courtland St, Spring Valley (55965). Phone 507/346-7659.* Victorian Gothic architecture (1876); 23 stained-glass windows. Laura Ingalls Wilder site. Lower-level displays include country store, history room, and military and business displays. Open June-Aug, daily; Sept-Oct, weekends; also by appointment. $$

MYSTERY CAVE. *Hwy 118, Spring Valley (55965). Phone 507/352-5111. www.dnr.state.mn.us/state_parks/forestville_mystery_cave/.* 60-minute guided tours. Bring a jacket; the cave is 48 degrees. Open Memorial Day-Labor Day, daily; mid-Apr-Memorial Day, weekends. Picnicking. Vehicle permit required (additional fee). $$

WASHBURN-ZITTLEMAN HOUSE. *220 W Courtland St, Spring Valley (55965). Phone 507/346-7659.* Two-story frame house (1866) with period furnishings, quilts; farm equipment, one-room school, toys. Open Memorial Day-Labor Day, daily; Sept-Oct, weekends; also by appointment. $$

PLACES TO STAY

Austin

The town of Austin is a few miles west of the Byway, off I-90.

★★ **HOLIDAY INN AND CONFERENCE CENTER.** *1701 4th St NW, Austin (55912). Phone 507/433-1000; toll-free 800/985-8850. www.holiday-inn.com.* 121 rooms, 2 story. Pet accepted, some restrictions. Check-out 11 am. TV; cable (premium). In-room modem link. Laundry services. Restaurant, bar; entertainment. Room service. In-house fitness room, sauna. Game room. Indoor pool, children's pool, whirlpool, poolside service. Airport transportation. ¢

La Crosse, WI

This city isn't on the Byway, but it's close to the eastern end, just across the river.

★★ **BEST WESTERN MIDWAY HOTEL.** *1835 Rose St, La Crosse, WI (54603). Phone 608/781-7000. www.bestwestern.com.* 121 rooms, 2 story. Check-out noon. TV; cable (premium), VCR available. In-room modem link. Restaurant, bar; entertainment Tues-Sat. Room service. In-house fitness room, sauna. Indoor pool, whirlpool. Downhill, cross-country ski 10 miles. On the river; dockage. ¢

Minnesota

✻ *Historic Bluff Country Scenic Byway*

★★ **RADISSON HOTEL.** *200 Harborview Plz, La Crosse, WI (54601). Phone 608/784-6680. www.radisson.com.* 169 rooms, 8 story. Pet accepted. Check-out noon. TV. Restaurant, bar; entertainment. In-house fitness room. Indoor pool. Downhill, cross-country ski 8 miles. Free airport transportation. Overlooks the Mississippi River. **$**

★ **ROADSTAR INN.** *2622 Rose St, La Crosse, WI (54603). Phone 608/781-3070; toll-free 800/445-4667.* 110 rooms, 2 story. Complimentary continental breakfast. Check-out noon. TV; cable (premium). Downhill, cross-country ski 8 miles. **¢**

PLACES TO EAT

A long day of driving is sure to make you hungry. At the end of your journey, take a table at one of the following restaurants, both just across the river in La Crosse, Wisconsin.

★★ **FREIGHTHOUSE.** *107 Vine St, La Crosse, WI (54601). Phone 608/784-6211. www.freighthouserestaurant.com.* Closed Easter, Thanksgiving, Dec 24-25. Dinner. Bar. Former freight house of the Chicago, Milwaukee, and St. Paul Railroad (1880). Outdoor seating. **$$$**

★★ **PIGGY'S.** *328 S Front St, La Crosse, WI (54601). Phone 608/784-4877.* Steak menu. Closed Memorial Day, Labor Day, Dec 24-25. Lunch, dinner. Bar. Children's menu. **$$$**

Minnesota River Valley Scenic Byway

✳ MINNESOTA

Quick Facts

LENGTH: 287 miles.

TIME TO ALLOW: 7 days.

BEST TIME TO DRIVE: Fall, due to changing leaf colors.

BYWAY TRAVEL INFORMATION: New Ulm Chamber of Commerce: toll-free 888/463-9856; Byway local Web site: www.mnrivervalley.com.

BICYCLE/PEDESTRIAN FACILITIES: The Byway's roadways are not conducive to quality bicycle travel because of the variety of surface types, widths, shoulder characteristics, and high commercial vehicle use in certain areas. The roadways are, however, generally low-volume roadways, and many of the local roads will be quite peaceful if you have a bicycle that accommodates both gravel and paved surfaces.

Winding through the countryside between Mankato and Faribault is the 39-mile, newly paved Sakatah Singing Hills Bike Trail. The trail passes through a state park of the same name and through the village of Elysian, where a train depot serves as a wayside rest for the trail. The lakes in the area around Waterville are popular for fishing, and there are several lakeside resorts and campgrounds.

The Minnesota River flows gently between ribbons of oak, elm, maple, and cottonwood trees. The Minnesota River Valley Scenic Byway, which follows the river, wanders past rich farmland and through towns steeped in Minnesota history. Passing through rolling farmland and woodlands bordering river between Belle Plaine and Browns Valley—300 miles of highways and gravel roads—is a scenic, historical, and cultural experience.

Several sites tell the story of the Dakota Indians who lived here, the pioneers who settled here, and the tragic conflict between the two communities. Overshadowed by the Civil War, the struggle that occurred along the river valley in 1862 resulted in thousands of lost lives. Although struggles of the past have not been entirely forgotten, the Minnesota River Valley has been able to move on to form a new culture and way of life. Today, the production of agricultural products and the distribution of these products help to feed the nation.

THE BYWAY STORY

The Minnesota River Valley Scenic Byway tells cultural, historical, natural, recreational, and scenic stories that make it a unique and treasured Byway.

Cultural

The Minnesota River Valley is a productive land, and as a result, people have been living here for hundreds of years. They thrive on the land and develop rich cultures to complement their successful lifestyle along the Minnesota River. In every town and stop along the Byway, travelers have the opportunity to discover the

Minnesota

Minnesota River Valley Scenic Byway

industrious society that has made this land thrive. Towns and historic districts provide a peek at past successes and failures. Festivals and monuments define each community and store the memories that make the culture along the Byway unique.

The Dakota people share their regional history and thriving modern-day culture through festivals, museums, restaurants, hotels, casinos, and trading posts. European pioneers left their mark as well, and their strong ties to their European heritage remain and are celebrated today. Residents of New Ulm welcome visitors to numerous celebrations, shops, restaurants, and lodging, all of which focus on their unique German heritage. Milan residents are likewise proud of their Norwegian heritage, and their community festivals focus on the food, crafts, and traditions of those ancestors. Many other communities on or near the Byway demonstrate their proud heritage through festivals, events, museums, and residents.

The development of the railroad, the river, and the highways allowed more access into the western frontier. It also brought new developments to improve traditional agriculture. Some farmers tend land today that has been passed down from generation to generation for over a century. Their success over the years can be seen in ornate historic buildings and today's thriving communities. In Le Sueur, the museum features an exhibit on the Green Giant Company, which started in the Minnesota River Valley. For a look at past agriculture, the Olof Swensson Farm Museum is a great place to examine the culture that influenced today's lifestyle on the Byway. The traditions of agricultural life made this region what it is today, and much of the economy is still based on the rich soils of the river valley.

Historical

Once a wild and untamed river valley, the land of the Minnesota River Valley Scenic Byway used to belong solely to the Dakotas. As with many places in the country, this area was a region of great conflict between the native people and the settlers dreaming of a new life in the Midwest. Despite conflict and tensions, the land allowed its inhabitants to prosper and become one of the main food-producing areas of the country.

The land was rich and fertile and seemed ripe for the picking, yet struggles erupted during the same years that the Civil War was raging. In 1862, the largest and bloodiest Indian war in the history of the United States occurred. The Dakota people who once lived along the river in peace were threatened by the coming of new settlers from the east. For ten years, the land was divided and the first reservations were developed, but the Dakota eventually wanted their land and their way of life back. The result was a six-week war in which many settlers and Dakotas were killed. When the war was over, Abraham Lincoln pardoned 262 of the 300 Dakota men who were going to be hanged. The remaining 38 men became part of the largest mass execution in United States history. Other Dakota men, women, and children were taken to a prison camp at Fort Snelling and later sent west.

Today, visitors can visit the historic sites and communities that were caught up in turmoil. The Lower Sioux Agency Historic Site and the Fort Ridgely State Historic Site provide a closer look into the changes that were occurring across Minnesota at that time. Museums and historic buildings provide further insight into the characteristics of early settlements on the Byway.

Natural

When you aren't driving through enchanting towns and pastoral fields, you see that the land is overtaken by nature and the wilderness that's native to Minnesota. Prairies and woodlands combine to form natural areas full of plants and animals. The natural areas along the Byway are perfect places to see the Minnesota River Valley's natural ecosystems.

The Minnesota River is now gentle and calm. However, the river valley was once filled by the Glacial River Warren. The glacial river carved the valley down into ancient bedrock and exposed outcrops of gneiss. Visitors can see this bedrock throughout the valley and at the Gneiss Outcrop scientific and natural area. The valley topography varies from 1 to 5 miles in width and from 75 to 200 feet deep. The Minnesota River flows from the Hudson/Mississippi Continental Divide in Browns Valley through the steep bluffs and low floodplain area that characterize one of the most impressive landscapes in Minnesota.

The array of landscapes in the river valley boasts a large variety of wildlife and plant life. Wooded slopes and floodplains of willow, cottonwood, American elm, burr oak, and green ash rise into upland bluffs of red cedar and remnant tall grass where preserved tracts of prairie remain for public use and appreciation. Prairie chickens, turkeys, white-tailed deer, coyotes, foxes, beavers, and many species of fish and birds live here, as does the bald eagle.

Recreational

The Minnesota River is one of the great tributaries to the Mississippi River. It flows across Minnesota, providing some of the best recreational opportunities in the upper Midwest. Outdoor recreation abounds, including bird-watching, canoeing, hiking, trail riding, fishing, hunting, camping, boating, snowmobiling, cross-country skiing, and golfing. State parks and recreation areas combine outdoor recreation with opportunities to discover history along the Minnesota River Valley. Monuments and historic buildings commemorate the events of the past. An excursion to a lake or a museum can be equally enjoyable on the Byway, which successfully showcases both types of attractions.

All of these activities can be found along the Byway at the six state parks, at scientific and natural areas, at wildlife management areas, or at the local parks or waysides. Camping along the Byway is one of the best ways to experience the nature of the river valley. Spending time near the river, you can better enjoy the beautiful scenery at one of the Byway's lakes or take a stroll along the Minnesota River itself. Big Stone Lake was one of the earliest places for travelers to visit. At the turn of the century, visitors came to enjoy steamboat rides on the lake, as well as to spot wildlife.

Uncovering the history of the valley at the numerous historic sites and museums is another activity that many travelers enjoy. The museums and monuments are preserved and displayed by each of the Byway communities as they tell visitors the story of the valley. Historic homes and districts reveal opportunities for touring and shopping. Browsing the many unique gift and specialty shops and attending festivals and special events in the small towns along the Byway should not be missed.

Scenic

The Minnesota River Valley Scenic Byway showcases a variety of scenic elements. The route brings travelers along the river and through prairies, farmland, cities, woodlands, and wetlands. Vistas from high upon the bluffs of the river let your eyes wander over the landscape, just as bald eagles, which are so commonly seen here,

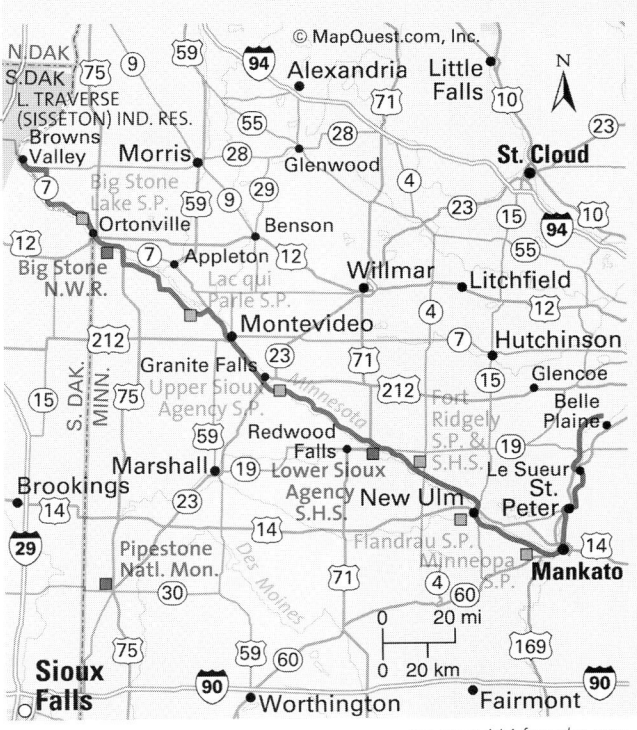

see page A14 for color map

Minnesota

❋ *Minnesota River Valley Scenic Byway*

would do. Some roads along the river bottom bring you right up to the river, where your senses enjoy the sights, sounds, and smells of the river valley. Along the way, you'll find magnificent farmsteads of today and yesterday, bustling communities, tremendous historical sites, wildlife, and some of the most unspoiled prairie in the Midwest—all of which entice you to stop and enjoy them.

THINGS TO SEE AND DO

Driving along the Minnesota River Valley Scenic Byway will certainly keep your senses engaged, but if you yearn to get out of the car and stretch your legs, or if you'd like to make a mini-vacation out of your trip, check out these attractions along the route.

BROWN COUNTY HISTORICAL MUSEUM. *2 N Broadway St, New Ulm (56073). Phone 507/354-2016.* Former post office. Historical exhibits on Native Americans and pioneers; artwork; research library with 5,000 family files. Open Mon-Fri, also Sat and Sun afternoons; closed holidays. **$**

EUGENE SAINT JULIEN COX HOUSE. *500 N Washington Ave, St. Peter (56082). Phone 507/931-2160. nrhp.mnhs.org.* Fully restored home (built in 1871) is the best example of Gothic Italianate architecture in the state. Built by the town's first mayor; late Victorian furnishings. Guided tours. Open June-Aug, Wed-Sun; May and Sept, Sat and Sun afternoons. **$$**

FLANDRAU. *1300 Summit Ave, New Ulm (56073). Phone 507/233-9800. www.dnr.state.mn.us/state_parks/flandrau/.* Comprised of 1,006 acres on the Cottonwood River. Swimming, cross-country skiing (rentals), camping, and hiking. Open daily 8 am-4 pm.

FORT RIDGELY. *Hwy 4, Fairfax (55332). Phone 507/426-7888 or 507/697-6321. www.mnhs.org/places/sites/fr/.* A 584-acre park. Fort partly restored; interpretive center. Nine-hole golf course (fee); cross-country skiing; camping; hiking; annual historical festival. June-Labor Day, Fri-Sat 10 am-5 pm, Sun noon-5 pm. **FREE**

THE GLOCKENSPIEL. *4th N and Minnesota sts, New Ulm (56073). Phone 507/354-4217.* A 45-foot-high musical clock tower with performing animated figures; carillon with 37 bells. Performances three times daily: noon, 3 pm, and 5 pm.

GUSTAVUS ADOLPHUS COLLEGE. *800 W College Ave, St. Peter (56082). Phone 507/933-8000. www.gustavus.edu.* On this campus of 2,300 students are Old Main (dedicated in 1876); Alfred Nobel Hall of Science and Gallery; Lund Center for Physical Education; Folke Bernadotte Memorial Library; Linnaeus Arboretum; Schaefer Fine Arts Gallery; and Christ Chapel, featuring door and narthex art by noted sculptor Paul Granlund. At various other locations on campus are sculptures by Granlund, sculptor-in-residence, including one depicting Joseph Nicollét, the mid-19th-century French explorer and cartographer of the Minnesota River Valley. Campus tours. In October, the college hosts the nationally known Nobel Conference, which has been held annually since 1965.

HARKIN STORE. *City Hwy 21, New Ulm (56073). Phone 507/354-8666 or 507/934-2160. www.mnhs.org/places/sites/hs/.* General store built by Alexander Harkin in 1870 in the small town of West Newton. The town died when it was bypassed by the railroad, but the store stayed open as a convenience until 1901, when rural free delivery closed the post office. The store has been restored to its original appearance and still has many original items on the shelves. Special programs in summer months. Open May, weekends 10 am-5 pm; June-Aug, Tues-Sun 10 am-5 pm; Sept, weekends 10 am-5 pm; first two weekends in Oct, Fri-Sun 10 am-5 pm. **$**

HERMANN'S MONUMENT. *Center Street Hill, Center and Monument sts, New Ulm (56073). Phone 507/354-4217.* Erected by a fraternal order, the monument recalls Hermann the Cheruscan, a German hero of AD 9. Towering 102 feet, the monument has a winding stairway to a platform with views of the city and the Minnesota River Valley. Open June-Labor Day, daily. Picnic area. **$**

HUBBARD HOUSE. *606 S Broad St, Mankato (56001). Phone 507/345-4154 or 507/345-5566.* Historic Victorian home (1871) with cherry woodwork, three marble fireplaces, silk wallcoverings, signed Tiffany lampshade; carriage house; Victorian gardens. **$$**

LAC QUI PARLE STATE PARK. *US Hwy 59 and State Hwy 7, Montevideo (56265). Phone 320/752-4736. www.dnr.state.mn.us/state_parks/lac_qui_parle/.* Approximately 529 acres. On Lac qui Parle and Minnesota rivers. Dense timber. Swimming, fishing; boating (ramps); hiking, riding, cross-country skiing; picnicking; camping. Open daily, hours vary.

LAND OF MEMORIES. *Land of Memories Parkway, Hwy 169/60, Mankato (56001). Phone 507/387-8649.* Picnicking, camping (hookups, dump station, rest rooms), fishing, boating (launch), and nature trails. **$$$**

MINNEOPA STATE PARK. *State Hwy 68 and US Hwy 169, Mankato (56001). Phone 507/389-5464. www.dnr.state.mn.us/state_parks/minneopa/.* 1,145-acre park. Scenic falls and gorge; historic mill site; fishing, hiking, picnicking, and camping. Memorial Day-Labor Day, daily 9 am-4 pm.

MOUNT KATO SKI AREA. *20461 Hwy 66, Mankato (56001). Phone 507/625-3363; toll-free 800/668-5286. www.mountkato.com.* Five quad, three double chairlifts; patrol, school, rental; snowmaking; cafeteria; bar. Open Nov-Apr, daily. **$$$$**

OLOF SWENSSON FARM MUSEUM. *151 Pioneer Dr, Montevideo (56265). Phone 320/269-7636. www.montechamber.com/cchs/swensson.htm.* A 22-room brick family-built farmhouse, barn, grist mill, and family burial plot on a 17-acre plot. Olof Swensson ran unsuccessfully for governor of Minnesota, but the title was given to him by the community out of respect and admiration. Open Memorial Day-Labor Day, Sun 1-5 pm or by appointment. **$$**

SCHELL GARDEN AND DEER PARK. *18th and Schell sts, New Ulm (56073). Phone 507/354-5528.* Garden with deer and peacocks (open all year). Brewery tours, museum, gift shop. Open Memorial Day-Labor Day, daily; rest of year, Sat. **$**

SIBLEY PARK. *800 Sibley Park Rd NE, Mankato (56001). Phone 320/354-2055. www.dnr.state .mn.us/state_parks/sibley/.* Fishing, picnicking; river walk, playground, zoo (open daily); beautiful gardens, scenic view of rivers. **FREE**

TREATY SITE HISTORY CENTER. *1851 N Minnesota Ave, St. Peter (56082). Phone 507/931-2160. www.tourism.st-peter.mn.us/treaty.php3.* County historical items relating to Dakota people, explorers, settlers, traders, and cartographers and their impact on the 1851 Treaty of Traverse des Sioux. Archives. Museum shop. Open Mon-Sat 10 am-4 pm, Sun 1-4 pm; closed major holidays. **$**

UPPER SIOUX AGENCY STATE PARK. *9805 Hwy 67, Granite Falls (56241). Phone 320/564-4777. www.dnr.state.mn.us/state_parks/upper_sioux_agency/.* 1,280 acres. Boating (ramps, canoe campsites). Bridle trails. Snowmobiling. Picnicking. Semi-modern campsite; rustic horse rider campground. Visitor center.

Minnesota

❄ *Minnesota River Valley Scenic Byway*

TOURTELOTTE PARK. *N Broad and Mabel sts, Mankato (56001). Phone 507/387-8649.* Picnicking, playground. Swimming pool (from early June-Labor Day, daily; fee), wading pool. **$**

W. W. MAYO HOUSE. *118 N Main St, Le Sueur (56058). Phone 507/665-3250 or 507/665-6965. www.mnhs.org/places/sites/wwmh/.* Home of Mayo Clinic founder; restored to 1859-1864 period when Dr. Mayo carried on a typical frontier medical practice from his office on the second floor. Adjacent park is the location of Paul Granland's bronze sculpture *The Mothers Louise.* Open mid-May-mid-June, Sat 1-4:30 pm; June-Aug, Tues-Sat 10 am-4:30 pm; other times by reservation. **$**

WILLIAMS MINNEOPA OUTDOOR LEARNING CENTER. *Hwy 68, Mankato (56001). Phone 507/625-3281.* Wide variety of native animals and vegetation; information stations; outdoor classroom. Open daily. **FREE**

YELLOW MEDICINE COUNTY MUSEUM. *726 Prentice St, Granite Falls (56241). Phone 320/564-4479.* Depicts life in the county and state dating from the 1800s. Two authentic log cabins and a bandstand are on site. Also here is an exposed rock outcropping estimated to be 3.8 billion years old. Open mid-May-mid-Oct, daily except Mon; mid-Apr-mid-May, Tues-Fri. **FREE**

PLACES TO STAY

If you choose to include an overnight stay in your trip along this Byway, Mobil Travel Guide recommends the following lodgings.

Granite Falls

★ **VIKING JR. MOTEL.** *1250 W Hwy 212, Granite Falls (56241). Phone 320/564-2411.* 20 rooms. Check-out 11 am. TV; VCR available (movies). ¢

Mankato

★★ **BEST WESTERN HOTEL & RESTAURANT.** *1111 Range St, Mankato (56003). Phone 507/625-9333; toll-free 800/780-7234. www.bestwestern.com.* 147 rooms, 2 story. Check-out noon. TV; VCR available (movies). Restaurant, bar. Room service. Sauna. Indoor pool, whirlpool. Downhill ski 6 miles, cross-country ski 1 mile. Free airport transportation. ¢

★★ **THE BUTLER HOUSE.** *704 S Broad St, Mankato (56001). Phone 507/387-5055. www.butlerhouse.com.* 5 rooms, 3 story. Offering turn-of-the-century elegance and regional cuisine in a restored Victorian mansion. **$**

★ **DAYS INN.** *1285 Range St, Mankato (56001). Phone 507/387-3332; toll-free 800/544-8313. www.daysinn.com.* 50 rooms, 2 story. Pet accepted, some restrictions; fee. Complimentary continental breakfast. Check-out 11 am. TV. In-room modem link. Indoor pool, whirlpool. Downhill, cross-country ski 5 miles. ¢

★★ **HOLIDAY INN.** *101 E Main St, Mankato (56001). Phone 507/345-1234; toll-free 800/465-4329. www.holiday-inn.com.* 151 rooms, 4 story. Check-out noon. TV. In-room modem link. Restaurant, bar. Room service. In-house fitness room. Indoor pool, whirlpool, poolside service. Downhill ski 4 miles, cross-country ski 1 mile. Civic Center nearby. ¢

★ **SUPER 8.** *Hwys 169 N and 14, Mankato (56001). Phone 507/387-4041; toll-free 800/800-8000. www.super8.com.* 61 rooms, 3 story. Check-out 11 am. TV. In-room modem link. Downhill ski 5 miles, cross-country ski 2 miles. ¢

New Ulm

★★ **THE BOHEMIAN.** 304 S German St, New Ulm (56073). Phone 507/354-2268; toll-free 866/499-6870. www.the-bohemian.com. 7 rooms, 3 story. Victorian. 7 rooms, each with private bath; romantic whirlpool and fireplace suites available. $$

★ **COLONIAL INN.** 1315 N Broadway St, New Ulm (56073). Phone 507/354-3128; toll-free 888/215-2143. 24 rooms. Check-out 11 am. TV; cable (premium). Cross-country ski 1 mile. ¢

★★ **DEUTSCHE STRASSE BED & BREAKFAST.** 404 S German St, New Ulm (56073). Phone toll-free 866/226-9856. 5 rooms, 3 story. Located in the historic district of New Ulm within blocks of unique German and antique shops. $

★★ **HOLIDAY INN.** 2101 S Broadway St, New Ulm (56073). Phone 507/359-2941; toll-free 800/465-4329. www.holiday-inn.com. 120 rooms, 2 story. Check-out noon. TV; cable (premium). In-room modem link. Restaurant, bar; entertainment. Room service. Sauna. Game room. Indoor pool, whirlpool. Cross-country ski 2 miles. Business center. ¢

★★ **JAVA RIVER CAFÉ.** 210 S 1st St, Montevideo (56265). Phone 320/269-7106. www.javarivercafe.com. American menu. Closed Sun. Breakfast, lunch, dinner. Featuring home-made specialty coffee and related foods with an emphasis on sustainable agriculture and land preservation. $

★ **D. J.'S.** 1200 N Broadway, New Ulm (56073). Phone 507/354-3843. German, American menu. Closed Dec 24-25. Breakfast, lunch, dinner. $$

★ **KAISERHOFF.** 221 N Minnesota Ave, New Ulm (56073). Phone 507/359-2071. German menu. Lunch, dinner. New Ulm's oldest and most venerable dining establishment. $

★ **VEIGEL'S KAISERHOF.** 221 N Minnesota St, New Ulm (56073). Phone 507/359-2071. German menu. Closed Dec 25. Lunch, dinner. Bar. Children's menu. $$

PLACES TO EAT

A long day of driving is sure to make you hungry. At the end of your journey, take a table at one of the following restaurants.

★★ **THE COUNTRY PUB.** 1179 E Pearl St, Kasota (56050). Phone 507/931-5888. www.countrypubmn.com. American menu. Dinner. Located in beautiful downtown Kasota, offering elegant white-tablecloth dining. $$

North Shore Scenic Drive

✤ MINNESOTA AN ALL-AMERICAN ROAD

Quick Facts

LENGTH: 154 miles.

TIME TO ALLOW: 1 day.

BEST TIME TO DRIVE: Late spring through late fall.

BYWAY TRAVEL INFORMATION: Minnesota Office of Tourism: 612/296-5027; Grand Portage Traveler Information Center: 218/475-2592; Byway local Web site: www.superiorbyways.org; Byway travel and tourism Web sites: www.northshorescenicdrive.com, www.lakecnty.com.

SPECIAL CONSIDERATIONS: The weather along the shore can be quite cool at times, so a jacket is recommended even in summer. Each town along the Byway offers plenty of visitor services, including gasoline and lodging.

RESTRICTIONS: The North Shore Scenic Drive can be traveled year-round. At times in winter, snowfall is heavy. However, because homes and businesses exist along the route, the plowing of this road is a priority.

BICYCLE/PEDESTRIAN FACILITIES: The North Shore Scenic Drive has sidewalks and 6- or 8-foot shoulders in places along the route that provide safe biking and pedestrian opportunities. Most pedestrian traffic concentrates around the various state parks. Trails and pedestrian warning signs and crossings are in place. Be aware that this route is a major transport route to and from the Canadian border, so large semi trucks and other similar vehicles are frequently seen along the Byway.

The North Shore of Lake Superior, the world's largest freshwater lake, is 154 miles of scenic beauty and natural wonders. It has what no other place in the Midwest can offer—an inland sea, a mountain backdrop, an unspoiled wilderness, and a unique feeling all its own.

The North Shore Scenic Drive runs from Duluth to the Canadian border. The drive along the lakeshore is rich with the history of Native Americans, French and British explorers, lumbering, and iron-ore mining. On one side of the highway, Superior National Forest, one of the great wilderness areas of the United States, carpets the hills of the Sawtooth Mountains with balsams, birch, and pine. On the other side, Lake Superior offers spectacular views from the rocky shore. Each of the eight state parks has beautiful trails, and the 200-mile Superior Hiking Trail provides the chance to experience this magnificent landscape firsthand. For those who want to get out on the lake, there are charter fishing, sailing, kayaking, and excursion boats.

The North Shore also has a rich history deeply rooted in its plentiful natural resources. The Grand Portage National Monument features a reconstructed NorthWest Company fur-trading post. Grand Marais is a quaint harbor town that is the entrance to the Gunflint Trail, a paved trail leading inland to the Boundary Waters Canoe Area Wilderness. Giant ore boats pull up to the docks at Two Harbors, and the much smaller, 100-year-old tugboat the *Edna G.* is displayed here. Small museums in Two Harbors and Tofte, as well as interpretive programs at the state parks, tell the story of this area. Whether you are looking for a wilderness expedition or the comforts of a modern lodge, the lofty pines

Minnesota

❋ *North Shore Scenic Drive*

of the Superior National Forest and crashing waves of Lake Superior seem to have a magic that whispers, "Come back."

THE BYWAY STORY

The North Shore Scenic Drive tells historical, natural, recreational, and scenic stories that make it a unique and treasured Byway.

Historical

The North Shore Scenic Drive includes the rich and colorful heritage of settlers who were attracted to the area's bounty of natural resources. Many examples of this heritage can be found along the route, from the Voyageur era at Grand Portage National Monument to more recent times at the Split Rock Lighthouse State Historic Monument or the *Edna G.* Steam Tug Boat National Monument in Two Harbors. A host of interpretive resources and festivals keep this heritage alive, allowing you to experience the story of the North Shore's legacy.

It is believed that the first people to settle the North Shore region arrived about 10,000 years ago. These Native Americans entered the region during the final retreat of the Wisconsin glaciation. Many waves of Native American people inhabited the North Shore prior to European contact.

The first Europeans, French explorers, and fur traders reached Lake Superior country around AD 1620. By 1780, the Europeans had established fur-trading posts at the mouth of the St. Louis River and at Grand Portage. During the fur-trading days, Grand Portage became an important gateway into the interior of North America for exploration, trade, and commerce. The trading post came alive for a short time each summer during a celebration called the Rendezvous, when voyageurs and traders from the western trading posts met their counterparts from Montreal to exchange furs and trade goods. Grand Portage is now a national monument and an Indian reservation. At this national monument, some of the original structures of the fur-trading days have been reconstructed.

In 1854, the Ojibwe signed the Treaty of La Pointe, which opened up northeastern Minnesota for mineral exploration and settlement. The late 1800s saw a rise in commercial fishing along the North Shore. Some of the small towns along the North Shore, such as Grand Marais and Little Marais, were first settled during this period. Many of the small towns still have fish smokehouses, thus keeping the fishing heritage alive. The North Shore Commercial Fishing Museum in Tofte interprets this period.

Lumber barons moved into the region between 1890 and 1910, and millions of feet of red and white pine were cut from the hills along the North Shore. Temporary railroads transported the logs to the shore, where they were shipped to sawmills in Duluth, Minnesota, and Superior, Bayfield, and Ashland, Wisconsin. Today, many of those old railroad grades are still visible, and some of the trails still follow these grades.

Miners digging for high-grade ore from the Iron Ranges in northeastern Minnesota established shipping ports like Two Harbors in 1884. With the rise of taconite in the 1950s, they developed shipping ports at Silver Bay and Taconite Harbors along Superior's shores. These harbors are still in use, and visitors driving the route may see large 1,000-foot-long ore carriers being loaded or resting close to shore. The shipping history and Lake Superior's unpredictable storms have left the lake bottom dotted with shipwrecks, which now provide popular scuba diving destinations.

With the completion of the North Shore highway in 1924, tourism became an important industry along the shore. However, even earlier (in the 1910s), Historic Split Rock Lighthouse attracted tourists who arrived by sailboat. Split Rock Lighthouse is still the best-known North Shore landmark. Visitors who explore the

see page A15 for color map

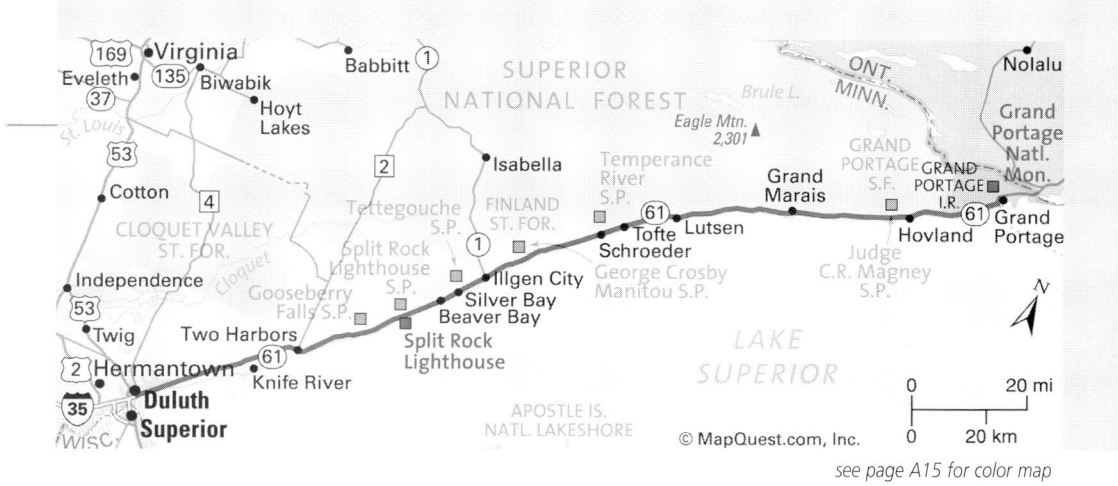

North Shore today will come to know not only this landmark, but also many others that point to past times in this area of Minnesota.

Natural

The North Shore Scenic Drive allows you to experience a number of unique natural and geological features. The Byway follows the shoreline of the world's largest freshwater lake, Lake Superior, which contains 10 percent of the world's fresh-water supply. This Byway is a marvelous road to travel, never running too far from the lakeside and at times opening out onto splendid views down the bluffs and over the blue water. This allows visitors a chance to experience a landscape that has seen little alteration from its original state. During the winter, the many parks and hundreds of miles of groomed trails draw outdoor enthusiasts for cross-country skiing, snowshoeing, and snowmobiling. Lutsen Mountain is the largest and highest downhill ski area in the Midwest.

The North Shore's spectacular topography originated a billion years ago, when molten basalt erupted from the mid-continental rift. The Sawtooth Mountains, which frame the North Shore, are remnants of ancient volcanoes. The glaciers that descended from Canada 25,000 years ago scoured the volcanic rock into its current configuration. Cascading rivers coming down from the highlands into Lake Superior continue to reshape the landscape today. Northeastern Minnesota is the only part of the US where the expansive northern boreal forests dip into the lower 48 states. The Lake Superior Highlands have been identified to be of great importance for biodiversity protection by the Minnesota Natural Heritage Program and the Nature Conservancy. This landscape contains significant tracks of old-growth northern hardwood and upland northern white cedar forest.

The natural environment along the corridor supports a number of wildlife species. Wildlife that may be observed from the road or at the state parks adjacent to the road include beaver, otter, timberwolf, white-tailed deer, coyote, red fox, black bear, and moose. Federally listed threatened species of bald eagle, gray wolf, and peregrine falcon also have populations here. Northeastern Minnesota is recognized as one of the better areas in the nation for viewing rare birds. Diversity of habitat, geography, and proximity to Lake Superior combine to attract a variety of bird life that draws bird-watchers from around the world. In the fall, hawks migrating along the shore of Lake Superior number in the tens of thousands. Winter is an excellent time to see northern owls, woodpeckers, finches, and unusual water birds.

Besides the state parks, a number of scientific and natural areas have been set aside along the route. These sites were selected because they contain excellent examples of the area's geologic history or harbor unique plant communities. Due to the unique geography of the landscape and climate changes during the ice ages, remnant species of arctic plants can be found along the shore. Examples of rare alpine plants that can also be found along the shore include butterwort, northern eyebright, alpine bistort, small false asphodel, and moonwort.

Minnesota

❋ *North Shore Scenic Drive*

Recreational

Imagine yourself nestled next to a fireplace in the lodge after a long day of skiing or drying out next to a campfire from an afternoon of kayaking on a Minnesota river. The North Shore is one of the primary destinations for recreational activities in the Midwest, as well as recreational driving. With well-developed facilities for outdoor activities that include camping, hiking, biking, skiing, snowmobiling, fishing, canoeing, and other water-related activities. The impressive scenery and natural beauty of the North Shore has been an attraction for tourists since the completion of Highway 61 in the early part of this century. The North Shore is dotted with innumerable points of interest, giving visitors a reason to come back again and again. The businesses along the corridor are set up to easily accommodate travelers from short weekend trips to multiple-week family vacations. They work together as partners to make this a quality recreational experience.

The North Shore Scenic Drive is home to one of the greatest trail systems in the nation. The Superior Hiking Trail stretches 200 miles from Two Harbors to the Canadian border, connecting the eight state parks and giving travelers an opportunity to enjoy the natural beauty and vistas from the highlands and the cliffs. The trail system is well signed and offers unlimited opportunities ranging from day hikes to multiple-week treks through a rugged wilderness setting. The trail is designated as a National Recreation Trail by the US Forest Service. A shuttle service offers transportation between trail heads. The Lake Superior Water Trail allows visitors to travel the coastline of Lake Superior by kayak. The trail uses public land for designated rest areas and will eventually be part of the Lake Superior Water Trail encircling all of Lake Superior. The Gitchi-Gami Trail is planned to accommodate bicyclists, pedestrians, and in-line skaters following the right of way of the North Shore Scenic Drive.

And when you're ready for a break from outdoor excitement, try touring some of the classic historical sites in the area. The Minnesota Historical Society takes care of the Split Rock Lighthouse site and provides interpretive programs. The Lake County Historical Museum, which is housed in the old Duluth & Iron Range Railway Depot, contains excellent exhibits on the region's history. In Two Harbors, you can tour an operating lighthouse and the *Edna G.*, which was the last steam tug to work the Great Lakes. So pack up your sled or your river raft and try them out in northeastern Minnesota, and be sure to bring your camera, because there are memorable sights to be seen!

Scenic

One of the main draws of the North Shore is its reputation as one of the most scenic drives in the United States The route offers splendid vistas of Lake Superior and its rugged shoreline, as well as views of the expansive North Woods. The road crosses gorges carved out by cascading rivers, offering views of waterfalls and adding diversity to the landscape. The attraction of the falls at Gooseberry State Park makes this the most visited state park in Minnesota.

Different seasons and changes in weather continually alter the landscape's appearance, keeping the route fresh and interesting. During the fall, the corridor displays bright-red sugar maples and warm gold birch and aspens. The vibrant colors attract many day and weekend travelers to the North Shore. After the leaves have fallen and the ground is snow covered, new views of the lake open up with opportunities to spy wildlife like deer, moose, and wolf that use the frozen lakeshore as a way to travel. Spring ice break-up offers another fascinating scene as glimmering mountains of ice are driven up the shore. In summer, the lake, with an average water temperature of 40 degrees, offers a welcome breeze to cool down the summer heat.

A few charming towns dot the shoreline. Most started out as small fishing and harbor towns, shipping ore and timber. These towns still have a distinct sense of connection with the lake. For many travelers, the quaint town of Grand Marais is as far as they will go; however, the scenery becomes even more spectacular as you continue driving north toward the Canadian border. For long stretches, there is nothing alongside the road but trees, cliffs, beaches, and the whitecap-crested lake. The Byway ends at Grand Portage. Here, a national monument marks the beginning of a historic 9-mile portage trod by Indians and Voyageurs. A few miles farther at Grand Portage State Park, you can view the High Falls. At 120 feet, these are the highest waterfalls in Minnesota. The falls' magnificence is a fitting end to a trip up the North Shore.

HIGHLIGHTS

The North Shore Scenic Drive is comprised of the northern 150 miles of Minnesota Trunk Highway 61, from Duluth to Two Harbors all the way to Grand Portage near the US/Canadian border.

- Begin your experience on the North Shore Scenic Drive in **Canal Park** in Duluth. Follow Canal Park Drive over Interstate 35 to Superior Street.

- Take a right on Superior Street and follow this road through a portion of historic **downtown Duluth** and past the refurbished **Fitger's Brewery Complex,** with its many fine shops and restaurants.

- To continue along the Byway, take a right onto London Road. This intersection is at 10th Avenue East and Superior Street. Follow London Road for the next 5.4 miles, driving past **Leif Erikson Park,** which contains the **Rose Garden,** a popular attraction for visitors.

- To continue on the route, follow the signs for the North Shore Scenic Drive. The route proceeds along the lake, passing through small settlements and stunning lake vistas. There are no less than six rivers that cross under the Byway during this 19-mile segment to Two Harbors.

- Connect with the Byway loop in downtown **Two Harbors** by turning right onto Highway 61 just west of town. Follow Highway 61 to 6th Street (Waterfront Drive), where you take a right and pick up the Byway once again.

- Follow 6th Street to South Avenue, where you take a left. Prior to taking a left, be sure to stop at the **3M Museum.** As you drive down South Avenue, you can begin to see the massive ore docks that extend from the waterfront.

- Continue on the route by taking a left onto First Street (Park Road) and following it to Highway 61. First Street takes you past hiking trails that line the shoreline and the peaceful setting of **Burlington Bay,** the second of the two harbors that gives the community its name.

- Taking a right on Highway 61 and continuing on that route will lead you along the remainder of the North Shore Scenic Drive, which ends approximately 122 miles later at the Canadian border.

Minnesota

❋ North Shore Scenic Drive

THINGS TO SEE AND DO

Driving along the North Shore Scenic Drive will certainly keep your senses engaged, but if you yearn to get out of the car and stretch your legs, or if you'd like to make a mini-vacation out of your trip, check out these attractions along the route.

AERIAL LIFT BRIDGE. *525 Lake Ave S, Duluth (55802). Phone 218/722-3119.* 138 feet high, 336 feet long, 900 tons in weight. The bridge connects the mainland with Minnesota Point, rising 138 feet in less than a minute to let ships through. **FREE**

DEPOT MUSEUM. *520 South Ave, Two Harbors (55616). Phone 218/834-4898.* Historic depot (1907) highlights the geological history and the discovery and mining of iron ore. Mallet locomotive (1941) and the world's most powerful steam engine are on display. Open June-Aug, Tues-Wed, Fri-Sat 10 am-4 pm; Sept-May, Wed, Fri-Sat 10 am-4 pm.

DEPOT SQUARE. *Duluth (55701).* Reproduction of 1910 Duluth street scene with an ice cream parlor, storefronts, gift shops, and trolley rides.

DULUTH LAKEWALK. *Duluth (55701). Phone toll-free 800/4-DULUTH.* The Duluth Lakewalk stretches for 4.2 miles along Lake Superior. An entrance point is just feet from the beginning of the North Shore Scenic Drive. The Lakewalk features a boardwalk and an adjacent trail for bikers, runners, and in-line skaters. Never more than a few feet from Lake Superior, the Lakewalk offers excellent views of the harbor. **FREE**

THE *EDNA G.* *Waterfront Dr, Two Harbors (55616). Phone 218/834-4898.* The *Edna G.* served Two Harbors from 1896 to 1981. It was designated a National Historic Site in 1974 as the only steam-powered tug still operating on the Great Lakes. Now retired, the *Edna G.* features seasonal tours where visitors can see her beautiful interior décor of wood paneling and brass fittings. Visitors get a look at the captain's quarters, the engine room, the crew's quarters, and the pilot house.

FITGER'S BREWERY COMPLEX. *600 E Superior St, Duluth (55802). Phone 218/722-8826; toll-free 888/FITGERS. www.fitgers.com.* Historic renovated brewery transformed into more than 25 specialty shops and restaurants on the shore of Lake Superior. Summer courtyard activities. Open daily; closed holidays. **FREE**

GLENSHEEN. *3300 London Rd, Duluth (55804). Phone 218/726-8910. www.glensheen.org.* Circa 1905-1908. A historic 22-acre Great Lakes estate on the west shore of Lake Superior; owned by the University of Minnesota. Tours. Grounds open daily. Mansion open May-Oct, daily; rest of year, weekends; closed holidays. **$$**

GOOSEBERRY FALLS STATE PARK. *3206 Hwy 61, Two Harbors (55616). Phone 218/834-3855. www.dnr.state.mn.us/state_parks/gooseberry_falls.* Known as the gateway to the North Shore Scenic Drive, Gooseberry Falls State Park entices visitors to stop and enjoy the area before continuing on the rest of the drive. The Gooseberry River plummets down a wall of rock, creating the scenic falls that make the area so popular. Learn more about the park's history and nature at the Joseph N. Alexander Visitor Center or explore it yourself on some of the hiking trails that are provided. 1,662 acres. Fishing, hiking, cross-country skiing, snowmobiling, picnicking, and camping (dump station). State park vehicle permit required. Spring and summer 8 am-4:30 pm.

GRAND PORTAGE NATIONAL MONUMENT. *315 S Broadway, Grand Portage (55604). Phone 218/387-2788. www.nps.gov/grpo/.* Once, this area was a rendezvous point and central supply depot for fur traders operating between Montreal and Lake Athabasca. Partially reconstructed summer headquarters of the North West Company includes stockade, great hall, kitchen, and warehouse. The Grand Portage begins at the stockade and runs 8 1/2 miles NW from Lake Superior to Pigeon River. Primitive camping at Fort Charlotte (accessible only by hiking the Grand Portage or by canoe). Buildings and grounds open mid-May-mid-Oct, daily. Trail open all year. **$$**

GUNFLINT TRAIL. *County Rd 12, Grand Marais (55604). Phone toll-free 800/338-6932. www.gunflint-trail.com.* Penetrates into an area of hundreds of lakes where camping, picnicking, fishing, and canoeing are available. Starting at the northwest edge of town, the road goes north and west 58 miles to Saganaga Lake on the Canadian border.

✪ ISLE ROYALE NATIONAL PARK. *800 E Lakeshore Dr, Houghton, MI (49931). Phone 906/482-0984. www.nps.gov/isro/.* This unique wilderness area, covering 571,790 acres, is the largest island in Lake Superior, 15 miles from Canada (the nearest mainland), 18 miles from Minnesota, and 45 miles from Michigan. There are no roads, and no automobiles are allowed. The main island, 45 miles long and 8 1/2 miles across at its widest point, is surrounded by more than 400 smaller islands. Moose, wolf, fox, and beaver are the dominant mammals. More than 200 species of birds have been observed, including loons, bald eagles, and ospreys. More than 165 miles of foot trails lead to beautiful inland lakes, more than 20 of which have game fish, including pike, perch, walleye, and, in a few, whitefish cisco. There are trout in many streams and lakes. Fishing is under National Park Service and Michigan regulations. Isle Royale can be reached by boat from Grand Portage in Minnesota. The park is open approximately May-Oct.

LAKE SUPERIOR MARITIME VISITORS CENTER. *600 Lake Ave S, Duluth (55802). Phone 218/727-2497. www.lsmma.com.* Ship models, relics of shipwrecks, reconstructed ship cabins; exhibits related to maritime history of Lake Superior and Duluth Harbor and the Corps of Engineers. Vessel schedules and close-up views of passing ship traffic. Open Apr-mid-Dec, daily; rest of year, Fri-Sun; closed Jan 1, Thanksgiving, Dec 25. **FREE**

LIGHTHOUSE POINT AND HARBOR MUSEUM. *520 South Ave, Two Harbors. Phone 218/834-4898.* Displays tell the story of iron-ore shipping and the development of the first iron ore port in the state. A renovated pilot house from an ore boat is located on the site. Shipwreck display. Tours of operating lighthouse. Open May-Nov 1, daily. **$$**

LUTSEN MOUNTAINS SKI AREA. *467 Ski Hole Rd, Lutsen (55612). Phone 218/663-7281. www.lutsen.com.* Seven double chairlifts, surface lift; school, rentals; snowmaking; lodge, cafeteria, bar. Longest run 2 miles; vertical drop 1,088 feet. Gondola. Cross-country trails. Open mid-Nov-mid-Apr, daily. **$$$$**

SPLIT ROCK LIGHTHOUSE STATE PARK. *Hwy 61, Two Harbors (55616). Phone 218/226-6377. www.dnr.state.mn.us/state_parks/split_rock_lighthouse/.* 1,987 acres. Lighthouse served as guiding sentinel for north shore of Lake Superior from 1910 to 1969. Also in the park is a historic complex (fee) that includes fog-signal building, keeper's dwellings, several outbuildings and the ruins of a tramway (mid-May-mid-Oct, daily). Waterfalls. Picnicking. Cart-in camping (fee) on Lake Superior; access to Superior Hiking Trail. State park vehicle permit required. **$**

Minnesota

❋ North Shore Scenic Drive

🟥 **SUPERIOR NATIONAL FOREST.** *Gunflint Trail (County Rd 12), Duluth (55808). Phone 218/626-4300. www.superiornationalforest.org.* With more than 2,000 beautiful clear lakes, rugged shorelines, picturesque islands, and deep woods, this is a magnificent portion of Minnesota's famous northern area. The Boundary Waters Canoe Area Wilderness, part of the forest, is perhaps the finest canoe country in the United States (travel permits required for each party, **$$** for advance reservations, phone toll-free 800/745-3399). Scenic water routes through wilderness near the international border offer opportunities for adventure. Adjacent Quetico Provincial Park is similar. Boating, swimming, water sports; fishing and hunting under Minnesota game and fish regulations; winter sports; camping (fee), picnicking, and scenic drives along the Honeymoon, Gunflint, Echo, and Sawbill trails.

WILLIAM A. IRVIN. *350 Harbor Dr, Duluth (55802). Phone 218/722-7876.* Guided tours of former flagship of United States Steel's Great Lakes fleet that journeyed inland waters from 1938-1978. Explore decks and compartments of restored 610-foot ore carrier, including the engine room, elaborate guest staterooms, galley, pilothouse, observation lounge, elegant dining room. Open May-mid-Oct. **$$**

PLACES TO STAY

If you choose to include an overnight stay in your trip along this All-American Road, Mobil Travel Guide recommends the following lodgings.

Duluth

★★ **BEST WESTERN EDGEWATER.** *2400 London Rd, Duluth (55812). Phone 218/728-3601; toll-free 800/780-7234. www.bestwestern.com.* 282 rooms. Pet accepted, some restrictions. Complimentary continental breakfast. Check-out noon. TV; cable (premium). Many rooms with balconies, views of Lake Superior. Refrigerators. Valet services. Restaurant adjacent. Playground. Exercise equipment, sauna. Game room. Indoor pool, whirlpool. Downhill ski 7 miles, cross-country ski 1 mile. Lawn games. Miniature golf. Business center. ¢

★★ **COMFORT SUITES.** *408 Canal Park Dr, Duluth (55802). Phone 218/727-1378; toll-free 800/228-5151. www.choicehotels.com.* 82 rooms. TV; cable (premium). Complimentary continental breakfast. Check-out 11 am. Indoor pool; whirlpools. Coin laundry. Restaurant adjacent. Business services available. In-room modem link. Bellstaff (in season). Downhill, cross-country ski 5 miles. Refrigerators; some in-room whirlpools. **$**

★★★ **FITGERS INN.** 🙂 *600 E Superior St, Duluth (55802). Phone 218/722-8826; toll-free 888/348-4377. www.fitgers.com/lodging/.* Housed in what was once a thriving brewery, this historic hotel pairs European style with modern facilities. Well-designed guest suites, luxurious amenities, a dinner theater, restaurants, and shopping make this a popular choice in Duluth. 62 rooms, 5 suites. Check-out noon. TV; cable (premium), VCR available. In-room modem link. Some fireplaces, in-room whirlpools. Bar; entertainment weekends. Health club privileges, in-house fitness room. Business services. Shopping arcade. Most rooms overlook Lake Superior. ¢

★★ **HOLIDAY INN HOTEL & SUITES.** *200 W First St, Duluth (55802). Phone 218/722-1202; toll-free 800/477-7089. www.holiday-inn.com.* 353 rooms. Check-out noon. TV. In-room modem link. Refrigerators. Microwave, wet bar in suites. Restaurant, bar. In-house fitness room, saunas. Health club privileges. Indoor pools, whirlpool. Downhill, cross-country ski 7 miles. Free garage parking. Adjacent to a large shopping complex. ¢

148

★★ **RADISSON HOTEL.** *505 W Superior St, Duluth (55802). Phone 218/727-8981; toll-free 800/333-3333. www.radisson.com.* 268 rooms. Pet accepted. Check-out noon. TV; VCR available. Restaurant, bar. Health club privileges, sauna. Indoor pool, whirlpool, poolside service. Downhill, cross-country ski 10 miles. Business services. ¢

★ **SUPER 8.** *4100 W Superior St, Duluth (55807). Phone 218/628-2241; toll-free 800/800-8000. www.super8.com.* 59 rooms. Complimentary continental breakfast. Check-out 11 am. TV; cable (premium). Coin laundry. Restaurant opposite. Sauna, whirlpool. Downhill, cross-country ski 5 miles. Business services. ¢

Grand Marais

★★ **BEARSKIN LODGE.** *124 E Bearskin Rd, Grand Marais (55604). Phone 218/388-2292; toll-free 800/338-4170. www.bearskin.com.* 4 kitchen units (1-3 bedrooms), 11 kitchen cottages (2-3 bedrooms). No A/C. Check-out 10 am, check-in 4 pm. Some balconies, screened porches. Fireplaces, microwaves. Coin laundry. Dining room (reservations required). Box lunches. Free supervised children's activities (June-Aug), ages 3-13. Playground. Whirlpool, sauna. Private swimming beach. Cross-country ski on site. Boats, motors, canoes. Fishing guides. Private docks. Grills, picnic tables. Hiking trails, mountain bikes. Business services. Grocery, package store 5 miles. Nature program. $

★ **DREAM CATCHER BED & BREAKFAST.** *2614 County Rd 7, Grand Marais (55604). Phone 218/387-2876; toll-free, 800/682-3119. www.dreamcatcherbb.com.* 3 rooms, 2 story. Situated in the woods of northeastern Minnesota, high above Lake Superior's rugged North Shore, this bed-and-breakfast offers unique rooms and a full breakfast each morning. $

★★ **EAST BAY HOTEL AND DINING ROOM.** *Wisconsin St, Grand Marais (55604). Phone 218/387-2800; toll-free 800/414-2807.* 36 rooms, 4 suites. No A/C. Pet accepted. Check-out 11 am. TV. Some in-room whirlpools, refrigerators. Microwaves, fireplaces, wet bars in suites. Restaurant. Bar; entertainment. Room service. Massage. Whirlpool. Cross-country ski 5 miles. On lake. ¢

★★ **NANIBOUJOU LODGE.** *20 Naniboujou Trail, Grand Marais (55604). Phone 218/387-2688. www.naniboujou.com.* 24 rooms. No A/C. No room phones. Check-out 10:30 am. Some fireplaces. Restaurant. On lake/river. Cross-country ski on site. Business services. Totally nonsmoking. ¢

Little Marais

★★ **STONE HEARTH INN BED & BREAKFAST.** *5698 Lakeside Estates Rd, Little Marais (55614). Phone 218/226-3020; toll-free 888/206-3020.* 8 rooms, 2 story. Romantic 1920s inn on Lake Superior. Old-fashioned porch, large stone fireplace in living room. Guest rooms are furnished with antiques; some have whirlpools, gas fireplaces, and kitchens. All rooms have private baths. $

★★ **LIGHTHOUSE BED & BREAKFAST.** *1 Lighthouse Point, Two Harbors (55616). Phone 218/834-4814; toll-free, 888/832-5606 www.lighthousebb.org.* 3 rooms, 2 story. Built in 1892, the Two Harbors Lighthouse is listed on the National Register of Historic Places and is a working lighthouse. Guests become assistant lighthouse keepers and become registered keepers of the light during their stay. $

Minnesota

❈ *North Shore Scenic Drive*

PLACES TO EAT

A long day of driving is sure to make you hungry. At the end of your journey, take a table at one of the following restaurants.

Duluth

★★★ **BELLISIO'S.** *405 Lake Ave S, Duluth (55802). Phone 218/727-4921.* The restaurant features wine racks from floor to ceiling and white tablecloths with full table settings to help guests enjoy a true Italian experience. Italian menu. Lunch, dinner. Bar. Wine cellar. $$
[D]

★★ **BENNETT'S ON THE LAKE.** *600 East Superior St, Duluth (55802). Phone 218/722-2829. www.bennettsonthelake.com.* American menu. Breakfast, lunch, dinner. Situated in a turn-of-the-century factory building, with views of Lake Superior and displays of local artwork. $$

★ **FITGER'S BREWHOUSE.** *600 E Superior St, Duluth (55802). Phone 218/726-1392. www.fitgers.com.* Bar food. Lunch, dinner. Duluth's oldest operating brewery and pub, established in 1857. $

★ **GRANDMA'S CANAL PARK.** *522 Lake Ave S, Duluth (55802). Phone 218/727-4192. www.grandmasrestaurants.com.* Closed some major holidays. Dinner. Children's menu. Under the Aerial Lift Bridge at the entrance to the harbor. $
[D] [SC]

★★ **PICKWICK.** *508 E Superior St, Duluth (55802). Phone 218/727-8901.* The wood-paneled dining area and antique-filled rooms of this historic bar and restaurant offer incredible views of Lake Superior. The menu is an eclectic mix of regional and internationally influenced dishes, from smoked white fish to Grecian lamb chops. Continental menu. Closed Sun; major holidays. Bar. Lunch, dinner. Children's menu. Open-hearth grill. Pub atmosphere. Family-owned. $$
[D]

★ **SCENIC CAFÉ.** *5461 North Shore Dr, Duluth (55804). Phone 218/525-2286. www.scenic-cafe.com.* International menu. Lunch, dinner. Creative, worldly cuisine with excellent desserts and an extensive beer and wine selection. $

★ **TOP OF THE HARBOR.** *505 W Superior St, Duluth (55802). Phone 218/727-8981. www.radisson.com/duluth/.* Specializes in steak, salmon, and trout. Breakfast, lunch, dinner. Children's menu. Reservations accepted. Panoramic view of the city, Duluth Harbor, and Lake Superior from this 16th-floor revolving restaurant. $$
[SC]

Grand Marais

★ **BIRCH TERRACE.** *Hwy 61, Grand Marais (55604). Phone 218/387-2215.* Dinner. Bar. Children's menu. North Woods mansion built in 1898; fireplaces. $$
[D]

Crowley's Ridge Parkway

❊ MISSOURI

Part of a multistate Byway that continues in AR.

Quick Facts

LENGTH: 14.2 miles.

TIME TO ALLOW: 30 minutes or more.

BEST TIME TO DRIVE: Each season offers unique opportunities along the Byway, which makes a for a pleasant visit year-round. High season is summer.

BYWAY TRAVEL INFORMATION: Tourism Advisory Council: 573/334-4142.

SPECIAL CONSIDERATIONS: Plenty of services are available along the route for any special needs; however, watch your engine temperature so that you notice if it begins to increase.

BICYCLE/PEDESTRIAN FACILITIES: This route was originally designed to be used primarily by vehicles; little thought was given to pedestrians or bicyclists. Many trails are now under development to provide better access for hikers and bikers.

Crowley's Ridge, characterized by sand, gravel, and deep gullies, has formed a primary route of transport and commerce for the people of this region for several centuries. During different time periods, the various inhabitants of different nationalities used the ridge to escape the swamps and wetlands. A combination of land travel on the ridge and connecting rivers provided a means of subsistence over a vast corridor of time and place.

THE BYWAY STORY

Crowley's Ridge Parkway tells archaeological, historical, and natural stories that make it a unique and treasured Byway.

Archaeological

The earliest recorded cemetery in the New World is found in Greene County on Crowley's Ridge. Excavated in 1974, the Sloan Site was both home and burial ground for a small group of Native Americans who lived here approximately 10,500 years ago. Living in small bands and in semi-permanent villages, they established the earliest documented cemetery in North America.

The Mississippian site of Parkin is now the Parkin Archaeological State Park in Cross County and provides a museum, visitor center, walking tour, and research station. This site features a rectangular planned village of 400 houses, with a plaza surrounding mounds and evidence of a large population that experienced interaction with the de Soto expedition.

Missouri

✤ *Crowley's Ridge Parkway*

Historical

Crowley's Ridge, with its heavy clay and gravel, proved to be a focal point for the historical and cultural events that shaped the region around it. From the unique plant life of the ridge to the lowland swamps, this region is a story waiting to be told. The earliest people in this region were Native Americans, and they were always few in number.

The first known Native Americans were named the Mound Builders. They date from AD 900 to 1500 and were agriculturists who lived in permanent villages. As permanent residents, they made ceramic pots and various utensils, some of which remain on and below the landscape surface today. Sizable collections of these artifacts can be found in the small museums throughout this region. Huge collections also exist at Arkansas State University and Southeast Missouri State University, some large collections are in private hands, and many artifacts exist in the hundreds of mounds that can still be found. However, these mounds are being challenged by the migration of the rice-planting culture and the disastrous raids of illegal collectors and artifact thieves. So the story of the Native Americans in this region continues as a largely untold story.

The early European immigrants to this region were German, Scottish-Irish, and English. Many of them came from Virginia, Tennessee, Kentucky, and Mississippi, and they brought African-American slaves with them. The Byway intersects many of these ethnic islands that have shaped the history and culture of the region. The migration of the cotton culture brought a lifestyle that continues to influence local food, religion, architecture, family values, and religion. Midwestern barns and Southern homes and churches can be found as well.

There continues to exist in this region what the residents call the Chalk Bluff Trail. This trail extends from Cape Girardeau, Missouri, to Helena, Arkansas, and runs along the high ground of the ridge. The trail was originally used by Native Americans as they moved through the region on hunting expeditions. European settlers used the trail as they migrated to the West: it was the only safe way to travel through the swamps in order to capitalize on the 19th-century opportunities that existed in what is now the great American Southwest. Crowley's Ridge was the only all-weather access for settlers to transport trade items. In the 1850s, a plank road was built from Gideon to Portageville, providing the first east-west avenue to Mississippi River ports. The ridge became a vital highway during the Civil War, used by both Union and Confederate troops. On April 30, 1863, an engagement known as the Battle of Chalk Bluff occurred on the banks of the St. Francis River where it crosses Crowley's Ridge in a narrow gap. Small by most Civil War measures, the event was nevertheless significant to people in this region. Local cemeteries contain graves of both Union and Confederate soldiers.

Following the Civil War, many changes occurred in this area, including a restructuring of agriculture and technology that provided for increased farm size. New plows, threshers, seeders, and shellers allowed farmers to work more acres and increase yields per acre. This put more pressure on transportation and further changed the region. The importation of cotton added to the region's transformation. Cotton farmers from the deep South came into the region as they tried to escape the boll weevil, which had temporarily devastated their lands and crops. Droughts in the 1930s withered those crops, however, and brought government programs to the area. In 1933, Congress passed the Agricultural Adjustment Act, which sought to stabilize agriculture. Soil conservation acts also impacted the region and reflected the changing nature of the times. Those benefits that came into the area generally went to the large planters, while little went to the growing number of sharecroppers and tenant farmers.

The land on and around Crowley's Ridge was generally worked well into the 20th century by sharecroppers who lived at the poverty level. The Southern way of life that existed in this area meant that there was little in the way of a middle class. African Americans in the region were almost entirely in the sharecropper class. This system was widespread across the South in the aftermath of the Civil War, but was especially rigid in the new cotton lands of southeast Missouri. This situation lead to one of the greatest stories to come out of Missouri in the 20th century, the story of the sharecropper strike of 1939.

The problems that plagued farmers in this region are still the same today in southeast Missouri. The large farm operator, through increased mechanization, has been able to survive and prosper, while the small farmer (and there are many in this area) has had little economic success. As farms continue to grow larger, owners of small farms have been forced to abandon both their land and their hope. Many have moved to larger cities or taken employment outside the farm. The area around the ridge continues to decline in population while land values increase.

Natural

Crowley's Ridge, characterized by sand, gravel, and deep gullies, has formed a primary route of transport and commerce for the people of this region over the past several centuries. The inhabitants used the ridge to escape the swamps and wetlands. A combination of land travel on the ridge and connecting rivers provided a means of subsistence over a vast corridor of time and place.

The ridge begins south of Cape Girardeau near Commerce, Missouri. It moves west in a great arc, coming back to the great river near Helena, Arkansas. The land period of glaciation, some 15,000 years ago, caused massive buildup and melting of the glaciers that resulted in great floodwaters beyond modern imagination and created the Ohio and Mississippi rivers as ice marginal streams. The Mississippi was west of Crowley's Ridge, the Ohio River to the east. The old route of the Mississippi may have followed the course of what is now the St. Francis River. Following the last ice melt, the Ohio joined the Mississippi River just south of the Thebes Gap at Cairo, Illinois. The narrow band of uplands known as Crowley's Ridge is the only remaining remnant of this story. The current river system moved to the east and left Crowley's Ridge.

Crowley's Ridge is a unique landform that is effectively isolated like an island in the middle of a small ocean of land. Plants and animals that inhabited the ridge were isolated and cut off from their natural migratory patterns, so a number of rare or endangered plants and animals are indigenous to the ridge. Of the various natural communities on the ridge, the rarest are the plants that occur along the seeps,

see page A17 for color map

Missouri

Crowley's Ridge Parkway

runs, and springs. Nettled chain fern, yellow fringed orchid, umbrella sedge, black chokeberry, and marsh blue violet are some examples of these rare and exquisite plants.

The natural changes that happened in the landscape affected not only plants and animals but humans as well. Some historians contend that Hernando de Soto was the first white explorer to reach Crowley's Ridge. If he did so, it remains unclear exactly where he made his first entrance into what is now Missouri. If and when he crossed the Mississippi River above Memphis, it is likely that he and at least some of his men followed Indian and animal trails to the high ridge into present-day Dunklin County, Missouri.

As other immigrants from all parts of northern Europe moved west, they generally moved down the Ohio, crossed the Mississippi, and landed on the Missouri shoreline. This landing usually occurred at Norfolk Landing, located south of Cairo, Illinois. From there, immigrants would cross the alluvial "prairie" swamps and head to the high ground of Crowley's Ridge. As early as the 1850s, there was discussion and even plans to drain the great wetlands. However, it was not until the early 20th century that the Little River Drainage District was formed and systematic and extensive drainage occurred.

Lumber companies moved into the region after 1913 when the massive log drives on the Mississippi were stopped by the Keokuk Dam. Soon, the region was a timberless, uninhabitable swamp. The Little River Drainage District did the rest. In what is the greatest land transformation in human history, southeast Missouri was moved from natural wetland to field and farm in less than half a century.

HIGHLIGHTS

The following is a suggested itinerary for traveling Crowley's Ridge Parkway in Missouri.

- **Malden:** Start your Byway tour in Malden. Downtown, you can visit the free **Malden Historical Museum** and enjoy the exhibits, which include the history of Malden and the Denis Collection of Egyptian Antiquities. Or spend some time at Malden's **Bootheel Youth Museum,** where adults and kids can both enjoy activities that explore the worlds of math, science, the arts, and more. Interactive exhibits include making (and standing inside!) a giant soap bubble, freezing shadows on the wall, and making music on sewer pipes. When you are ready to continue your drive, leave Malden on County Route J to the west. Enjoy the rural scenery. You may even spot the Military Road, but the easiest place to see the Military Road is at the next stop.

- **Jim Morris State Park and the Military Road:** Shortly after you turn south on County Route WW, you find **Jim Morris State Park.** In addition to showcasing many varieties of trees and other vegetation unique to the ridge, the park is the most practical way to see the **Military Road.** Access the road by the bicycle or pedestrian paths here. This road was used by military troops in the Civil War and by settlers as they moved across the Missouri bootheel. The entire road runs from 1 mile north of Campbell to the intersection of Routes J and WW, with a total length of approximately 6 miles, and can be reached at various points along the route.

- **Peach orchards:** As you drive south on County Route WW, keep an eye out for the more than 800 acres of peach orchards. In the spring, the scent of peach blossoms fills the air. If you are lucky enough to be driving this tour in the summer, be sure to stop at one of the peach stands and buy a tasty treat for the drive ahead.

- **Billy DeMitt Grave:** About 3 miles from the junction of Routes J and WW is the famous site where 10-year-old Billy DeMitt was killed during the Civil War for refusing to tell a group of guerrillas where his father was. A marker at the spot commemorates this tragic story, a true legend in southeast Missouri.

- **Beachwell Gullies:** Also in this area are the visible remnants of erosional forces that scoured the ridge about 200 feet deep from the top of the ridge to the bottom of the gully. The sand here is pure ocean sand (in a landlocked state!), and saltwater shells have been found in this area, evidence that this area was once part of a vast ocean. The gullies can be accessed by a pedestrian trail that is approximately 1 mile long.

- **Campbell:** Next, visit the historic city of Campbell, which has roots dating back to before the Civil War. Several buildings in the Historic District, many of which are on the National Historic Register, are open to visitors. Dine in the quaint local restaurants and visit the antique shops in the Historic District.

- **Chalk Bluff Park:** As you leave Campbell, turn southeast onto US Hwy 62. Shortly before the St. Francis River and the Arkansas border, stop and visit the **Chalk Bluff Conservation Area.** This area is full of Civil War history. The Military Road and trenches from the Civil War can still be seen today, especially along the river. Continue on across into the tip of Arkansas to see a related Civil War historical site, Chalk Bluff Battlefield.

- **Chalk Bluff Battlefield Park:** This park will be of interest if you are seeking Civil War history. It is located at a Civil War battlefield site where several skirmishes were fought. The history of the area is interpreted on plaques along a walking trail. Picnic tables are available as well.

THINGS TO SEE AND DO

Driving along the Crowley's Ridge Parkway will certainly keep your senses engaged, but if you yearn to get out of the car and stretch your legs, or if you'd like to make a mini-vacation out of your trip, check out these attractions along the route.

BOOTHEEL YOUTH MUSEUM. *700A N Douglas, Malden (63863). Phone 573/276-3600. www.bootheelyouthmuseum.org.* Hands-on children's museum offers more than 50 fun and educational exhibits, including a visit to Mars, a construction zone, and an earthquake exhibit. Exhibit hall and theater. $$

CHALK BLUFF BATTLEFIELD PARK. *County Rd 368, 2 miles north of St. Francis, AR (72454).* This park, listed on the National Historic Register, was the site of several Civil War clashes, including a major battle in 1863. Interpretive markers along a trail tell of the battle and the history of the town of Chalk Bluff. Open daily.

CROWLEY'S RIDGE STATE PARK. *2092 Hwy 168, Walcott, AR (72474). Phone 870/573-6751; toll-free 800/264-2405 for fees for cabins and lodges.* Named for pioneer settler Benjamin Crowley, this state park encompasses his homestead. Log and stone structures there were constructed by the Civilian Conservation Corps (CCC) in the early 1900s. Visitors will find trails, campsites, and 31-acre lake for swimming, fishing, and boating. Trails throughout the park offer a closer look at the plants and animals that make their home here. Open year-round.

MALDEN HISTORICAL MUSEUM. *201 N Beckwith, Malden (63863). Phone 573/276-5008. www.maldenmuseum.com.* Started nearly 50 years ago by a local resident who collected historical items, the museum now offers rotating displays of local historical items, including clothing, maps, photos, and other objects. Open Wed and Sat 1:30-4:30 pm or by appointment.

PLACES TO STAY AND EAT

If you're planning to drive this entire Byway, including the Arkansas portion (the total route is 212 miles long), your best bet is to use Memphis, Tennessee, as a home base. The Byway itself is fairly remote.

Little Dixie Highway of the Great River Road
❄ MISSOURI

Quick Facts

LENGTH: 30 miles.

TIME TO ALLOW: 1 hour or more.

BEST TIME TO DRIVE: Spring through autumn.

RESTRICTIONS: Snowfall in January and February can close the highway until snow plowing can occur.

BICYCLE/PEDESTRIAN FACILITIES: Bicycle lanes connecting Clarksville and Louisiana (10 miles) offer travelers the opportunity to see the Byway via means other than the automobile. Both communities have also developed pedestrian facilities, and hiking trails exist in several parks and refuge areas along the Byway.

The Little Dixie Highway (also called the Mississippi Flyway Byway) is home to a culture expressing a unique Southern flair. Accents of the South are revealed in the area's Victorian-era streetscapes and plantation-era mansions. This touch of the South has earned the region its nickname, Little Dixie. The area reminds travelers of the far-reaching effects of slavery, the Civil War, and the nation's reconstruction.

The Mississippi River played an enormous role in the shaping of the physical and cultural features of the area. The Little Dixie Highway stretches out alongside the river. Limestone bluffs along the route offer stunning views of the mighty Mississippi, and the towns of Clarksville and Louisiana allow access to the river. The Little Dixie Highway is part of the Great River Road, a path of highway that runs alongside the river through several states.

THE BYWAY STORY

The Little Dixie Highway of the Great River Road tells cultural, historical, natural, recreational, and scenic stories that make it a unique and treasured Byway.

Cultural

As you may expect from a region with such a diverse history, the Little Dixie Highway can boast equally varied and rich cultural qualities. Although the issue of slavery and the Civil War divided the nation, the cultural heritage of the United States proved remarkably resilient. The same can be said about Pike County, where churches and fraternal organizations helped to create a close-knit society that rebuilt itself after the war.

Missouri

✣ Little Dixie Highway of the Great River Road

In his novels, Mark Twain describes a lifestyle that so many people recognize as an essential part of the American historical experience. For many of the residents who grew up along the Little Dixie Highway, Twain's portrait of the Mississippi River corridor is part of their own life experiences. Because the Mississippi River enjoys an essential and dominant role in life along the Little Dixie Highway, its influence is celebrated during the Big River Days festival in Clarksville. Storytelling, music, and displays of the type of craftwork essential to life along the river offer an accompaniment to the festival's main attractions: a giant aquarium filled with the underwater life of the river and the opportunity to travel on the muddy waters of the Mississippi on a majestic riverboat.

Clarksville's craftsmen have taken an active role in preserving this town's cultural heritage by restoring its historic commercial district, which is listed on the National Register of Historic Places. These artisans have constructed the signs, forged the sign brackets, restored the storefronts, and performed the general contracting work necessary for the historically accurate restoration of this Victorian era commercial district. A 50-mile network of artists, artisans, and galleries, called the Provenance Project, also takes an active role in promoting towns from Clarksville north to Hannibal. These artists not only enrich the area aesthetically, but also attract visitors and new resident artists.

Historical

The Little Dixie Highway travels through an area bursting with history. The Mississippi River played an important and influential role in the area's history, both as a natural resource and as a geographical feature. In addition, the state of Missouri's geographical location between the North and the South left it in a volatile yet influential position.

For centuries, the Mississippi River has provided extremely fertile ground: wildlife for furs and food, an endless supply of water, and transportation. The flowing waters of the Great River have carried canoes, rafts, 19th-century steamboats, tugboats, and barges both to and from the markets of this nation, and to ships that sail to the rest of the world. But perhaps more than anything else, the Mississippi River has provided this entire nation with stories. Mark Twain probably would have agreed that no single geological feature in the country can claim to be the father of an historical and cultural offspring as rich and colorful as the Mississippi River. During the early 1800s, the river served as the edge of the West, the barrier dividing civilization from the wild frontier. As the young nation continued its growth and expansion, clashes continued between new settlers and the native peoples. Eventually, the river proved no longer to be a barrier to the frontier, but a starting line for settlers' hopes and dreams.

Slavery left marks on the history of the region traversed by the Little Dixie Highway. Missouri was a state whose role in the slavery debate proved pivotal. Within the state lay a region that came to be known as Little Dixie because of its links to the cultures and institutions of the American South. Pike County, home of the Little Dixie Highway, was one of eight counties within Little Dixie. The plantation homes and Southern-style architecture that marks the landscape throughout the length of the Byway stands as a testament to that time. As travelers look out over the Mississippi River and see the shores of Illinois just across the mighty river, they are also reminded of the fact that freedom for the slaves always lay tantalizingly close. Little wonder, then, that the region also included several stops on the Underground Railroad.

The stories of the struggle for justice and freedom soon turned into tales of triumph and renewal. New struggles for a new kind of justice and freedom accompanied the hope ushered in by the Emancipation Proclamation. As the nation attempted to rebuild, so, too, did the residents of Pike County work to repair the damage caused by slavery and the Civil War. The Freedmen's Bureau opened in the town of Louisiana and began fulfilling its mission to uplift the region's ex-slaves. Newly opened schools sought to fulfill the same mission for the children of these former slaves. Fraternal organizations and houses of worship dotted the social landscape, and many of the buildings built by these organizations still influence the physical landscape along this Byway today. These buildings provide evidence of the prosperous society that overcame the ravages of division.

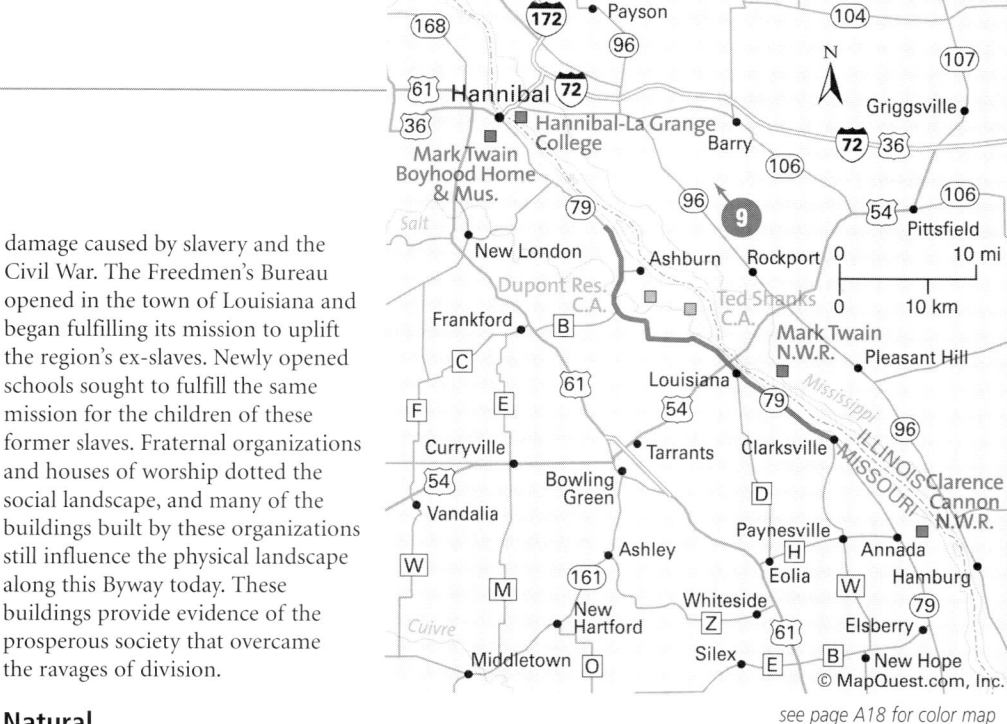

see page A18 for color map

Natural

There is nothing quite like strolling along the banks of the Mississippi. The calm waters of the giant stream seem to stretch across for miles and provide habitat for an abundance of plant life and wildlife. These living treasures of the river can be witnessed from the Byway itself, or from many parks and observation points along the Byway, which offer exceptional closeness with the river and its resources. This particular region of the river is an important bird habitat.

A flyway is simply a path for migrating birds. The Mississippi River provides a flight corridor for approximately 40 percent of all North American waterfowl, making this a prime spot for bird-watching. In addition, the Missouri Audubon Society has documented about 39 species of migratory shorebirds in the area, such as the killdeer and the sandpiper. As if that weren't enough, the area also serves as habitat for eagles. Walter Crawford, executive director of the World Bird Sanctuary in St. Louis, has called Clarksville the eagle-viewing capital of the United States. This "fowl" life along the Little Dixie Highway is celebrated with a number of annual festivals. The biggest and best of these festivals, Eagle Days, pays tribute to the area's most notable wildlife resident and brings nearly 10,000 visitors to Clarksville the last weekend of each January. The weekend prior to Eagle Days, the region's other raptors take center stage at Clarksville's Masters of the Sky festival.

The landscape surrounding the Little Dixie Highway reflects the impact made not only by the river itself, but also by the progression and retreat of two ice advances. The Byway is marked by steeply sloping uplands and prominent knobs and ridges dotted with limestone and shale outcroppings. Deciduous vegetation dominates the landscape in those areas not being used for agriculture.

Because this part of the country experiences the effects of four distinct seasonal changes, nature paints an ever-changing picture. The fall colors are particularly significant to local residents, who justifiably take pride in the beauty the fall brings to the region.

Recreational

The Little Dixie Highway runs parallel to the area's major source of recreation, the Mississippi River. Recreational opportunities related to the river include some river islands open to the public; great opportunities for

159

Missouri

Little Dixie Highway of the Great River Road

fishing, hiking, and cycling; a few wildlife reserves; two boat launches; the Eagle Center in Clarksville; a pair of riverfront municipal parks; Lock and Dam 24 at Clarksville; and a variety of scenic overlooks.

Upon traveling the Little Dixie Highway, some travelers may get the overwhelming urge to kick off their shoes, roll up their pants or overalls, pull out a corncob pipe, and set off down the river on a log raft for some remote camping site. Well, this Byway allows you to relieve that urge. The river contains many small islands, several of which are accessible to the public by log raft, canoe, or any other boat. Many recreationalists canoe over to an island for peaceful camping and great fishing, from launching points such as Silo Park. The Army Corps of Engineers manages the islands and should be contacted to determine which of the islands in Pike County are open to the public.

The Mississippi River has long been a prominent source of fish—catfish and crappie are the most common types found in the river. The parks and other points along the Byway offer great opportunities to land that 20-pound Mississippi River catfish. The reserves along the Byway also offer exceptional fishing. Clarksville and Louisiana each have a boat dock.

The Little Dixie Highway features bike trails or wide shoulders for cyclists and hikers. Cycling along the Byway offers you a chance to really take in the serenity and uniqueness of the Byway and adjacent river. Additionally, the parks and conservation areas along the Byway allow for leisurely walks right next to the river.

Scenic

You need only capture a view from one of the many scenic overlooks along the Byway to understand a little better why the Mississippi has played such an important role in the lives of Americans. The scenic overlooks from high bluffs north of Louisiana and at the Louisiana Cemetery, in particular, offer a bird's-eye view of the river and of the lush Mississippi Valley.

The valley is a deep shade of green most of the time, full of dark, fertile soil that has provided food for Americans for centuries. The Pinnacle, at 850 feet above sea level, is the highest point on the Mississippi River. Barges, other boats, and occasionally even riverboats can often be seen from these various overlooks, reminding you of the important role the river has played in transportation and commerce. For those who simply want an aesthetic treat, autumn provides an abundance of colors to please the eye.

THINGS TO SEE AND DO

Driving along the Little Dixie Highway of the Great River Road will certainly keep your senses engaged, but if you yearn to get out of the car and stretch your legs, or if you'd like to make a mini-vacation out of your trip, check out these attractions about 15 miles north of the Byway in Hannibal.

ADVENTURES OF TOM SAWYER DIORAMA MUSEUM. *323 N Main St, Hannibal (63401). Phone 573/221-3525. www.artcom.com/museums/nv/sz/63401-35.htm.* The Adventures of Tom Sawyer in three-dimensional miniature scenes carved by Art Sieving. Open daily; closed major holidays. **$**

BECKY THATCHER HOUSE. *209-211 Hill St, Hannibal (63401). Phone 573/221-0822. www.marktwainmuseum.org/BeckyThatcher.html.* House where Laura Hawkins (Becky Thatcher) lived during Samuel Clemens' boyhood; upstairs rooms have authentic furnishings. Open daily; closed Jan 1, Thanksgiving, Dec 25. **FREE**

HANNIBAL TROLLEY. *220 N Main St, Hannibal (63401). Phone 573/221-1161.* Narrated tour aboard trolley. Open-air in summer months, enclosed the rest of the season. Open mid-Apr-Oct, daily; rest of year, by appointment. **$$**

HAUNTED HOUSE ON HILL STREET. *215 Hill St, Hannibal (63401). Phone 573/221-2220.* Life-size wax figures of Mark Twain, his family, and his famous characters. Open Mar-Nov, daily. **$**

MARK TWAIN CAVE. *Hwy 79, Hannibal (63401). Phone 573/221-1656. www.marktwaincave.com.* This is the cave in *The Adventures of Tom Sawyer* in which Tom and Becky Thatcher were lost and where Injun Joe died. Bring a jacket; the cave's temperature is a constant 52 degrees. One-hour guided tours (Apr-May 8 am-8pm, Sept-Oct 9 am-6 pm, Nov-Mar 9 am-4 pm; closed Thanksgiving, Dec 25). Lantern tours to nearby Cameron Cave (Memorial Day-Labor Day, daily). Campground adjacent. **$$$**

★ **MARK TWAIN MUSEUM AND BOYHOOD HOME.** *208 Hill St, Hannibal (63401). Phone 573/221-9010. www.marktwainmuseum.org.* Museum houses Mark Twain memorabilia, including books, letters, photographs, and family items. Two-story white-frame house in which the Clemens family lived in the 1840s and 1850s, restored and furnished with period pieces and relics. Gift shop. Open daily; closed holidays. **$$**

★ **MARK TWAIN RIVERBOAT EXCURSIONS.** *Hannibal (63401). Phone 573/221-3222; toll-free 800/621-2322.* One-hour cruises on the Mississippi River; also two-hour dinner cruises. Open early May-Oct, daily at 11 am, 1:30 pm, and 4 pm.

MOLLY BROWN BIRTHPLACE AND MUSEUM. *600 Butler St, Hannibal (63401). Phone 573/221-2100. www.mollybrownmuseum.com.* Antique-filled home has memorabilia of the "unsinkable" Molly Brown, who survived the *Titanic* disaster. Open Apr-May, Sept-Oct, weekends 10 am-5 pm; Jun-Aug, daily 9:30 am-6 pm. **$**

MUSEUM ANNEX. *415 N Main, Hannibal (63401). www.marktwainmuseum.org/annex.html.* Audiovisual presentations and displays on Mark Twain and Hannibal.

OPTICAL SCIENCE CENTER AND MUSEUM. *214 Main St, Hannibal (63401). Phone 573/221-2020.* Hands-on learning stations provide first-hand optical demonstrations. Computerized light show, wrap-around theater. Open Apr-Oct, daily. **$$**

PILASTER HOUSE AND GRANT'S DRUGSTORE. *Hill and Main sts, Hannibal (63401).* The Clemens family lived in this Greek Revival house (built 1846-1847), which now contains a restored old-time drugstore, pioneer kitchen, doctor's office, and living quarters where John Clemens, Twain's father, died.

RIVERVIEW PARK. *Hannibal (63401). Phone 573/221-0154.* This 400-acre park on bluffs overlooking the Mississippi River contains a statue of Samuel Clemens at Inspiration Point. Nature trails. Picnicking; playground. Open Mon-Fri. **FREE**

ROCKCLIFFE MANSION. *1000 Bird St, Hannibal (63401). Phone 573/221-4140. www.hannibal.net/rockcliffe/.* Restored Beaux Arts mansion overlooking the river; 30 rooms, many original furnishings. Samuel Clemens addressed a gathering here in 1902. Guided tours. Open daily; closed Jan 1, Thanksgiving, Dec 25. **$$**

TOM AND HUCK STATUE. *Main and North sts, Hannibal (63401).* Life-size bronze figures of Huck Finn and Tom Sawyer by F. C. Hibbard.

Missouri

✵ Little Dixie Highway of the Great River Road

TWAINLAND EXPRESS. *400 N 3rd St, Hannibal (63401). Phone 573/221-5593. www.twainlandexpress.com.* Narrated tours past points of interest in historic Hannibal, some aboard open-air, train-style trams. Open May, Sept-Oct, weekends; June-Aug, daily. $$

PLACES TO STAY

If you choose to include an overnight stay in your trip along this Byway, Mobil Travel Guide recommends the following lodgings.

★★ **BEST WESTERN HOTEL CLEMENS.** *401 N 3rd St, Hannibal (63401). Phone 573/248-1150; toll-free 800/780-7234. www.bestwestern.com.* 78 rooms, 3 story. Complimentary continental breakfast. Check-out 11 am. TV; cable (premium), VCR available. Laundry services. Game room. Indoor pool, whirlpool. Free airport transportation. In a historic district near the Mississippi River. ¢

★★★ **GARTH WOODSIDE MANSION B&B.** *11069 New London Rd, Hannibal (63401). Phone 573/221-2789; toll-free 888/427-8407. www.garthmansion.com.* This second-empire/Victorian mansion (1871) is a great bed-and-breakfast for a romantic weekend getaway. Mark Twain was often a guest. 8 rooms, 3 story. No room phones. Complimentary full breakfast. Check-out 11 am, check-in 4 pm. On 39 acres with pond, woods. Totally nonsmoking. $

★★★ **HANNIBAL INN AND CONFERENCE CENTER.** *4141 Market St, Hannibal (63401). Phone 573/221-6610; toll-free 800-325-0777.* 241 rooms, 2 story. Pet accepted. Complimentary continental breakfast. Check-out 11 am. TV; cable (premium). In-room modem link. Restaurant, bar. Room service. Sauna. Indoor pool, whirlpool, poolside service. Game room. Tennis. $

★ **SUPER 8.** *120 Huckleberry Heights Dr, Hannibal (63401). Phone 572/221-5863; toll-free 800/800-8000. www.super8.com.* 59 rooms, 3 story. No elevator. Complimentary continental breakfast. Check-out 11 am. TV. Pool. City park adjacent. ¢

PLACES TO EAT

A long day of driving is sure to make you hungry. At the end of your journey, try the following restaurant.

★ **LOGUE'S.** *121 Huckleberry Heights Dr, Hannibal (63401). Phone 573/248-1854.* Closed holidays. Breakfast, lunch, dinner. Children's menu. $

Sheyenne River Valley Scenic Byway
❋ NORTH DAKOTA

Quick Facts

LENGTH: 63 miles.

TIME TO ALLOW: 9 hours.

BEST TIME TO DRIVE: Spring and summer offer wonderful roadside wildflower viewing, while late-summer travelers can enjoy vast fields of sunflowers.

BYWAY TRAVEL INFORMATION: Valley City Chamber of Commerce & Convention Visitor Bureau: 701/845-1891; Fort Ransom State Park: 701/973-4331.

SPECIAL CONSIDERATIONS: Part of the road from Kathryn to Lisbon is a gravel surface (about 15 miles). The Byway contains a total of about 26 miles of gravel roads. During the growing season, you may encounter minor changes in travel speeds when meeting or passing agricultural equipment.

RESTRICTIONS: Spring sometimes brings closures due to spring flooding and excessive rainfall. In winter, the Byway is reliable, with only occasional closures due to seasonal blizzards.

BICYCLE/PEDESTRIAN FACILITIES: Even with an increase in the number of off-road trails and facilities for bicycles in the Scenic Byway Corridor, most bicycling in the Sheyenne River Valley still takes place on ordinary roads and highways. Although the entire Sheyenne River Valley Scenic Byway has been used as a bicycle path, six bicycle trails have been identified in the Sheyenne River Valley Byway corridor. The trails were developed to provide safe and enjoyable routes and associated facilities in the Scenic Byway corridor for bicyclists of all abilities.

The Sheyenne River Valley Scenic Byway is distinguished by its small-town hospitality and a mixture of scenic hills and grassy flatlands. Panoramic pastoral scenery catches travelers' attention, featuring wildflowers, wild grasses, quaint farms, grassy hills, and wildlife. Old prairie churches, one-room schoolhouses, quaint farms, and historical towns allow the traveler to experience a little of the old frontier.

The Sheyenne River Valley Scenic Byway stretches past one historical site after another. The main streets of the towns are lined with historic buildings, old general stores, and brick banks. It isn't hard to imagine a brick bank being robbed by desperadoes; an old frontier jail reveals where the desperadoes would usually end up. Fort Ransom, on the southern end of the Byway, provides a glimpse into the lives of the frontier army. The Byway also contains significant Native American sites.

If the history and scenery of the Sheyenne River Byway aren't enough, you can't go wrong with the area's abundance of home cookin'. The mashed potatoes are real, and the desserts are homemade. Homemade ice cream can be sampled at Sunne Demonstration Farm located in Fort Ransom State Park, where, during the second weekend in July and September, the farm comes alive with demonstrations of life on the farm as it was in the late 18th and early 19th centuries.

THE BYWAY STORY

The Sheyenne River Valley Scenic Byway tells archaeological, cultural, historical, natural, recreational, and scenic stories that make it a unique and treasured Byway.

Sheyenne River Valley Scenic Byway

Archaeological

Archaeological studies show that the Sheyenne River Valley may have been intermittently occupied for over 11,000 years. The archaeological record of the Sheyenne River Valley has been categorized into a series of cultures, namely Paleo-Indian or Early Prehistoric (11,500 to 7,500 years before present), Archaic (7,500 to 1,600 years before present), Woodland (1,600 to 1,400 years before present) and the Late Prehistoric (1,400 years before present to the time of Euro-American settlement around 1850). Natural resources used to sustain populations, as well as the tools and methods of human organization required to utilize these resources, are the basis for differentiating these cultures. Significant Native American sites found along the Byway are described in this section.

Located in Valley City, Medicine Wheel Park includes a reproduction of a medicine wheel (a stone solar calendar) and an extensive Native American burial mound complex dating from the Woodland period. The Standing Rock, known as Inyun Bosndata by the Sioux Indians who consider it sacred, is also found at the park. The 4-foot-tall Standing Rock is an inverted cone shape that stands on a complex of prehistoric burial mounds dating from the Woodland period. This state historic site is marked and is located just east of Little Yellowstone Park off Highway 46.

Hosting the Viking sculpture at Fort Ransom, Pyramid Hill measures 650 feet long, 520 feet wide, approximately 100 feet high and is level on top, with the north, west, and south sides of uniform shape. Although geologists consider this to be a natural geologic formation, some think that about one-third of the mound is man-made, being built an estimated 5,000 to 9,000 years ago by an ancient civilization. Native American tradition holds this site to be a place of emergence. Additionally, other mounds, stone circles, and rock features are well represented in the Sheyenne Valley, though they have yet to be attributed to any specific cultural affiliation. Thirteen mound sites have been attributed to Middle Woodland, Sonota/Besant, Devils Lake-Sourisford, and Late Prehistoric periods.

Cultural

Diversity characterizes the cultural resources of North Dakota's Sheyenne River Valley. The cultural heritage of the Scenic Byway corridor has been strongly influenced by a mixture of customs and traditions passed on from the region's first immigrant settlers. The unaltered natural settings and rustic outdoor facilities of the region's most remote areas are a short distance from modern and sophisticated art centers, museums, restaurants, and hotels.

The cultural quality found in the Sheyenne River Valley Scenic Byway corridor is a result of this surprising diversity and contrast, as well as the number of cultural attractions and events. More than 60 attractions in the Scenic Byway corridor highlight the region's many cultural resources. Among these cultural attractions are pioneer cemeteries, historic churches, antique shops, guest inns, museums, monuments, and community concerts and theaters. Complementing these are close to 50 cultural events that are held regularly in communities throughout the Byway corridor. These events include farmers' markets, horse shows and rodeos, county fairs, community days, antique and art shows, vintage tractor and car shows, and ethnic festivals such as Syttende Mai (May 17, Norwegian Independence Day) and Fiesta Mexicana. Sodbuster Days is held annually on the second weekends in July and September. The event transports visitors back to pioneer days with demonstrations of pioneer farming methods at Sunne Farm at Fort Ransom State Park. The Sheyenne Valley Arts and Crafts Association Fall Festival, on the other hand, is held on the last weekend of September in Fort Ransom, during the peak of the fall color season. This two-day cultural event brings several hundred vendors and around 8,000 annual visitors.

Historical

Sheyenne Valley's history really began about 15,000 years ago, as tribes began to filter from the north and west into the area east of the Rocky Mountains, foraging for food. The region's earliest inhabitants were members of these prehistoric and early historic cultures, including the Paleo, Archaic, and Woodland Indians. The first written records for the region surrounding the Sheyenne River Valley came from the explorers Joseph Nicolas Nicollet and John Fremont, who noted the landmark Standing Rock on their maps during their exploration of 1839. They also recorded their experiences with the local Native Americans while camping in the area now known as Clausen Springs. These early histories preclude events that have shaped and changed the Sheyenne Valley.

The next written account from the area does not take place until James L. Fisk passed through the area, leading an emigrant train to the gold fields of Montana in 1862 and 1863. Also in 1863, Henry H. Sibley led his punitive expedition through the area after the Minnesota Uprising of the Sioux. Some of his campsites are state historic sites and are marked with monuments. Many of his soldiers kept diaries of their experiences on this expedition, a few of which can be accessed at the Barnes County Museum in Valley City. In 1867, the Fort Ransom military post was established on the Fort Abercrombie to Fort Totten Trail, to protect wagon trains on their way to the gold fields of Montana and to protect pioneer settlers and railroad workers passing through. Today, the Fort Ransom Historic Site recalls the importance of the original outpost.

With the advance of the Northern Pacific Railroad in 1872 came the first permanent white settlers to the valley. A depression in the following years kept the number of new settlers to a minimum, but then the Great Dakota Boom, beginning in the year 1878, marked the influx of a steady stream of mostly Norwegian immigrants to the valley. One such Norwegian was Theodore P. Slattum, whose story is probably typical of the settlers of the time. Slattum immigrated to Minnesota from Norway in 1870, and then moved to the Sheyenne Valley in 1879. He built a dugout for temporary living quarters until his cabin could be built. Seven children were raised in the cabin, with two dying in early childhood. The Byway reveals histories like Slattum's as it follows Native American footpaths, pioneer wagon trails, and military fort supply routes, segments of which are still visible.

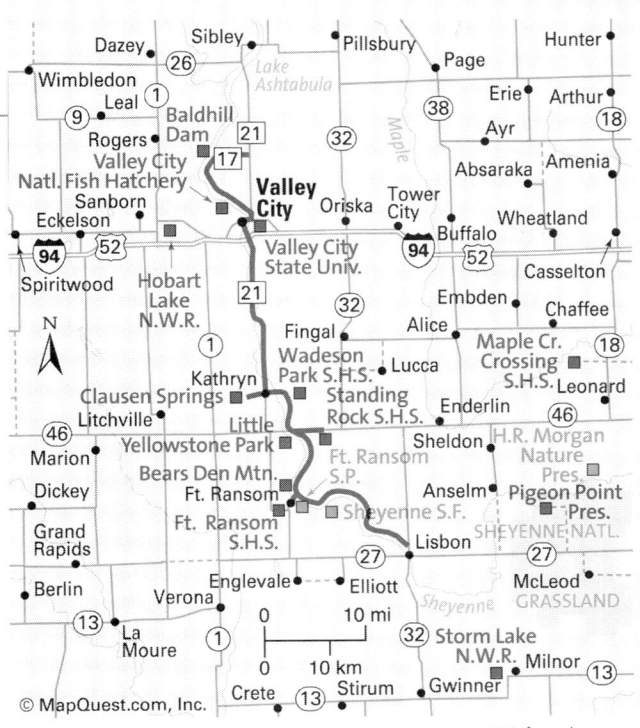

Dramatic changes to the Sheyenne River Valley landscape resulted from the agricultural activities, urban development, and construction of transportation routes that accompanied the region's settlement. The territories and numbers of many plant and animal species native to the Sheyenne River Valley were greatly reduced. Sections of the river were dammed for mills, water supply regulation, pollution abatement during low-flow periods, and recreational purposes; and dikes were constructed to reduce flooding in river valleys. But through it all, the Sheyenne River Valley, its many historical

N. Dakota

❋ Sheyenne River Valley Scenic Byway

structures, and its citizens have continued to share the valley's rich history and to provide countless visitors with memorable experiences.

Natural

Driving across the vast plains of North Dakota, approaching the Sheyenne River Valley, you will be pleasantly surprised as the interstate crests a hill, and the beautifully green and tree-lined Sheyenne River Valley vista stretches before you. Many travelers likely picture a flat, treeless North Dakota. Some may picture the state as rugged badlands. The fact is, North Dakota contains quite a variety of natural landscapes, with the Sheyenne River Valley representing some of the most accommodating and pleasant land in the state. Of course, to arrive at the Byway, you will likely travel through the wild and rugged badlands to the west or the flat and fertile Red River Valley to the east. Perhaps you will have to traverse the plains of South Dakota to the south or the vast prairies from the north. From whichever direction you come, the Sheyenne River Valley, with its rivers and forests, is a welcome and refreshing natural retreat.

Much of the valley's uniqueness can be attributed to the oak savannah riparian forest, a zone whose wide variety of vegetation supports the existing wildlife diversity. This rolling sand-dune landscape of oak savannah along the Sheyenne River is perhaps the most remarkable scenic quality of the area, but also allows for great camping in the cool, shady forests. Other unique areas include Clausen Springs Recreational Area, Little Yellowstone Park, Mooringstone Pond at the Writing Rock site, and the Sheyenne State Forest. These areas are beautifully wooded and sprinkled with wildflowers and flowing natural springs. For the angler, the Sheyenne River offers an abundance of game fish, from catfish to walleye. The river contains around 50 different fish species.

Although forests and river are the defining features along the Byway, the traveler longing to see undisturbed prairie lands will not be disappointed. Tall grass prairie once covered nearly 400,000 square miles of North America. Today, less than 1 percent of this ecosystem remains. However, a fine sample can be found within the boundaries of Fort Ransom State Park, a 900-acre site with a 1.5-mile segment of certified North Country Scenic Trail. If that's not enough, the Sheyenne National Grassland area is only about 10 miles east of the southern portion of the Byway and represents some of the most pristine, untouched grasslands in the nation.

Recreational

Recreational activities abound in the Sheyenne River Valley and are limited only by your imagination. You can participate in hiking, biking, and horseback riding all along the Byway in spring, summer, and fall. The area's river and lake provide great opportunities for canoeing, boating, and kayaking. In the winter, these activities are exchanged for cross-country and downhill skiing, snowshoeing, and snowmobiling. Fishing is a year-round pastime that can be enjoyed all along the corridor, as is the ever-popular bird-watching. The Byway is also a popular location for hunting, whether it be for waterfowl, upland bird, deer, moose, or fur-bearing species.

Standing out on the basically lakeless landscapes of the North Dakota plains, Lake Ashtabula in the Sheyenne River Valley has several resort areas located along its 27-mile length, providing a plethora of recreational opportunities. There are resorts, camping areas with a total of about 140 campsites, swimming areas, fish-cleaning stations, boat-rental locations, restaurants, and cabin rentals.

Fort Ransom State Park provides a wealth of outdoor activities, including educational programs, camping, fishing, picnicking, horseback riding, canoeing, cross-country skiing, hiking, birding, and nature photography. Amenities include an outdoor amphitheater, canoe landing, campsites and rentals, nature trails, playground, picnic shelters, visitor interpretive center, horse corrals, and pioneer farming methods demonstration farm, the Sunne farm, home to the Fort Ransom Sodbusters Association.

A ski resort, Bears Den Mountain, is not something you would expect to see on the flat plains of North Dakota, but you can find it at Fort Ransom. This winter seasonal recreation site offers a chairlift and T-bar, two beginner rope tows, snowmaking, grooming, a cafeteria, and rental equipment. The Sheyenne Valley Snowmobile Trail system has a wide variety of riding areas, such as nice flat ditches, shelter belts, national grasslands, and the Sheyenne River bottom. These 285 miles of trails cross many private lands, with access being provided by a lease through local organizations.

Scenic

The scenic qualities of the Sheyenne River Valley are widely recognized as one of this Byway's most distinguishing features. Viewing the fall colors of native prairie and hardwood forests in the Sheyenne River Valley is one of the most popular attractions for travelers. Seasonal changes provide for a pleasant drive at any time of the year. This beautifully forested river valley; its associated panoramic views; tumbling, spring-fed creeks; and resident wildlife make this a truly scenic Byway. The river is lined by riparian forests of basswood, American elm, green ash, and bur oak that are home to several rare species of plants and animals.

In addition to remarkable natural features, human activities are important components of the scenery of the Sheyenne River Valley. Picturesque farmsteads nestle in the valley, while historic churches, one-room schoolhouses, and pioneer cemeteries dot the countryside. Fields of bright sunflowers, small grains, hay, and pastureland form intricate patterns on the land, and winding roadways connect small rural communities. This colorful mixture of man-made resources contributes to the scenic quality that is so popular in the Sheyenne River Valley.

The Valley City Historic Bridges Tour also highlights man-made developments that contribute to the quality of the corridor's visual environment. The most spectacular of these is the Highline Bridge, which is 3,860 feet long and hangs 162 feet above the riverbed. It is one of the highest and longest single-track railroad bridges in the nation—not something you would expect to see on the plains of North Dakota.

Valley City has also designated a new city park featuring a re-creation of a Native American medicine wheel (an ancient solar calendar) that includes a bike path, five information panels, a solar system walk, and a breathtaking overlook of the city, located next to an extensive Native American burial mound complex.

THINGS TO SEE AND DO

Driving along the Sheyenne River Valley Scenic Byway will certainly keep your senses engaged, but if you yearn to get out of the car and stretch your legs, or if you'd like to make a mini-vacation out of your trip, check out these attractions along the route.

BALDHILL DAM AND LAKE ASHTABULA. *2630 114th Ave SE, Valley City (58072). Phone 701/ 845-2970.* Impounds the Sheyenne River and Baldhill Creek in a water supply project, creating the 27-mile-long Lake Ashtabula; eight recreational areas. Swimming, water-skiing, fishing, boating (ramps); picnicking (shelters), concessions, camping (hookups; fee). Fish hatchery. Open May-Oct.

N. Dakota

✱ Sheyenne River Valley Scenic Byway

CLAUSEN SPRINGS PARK. *4 miles W on I-94, then 16 miles S on ND 1, then 1 mile E and 1/2 mile S following signs; Valley City (58072).* A 400-acre park with a 50-acre lake. Fishing, canoeing, boating (electric motors only); nature trails, bicycling, picnicking, playground, camping, tent and trailer sites (electric hookups, fee). Open May-Oct, daily.

FORT RANSOM STATE PARK. *5981 Walt Hjelle Pkwy, Ft. Ransom (58033). Phone 701/973-4331; toll-free 800/807-4723 (reservations). www.ndparks.com/Parks/frsp.htm.* Former frontier Army post. The park also preserves a historic farm that re-creates the life of early Minnesota homesteaders. Sodbuster Days festival is held the second weekends in July and Sept. Arts and crafts festival is the last weekend in Sept. Downhill ski resort. Camping, picnic areas. Hiking, cross-country skiing, and snowmobiling.

LITTLE YELLOWSTONE PARK. *4 miles W on I-94, then 19 miles S on ND 1, then 6 miles E off ND 46; Valley City (58072).* Picnicking, camping (electric hookups, fee) in a sheltered portion of the rugged Sheyenne River Valley; fireplaces, shelters, rustic bridges. Open May-Oct, daily.

PLACES TO STAY

If you choose to include an overnight stay in your trip along this Byway, Mobil Travel Guide recommends the following lodgings.

★ **SUPER 8.** *822 11th St SW, Valley City (58072). Phone 701/845-1140; toll-free 800/800-8000. www.super8.com.* 30 rooms, 2 story. Check-out 11 am. TV, VCR available (movies). ¢

★★ **WAGON WHEELS INN.** *I-94, exit 292, Valley City (58072). Phone 701/845-5333.* 88 rooms, 1 story. Comfortable accommodations including Jacuzzi suites, cable television, an indoor pool, and a whirlpool. $

PLACES TO EAT

A long day of driving is sure to make you hungry. At the end of your journey, try the following restaurant.

★ **DUTTON'S PARLOUR.** *256 Central Ave N, Valley City (58072). Phone 701/845-3390.* American menu. Breakfast, lunch. Located in downtown Valley City, this turn-of-the-century ice cream parlor offers homemade lunch items and sandwiches. $

Amish Country Byway

OHIO

Quick Facts

LENGTH: 76 miles.

TIME TO ALLOW: 4 to 6 hours.

BEST TIME TO DRIVE: Early autumn means harvest season (which brings produce stands) and stunning fall foliage.

BYWAY TRAVEL INFORMATION: Chamber of Commerce/Travel and Tourism Bureau of Holmes County: 330/674-3975; Byway local Web site: www.visitamishcountry.com.

SPECIAL CONSIDERATIONS: Please respect the privacy and religious beliefs of the Amish and don't take pictures of them. Because of the unique agriculture and culture of Amish Country, you must share the road with Amish buggies, agriculture equipment, cyclists, etc. The two-lane state routes and US 62 should be traveled at a somewhat slower pace than most paved roads. Keep in mind that while rest areas, public rest rooms, and some gas stations remain open on Sundays, many services are not available on that day.

RESTRICTIONS: Roads are sometimes bad in the winter because of ice or snow. Every ten years or so, spring and summer bring flooding, but rarely are the major highways closed.

BICYCLE/PEDESTRIAN FACILITIES: The Great Ohio Bicycle Adventure has included the Amish Country Byway on its route numerous times. Many bicycling groups meander the Byway. In most areas frequented by bicycles and buggies, the curbs or road berms have been widened enough to allow slower travelers, including buggies, to move to the side of the road and let vehicles pass by.

In this 21st century of cell phones, computers, fast cars, multiple and demanding appointments, and time commitments, there is a community within America that holds steadfast to its traditional beliefs and customs. The Amish people in Holmes County, Ohio, make up the largest concentration of Amish communities in the world, and they provide a unique look at living and adapting traditional culture. Traditional but not old-fashioned, the Amish continue to live simply, the way they always have.

When the Amish and others first settled and explored this northern Appalachian region, many depended on agriculture-based professions, and this profession continues today. The growing and harvest season in this area is particularly exciting because of the large produce auction held along the Byway. Holmes County is a strong dairy-producing region, and cheese and specialty meat products are made locally by both the Amish and non-Amish residents.

Part of this simple living is inherent in the religious tradition of the Amish people. This community is a living reminder of the principles of religious freedom that helped shape America. With a devout sense of community and adherence to beliefs, the Amish Country is a rare opportunity to witness a different way of life that will remind you that life can be simple.

THE BYWAY STORY

The Amish Country Byway tells cultural, historical, natural, recreational, and scenic stories that make it a unique and treasured Byway.

Ohio

❋ Amish Country Byway

Cultural

One of the most important features of the Amish Country Byway is seen in the visible aspects of culture along the Byway. The Amish have established themselves in the Holmes County area, and it is estimated that one in every six Amish in the world live in this area. The Amish choose to live a simple way of life, which is clearly evident by the presence of horses and buggies, handmade quilts, and lack of electricity in Amish homes. Entrepreneurial businesses owned by the Amish add to the friendly atmosphere along the Byway while creating a welcomed distance from the superstores of commercial America. In the 21st century, the Amish Country Byway is an important example of a multicultural community, as both the Amish and English (that is, non-Amish) traditions are strong in the region. These two cultures have built on similarities while still respecting differences. By working together, they have created a thriving, productive community that is a wonderful experience for all who visit it.

The Amish, as a branch of the Anabaptist people, are traditionally devout and religious. Like so many other immigrants, they came to America in search of religious freedom. In Europe, the Anabaptists had been persecuted for their beliefs, but today, Amish beliefs are more accepted and laws have been passed protecting their rights in regard to education, Social Security, and military service. Horses and buggies, plain dress, independence from electricity, plain window curtains, homemade quilts, spinning tops, and lots of reading materials are some of the things you might find in an Amish home—all evidence of their simple living. A community event, such as a barn raising, helps build relations among neighbors and is an efficient way to get work done.

Another important aspect of the Amish Country Byway is the influence of early Native American Indians and Appalachian folklore. The presence of both is felt along the Byway, as festivals and parades, such as the Killbuck Early American Days Festival, celebrate these early settlers. Coal fields and stone quarries drew settlers from the east, and today, this influence is manifested in the strong mining and manufacturing industries in the area.

Agriculture is the economic heart of Amish Country, and visitors to the area are likely to see rows of haystacks or fields being plowed. Holmes County boasts the second largest dairy production in the state, the largest local produce auction during the growing season, and weekly livestock auctions in the communities along the Byway. The Swiss and German heritage of the early settlers in the county is evident in the many specialty cheese and meat products and delicious Swiss/Amish restaurants. A variety of festivals and local produce stands along the Byway allow visitors to taste a part of Amish Country. Agriculture-based weekly auctions are held at the Mount Hope, Farmerstown, and Sugarcreek sale barns, and specialty sales are held throughout the year at various times.

When the Amish settled in the area, most depended on agriculture as their profession, but others who were not farmers worked instead in blacksmith shops, harness shops, or buggy shops. In addition, many specialties sprang up, such as furniture-making. Today, shops are scattered throughout the Byway, specializing in everything from furniture to gazebos. Artists and craftsmen have made the Amish Country Byway home, and travelers can see cheese-makers, bakers, quilters, potters, and a variety of other artisans at work.

Historical

The story of the Amish Country Byway is the story of the movement and settlement of people. The Byway serves as a reminder of why people came to America and the struggles that many had in settling new and uncharted lands. Holmes County settlers came from the east, but even before that, many had come from Europe.

Today, roads forged by the early settlers in the area have been upgraded to highways, and while Amish farmers still use horses and buggies as transportation, the roads have improved their journey. The historic nature of the Byway is felt from these roads to the numerous buildings that stand on the National Historic Register.

The Amish branch of the Anabaptist faith began in Switzerland in the 16th century with the Swiss Brethren movement. The Reformation in Europe had created a new Europe in which the Catholic Church was not the only church. As a result, persecutions developed as people of different faiths conflicted with one another. To escape some of these persecutions, many Anabaptists immigrated to America and settled primarily in southeastern Pennsylvania. Because many who settled in the area were of the Swiss Brethren and Mennonite movements, there remains a strong influence of Swiss and German food and culture in the area.

In 1807, three Amish men began the uncertain and dangerous task of scouting out additional lands, yet unknown to them. Long before the Amish came to call Holmes County home, bison herds crisscrossed the state, led by instinct down the valleys and along the terminal moraine. Indians later used the trails left by the bison. Eventually, these trails became the main paths of the Amish. Today, those paths make up State Route 39, one of the main arteries of the Amish Country Byway. In the 1830s, before the railroad, Amish and English farmers would drive their fattened pigs along well-worn paths to the Ohio Erie Canal at Port Washington from Millersburg. This walk to the canal was referred to as a three-day drive. Today, along the Byway, Amish and English neighbors continue to work together, making Holmes County an important agricultural, furniture-manufacturing, and cheese-producing region of Ohio and the nation.

see page A20 for color map

Visiting any of the places of interest along the Byway is only one way to experience the historical nature of the Byway. The majority of the historic buildings are located in Millersburg, and the entire downtown district has been designated a historical area. The entire Byway speaks of the past, and the many people living along the Byway still engage in the activities and livelihoods in which their ancestors participated.

Natural

The Amish Country Byway may be known primarily for its distinctive cultural and historical aspects; however, many natural features in the area make this a place where people would naturally choose to settle. The area is diverse in its natural features, and you can enjoy them from your car or by exploring various regions along the Byway. From natural passageways created by the earth forces to forests and wildlife, the area is rich in resources and natural beauty.

The Amish Country Byway is literally the product of being at the upper edge of the terminal moraine, making it the northwest gateway to Appalachia. (A terminal moraine marks the farthest point to which a glacier has advanced.) Because of this terminal moraine, natural paths developed, one of which has become the Byway

Ohio

❋ Amish Country Byway

as it is known today. This activity causes arc- or crescent-shaped ridges to form, and this is what has happened along the Byway. The terminal moraine runs west to east, paralleling State Route 39 with visible formations along the road. Large rock cliffs and small, deep lakes in the northwest portion of the Amish Country Byway are the products of this geological process. Briar Hill Stone Quarry is the largest sandstone quarry in operation in the United States and is located just off the Byway near Glenmont. This quarry has provided an important natural resource, and many Amish farms and schools have been established in this area.

Another important industry along the Byway is drawn from the forests of large oak and cherry trees. Rich soil and available water tables underneath the ground have made this a rich area for timber to grow. There has been an increase in the demand for hardwood over the past few years, and the Amish Country Byway is known for its good timber. The Amish often use this timber as they make their furniture, and you can see the finished products for sale along the Byway.

The natural wonders found at the Killbuck Marsh Wildlife Area are mainly the wildlife of the wetland region. This area is home to birds and other wildlife. Most notably, the American bald eagle has roosting nests in the marsh area. Local bird-watchers, especially the Amish, have formed groups and organizations to document the birds and provide educational activities to inform others about the value of birds and wildlife as a resource in the area.

Recreational

From leisurely drives to hiking and biking, recreational opportunities exist for everyone traveling the Amish Country Byway. Every season can be enjoyed, with hiking, biking, rodeos, horseback riding, tennis, golfing, and hunting in the summer months. You can also enjoy numerous water activities, such as canoeing, swimming, fishing, and boating. Winter can be a memorable experience, with activities such as cross-country skiing and snowmobiling.

The Amish Country Byway is not one for speed demons. By slowing down, you get to experience the many recreational opportunities that are unique to this Byway. There are carriage rides, hay rides, and sleigh rides that reflect the agricultural traditions of the area, while unique activities, such as hot air balloon rides and airplane rides, may also be enjoyed. Because of the high aspects of culture and history along the Byway, one of the most popular activities is visiting Amish homesteads and farms, antique shops, and museums. In addition, you can find many places to stop and enjoy some good cooking or shopping.

More traditional recreational activities abound along the Amish Country Byway. The Holmes County Trail goes through Millersburg and links the Byway with the northern part of the county and state. This trail is open to bicycles, hikers, and buggies. The local Amish citizens who sit on the Rails to Trails board provide valuable insight into how to make this a success for the Amish, their English neighbors, and visitors. This trail travels through beautiful Amish Country and is a good way to get off of the main Byway route.

Another way to get off the main Byway route is to use the area's river and creek network. These rivers and creeks were critically important to the transportation and commerce of the past, and today, they provide a great opportunity for visitors to go canoeing, swimming, boating, or fishing. The Killbuck Creek feeds into Killbuck Marsh, which also provides excellent bird-watching opportunities. Tucked away on the western edge of the Byway, the Mohican River is the basis for making this one of the most popular recreational retreats in the state of Ohio. Canoeing is especially popular along the river—this area has been coined the Camp and Canoe Capital of Ohio.

Scenic

The view on the Amish Country Byway is one of rolling hills; undisturbed marshland and forests; beautiful trees and landscapes; well-kept farmhouses, barns, and ponds; neat rows of agricultural crops and vegetables; brilliant displays of flowers; and bucolic scenes of Amish farmers/laborers with their families and children.

The simple living of the Amish and the gentle hospitality of the residents of Holmes County make the Amish Country Byway a scenic trip indeed. The gently rolling farmlands of the Byway give you a chance to experience the area's grand agricultural tradition. Bales of hay, freshly plowed fields, barn raisings, and locally grown produce sold at the roadside are some of the scenes that will greet you.

Red barns, buggies, and laundry hung out neatly on the line to dry are some of the unique scenes along the Byway. The Amish idea of simple living is reflected in the balance and harmony with nature. The friendly and hospitable nature of those who live along the Byway may be lacking in other big cities and thus makes this Byway unique.

HIGHLIGHTS

The following itinerary gives you an idea for spending a day on the eastern half of State Road 62.

- Enjoy the peace of the early morning at the **Killbuck Marsh Area.** Fish or hike while the sun rises. Have a picnic breakfast while enjoying the birds and scenery.
- Head north to **Millersburg.** If you're interested in historic buildings, you're in luck. Downtown Jackson, Clay, and Washington streets are a Historic District. There are plenty of National Register of Historic Places homes to see, like the G. Adams House, the G. W. Carey House, and the Victorian House and Museum. There are also a lot of neat shops downtown, like Maxwell Brothers and the Three Feathers Pewter Studio/Gallery. Take time to look around and see what else you can discover. You can find plenty of places to eat in Millersburg, so this a perfect place to have lunch before you head off to the Rolling Ridge Ranch.
- The **Rolling Ridge Ranch** is a good place for families to visit. Kids love the petting zoo and playground. You can take wagon rides and see many kinds of animals there.
- **Behalt-Mennonite Information Center** is an essential next stop, because you can learn about the Amish and Mennonites and why they live the way they do. See the historical mural. This is interesting, free education.
- Just about the time you're done at the Information Center, you'll be ready for dinner. The **Alpine Alpa Restaurant** is just up the road from the Information Center. It's famous not only for its food but also for the interesting things you can see and buy there. You can also experience a bit of Swiss culture.

THINGS TO SEE AND DO

Driving along the Amish Country Byway will certainly keep your senses engaged, but if you yearn to get out of the car and stretch your legs, or if you'd like to make a mini-vacation out of your trip, check out these attractions along the route.

Ohio

❋ *Amish Country Byway*

BEHALT-MENNONITE INFORMATION CENTER. *5798 County Rd 77, Berlin (44610). Phone 330/893-3192.* Behalt means "to keep or to remember," and this center remembers the history of the Amish and Mennonites in the US. Heavily persecuted in Europe, the Anabaptists from which both Amish and Mennonites trace their heritage fled into Russia and North America. A beautifully illustrated cyclorama depicts scenes from Amish and Mennonite history; volunteers are on hand to answer questions and help you understand this peace-loving culture and faith. Open Mon-Sat 9 am-8 pm; until 5 pm Nov-May. **FREE**

KILLBUCK MARSH WILDLIFE AREA. *Off SR 83 near Shreve (44676). Phone 330/567-3390.* This 5,000-acre marsh, fed by Killbuck Creek, offers unparalleled opportunities for viewing wildlife, including endangered trumpeter swans and sandhill cranes. Some hunting is allowed in season. Trails. Open daily; some areas off-limits during nesting seasons.

ROLLING RIDGE RANCH. *3961 County Rd 168, Millersburg (44654). Phone 330/893-3777.* Rolling Ridge is a terrific place for kids and adults to view 350 different species of animals from the comfort of a car; 80 acres. Wagon rides, petting zoo, playground. Open Mon-Sat 9 am-7 pm. **$$**

PLACES TO STAY

If you choose to include an overnight stay in your trip along this Byway, Mobil Travel Guide recommends the following lodgings.

★★ **LAMPLIGHT INN BED & BREAKFAST.** *5676 TR 362, Berlin (44610). Phone 330/893-1122; toll-free 866/500-1122. www.bbonline.com/oh/lamplight/.* 5 rooms, 3 story. Located within walking distance of historic downtown Berlin. All rooms have private entrances and baths. **$**

★ **THE BARN INN BED & BREAKFAST.** *6838 County Rd 203, Millersburg (44654). Phone 330/674-7600; toll-free 877/674-7600. www.bbonline.com/oh/thebarn/.* 7 rooms, 2 story. Located in the picturesque Honey Run Valley, this bed-and-breakfast was once home to a dairy farm and has been restored to its former splendor. **$$**

★★ **BIGHAM HOUSE BED & BREAKFAST.** *151 S Washington St, Millersburg (44654). Phone toll-free 866/689-6950. www.bbonline.com/bighamhs/.* 4 rooms, 2 story. Located in the heart of Amish country with private baths and a TV in every room. **$$**

★★ **INN AT HONEY RUN.** *6920 County Rd 203, Millersburg (44654). Phone 330/674-0011; toll-free 800/468-6639. www.innathoneyrun.com.* 39 rooms, 2-3 story. Check-out noon, check-in 4:30 pm. TV; VCR. In-room modem link. Hiking; shuffleboard. **$$**

★★ **MILLER HAUS BED & BREAKFAST.** *3135 County Rd 135, Walnut Creek (44687). Phone 330/893-3602. www.millerhaus.com.* 9 rooms, 2 story. Situated on a 23-acre farm in the middle of Amish Country, within walking distance to sights and attractions. **$**

PLACES TO EAT

A long day of driving is sure to make you hungry. At the end of your journey, take a table at one of the following restaurants.

★★ **CHALET IN THE VALLEY.** *5060 State Rd 557, Millersburg (44654). Phone 330/893-2550. www.chaletinthevalley.com.* Austrian menu. Closed Mon. Lunch, dinner. Beautiful farm setting serving traditional Viennese dishes. **$**

★ **ALPINE ALPA.** *1504 State Rd 62, Wilmont (44689). Phone 330/359-5454. www.alpine-alpa.com.* Swiss menu. Lunch, dinner. Beautiful Swiss chalet on the Byway serving traditional specialties. **$**

CanalWay Ohio Scenic Byway

❋ OHIO

Quick Facts

LENGTH: 110 miles.

TIME TO ALLOW: 4.5 hours.

BEST TIME TO DRIVE: Spring and summer.

BYWAY TRAVEL INFORMATION: Byway local Web site: www.canalwayohio.com.

SPECIAL CONSIDERATIONS: Be sure to allow plenty of time to savor each of the many points of interest found along the Byway.

RESTRICTIONS: Riverview Road in south Cuyahoga County is occasionally closed due to flooding. This happens only during major storms and doesn't last for very long. Fort Laurens, Towpath, and Dover Zoar roads also lie in the flood path.

BICYCLE/PEDESTRIAN FACILITIES: The Towpath Trail is excellent for bicycle and pedestrian travel. It runs parallel to the Byway in many places. The Towpath Trail is currently 44 miles long and will eventually be extended to 88 miles in length.

The construction of the Ohio and Erie Canal in 1825 drastically changed the people and pace of this region. Many Byway visitors travel this route to learn about the development of the pre- and post-canal eras. The route also offers an impressive display of historical sites, along with many opportunities for hiking, biking, and water sports.

THE BYWAY STORY

The CanalWay Ohio Scenic Byway tells archaeological, cultural, historical, natural, recreational, and scenic stories that make it a unique and treasured Byway.

Archaeological

The CanalWay Ohio Scenic Byway boasts more than 500 archaeological sites; seven are listed on the National Register of Historic Places. With all possibility, many more sites exist, because a systematic, corridor-wide archaeological survey has not yet been made.

One site, Irishtown Bend, is located in Cleveland's Flats. This is where an early settlement of Irish canal workers lived. While visitors can't see very much at this site because the house foundations that have been semi-excavated have been covered up for protection, the findings at this site are significant. They show that people in this particular settlement were drinking alcohol less than those in surrounding Irish settlements, thus dispelling in part the working-class Irish myth. Artifacts from Irishtown Bend are displayed periodically.

The locations of these archaeological sites are not advertised in order to protect them from ransacking and vandalism.

Ohio

✽ CanalWay Ohio Scenic Byway

Cultural

The influences on area culture can be classified into four elements: the resilience of farming communities, life during the canal period, the gumption of immigrants, and the accomplishments of wealthy industrialists.

Farming is still important to area culture, and the strength of past and present farmers is reflected in the sturdiness of the farms themselves: many farms from the mid-19th century are still standing and working. The village of Zoar, which was founded in 1817 by a German religious sect, the Zoarites, is still working according to many of its old agricultural practices. Burfield Farm (near Bolivar) is another durable example of early area farming and its influences.

The sway of the canal era on area culture is apparent because its flavor is still potent in the historic cities of Clinton, Canal Fulton, Boston, Everett, and Peninsula.

The neighborhoods that immigrants developed retain their special foreign flavor, creating small pockets of spice that add verve to the Byway. Particularly, these National Register historic neighborhoods near Cleveland are based on Archwood-Dennison Avenue, Broadway Avenue, Ohio City, and Warszawa.

The ambitions of early wealthy industrialists are represented in the homes they built. Some of the more notable are the Stan Hywet Hall (Goodyear founder Frank A. Seiberling's home) and the Anna Dean Farm (Diamond Match founder O. C. Barber's estate).

Replete with continuous high-culture events, the larger cities on the route (Cleveland, Akron, Barberton, and Strongsville) retain several historic and well-known venues. The area's local culture is manifested in locally produced festivals and entertainment.

One of Cleveland's most prominent venues is Playhouse Square Center, which recently celebrated its 75th anniversary and is the hub of the region's performing-arts scene. Akron, Barberton, and Strongsville are similarly awash in cultural activities.

Historical

People have lived in the CanalWay Ohio area for nearly 12,000 years. The area was an important transportation route for American Indians, and it was deemed neutral territory so that all might travel safely from the cold waters of the Great Lakes to warmer southern waters. European explorers and trappers arrived in the 17th century, and immigrants slowly moved in to farm over the centuries. The modern catalyst for area development was the Ohio and Erie Canal, which gave way to the area's industrial era. The CanalWay Ohio Scenic Byway takes you through places that still show evidence of the early inhabitants and the effects of industrial development.

The countryside north of Dover consists of wooded ravines and hillsides. It is separated by tilled croplands and isolated farms. This setup comes from early settlement of the Pennsylvania Germans and the Moravians. The Zoar Historic Village represents the Zoarites, the religious group that helped settle this former western frontier.

The newly constructed Indigo Lake Visitor Shelter is a stop on the old railroad. Here, visitors can fish in the adjacent lake, access the canal towpath, or walk to Hale Farm. This farm is a 19th-century farm, presented to visitors through a full program of living history.

Towns such as Dover, Bolivar, Navarre, Barberton, Canal Fulton, and Clinton show the influence of the canal with their early 19th-century architecture, original canal-oriented street patterns, locks and spillways, and towpath trails along the historic canal route. Tours on the *St. Helena II* Canal Boat in Canal Fulton give visitors firsthand experience of canal life. The Canal Visitors Center is housed in an 1852 canal tavern.

Waters from the Summit Lakes provided the cooling needed for the industrial rubber boom, which created the prosperous Akron of the 1920s. Brick factories that were four and five stories high illustrate life in industrial-era Akron.

These buildings were capped by tall clock towers rimmed by rail lines, and they were surrounded by neighborhoods for industry workers.

Close to the Byway is the National Historic Landmark Stan Hywet Hall. This used to be the home of F. A. Seiberling, the founder of the Goodyear Rubber Company. Visitors will enjoy viewing the home's superb landscape (designed by Warren Manning), as well as taking tours of the house itself. These modest homes are distinct from the large Tudor Revival mansions created by the wealth of the industrial boom economy of the 1920s.

Overhead train trestles, interstate pipelines, fields of oil tanks, and aluminum and steel works tell the story of industrial might created by the marriage of Great Lakes iron ore and Appalachian coal—brought together by the Ohio and Erie Canal.

Natural

The CanalWay Ohio Scenic Byway is a biological crossroad because it transects three regions: lake plains, glaciated plateau, and unglaciated plateau. This results in a great diversity of plants and animals, textbook examples of forest communities and habitats.

The primary trees in the glaciated plateau region are beech-maple (the most common), oak-hickory, and hemlock-beech. Ice Age hemlock-beech forests are found in ravines, while oak-hickories are found atop ridges and in drier areas. The Tinkers Creek Gorge National Landmark has a rare settlement of hemlock-beech on the moist valley floor.

Rolling hills and steep valleys characterize the unglaciated plateau region. Oak-hickory is common in this southern part of the route. Many of these lands are public and are being preserved by public agencies.

Recreational

There is plenty to do along the CanalWay Ohio Scenic Byway. Towpath Trail is extremely popular; over 3 million users enjoy the trail in a typical year. Other recreation includes skiing, golfing, picnicking, hiking, fishing, and boating, and water excursions are available along the Cuyahoga and Tuscawaras rivers as well. Also, the route passes through several entertainment districts.

see page A21 for color map

Scenic

The scenery found along the CanalWay Ohio Scenic Byway is interestingly diverse. It ranges from the heavy industry of the Cleveland Flats (and the resulting immigrant neighborhoods) to rolling hills and farmland. You can also see the remnants of the towns and villages associated with the canal, as well as samples of 200-year-old architecture. You will find hundreds of varied vistas along the route.

Ohio

✽ CanalWay Ohio Scenic Byway

HIGHLIGHTS

This itinerary takes you through Cuyahoga Valley National Park and into Akron. Although it does not cover the entire distance of the Byway, it gives you a taste of how you can drive a portion of this road.

- Start with a dawn picnic breakfast at the **Cuyahoga Valley National Park** (**CVNP**). The CVNP is on the Byway, and its entrance is about 8 miles south of Cleveland. In this gorgeous 33,000-acre region, you can do just about anything outdoorsy (hiking, golfing, bicycling, horseback riding, and so on). Admittance is free. After breakfast, go biking along the Towpath Trail, which runs through the park for about 44 miles. Take an hour or two to enjoy a bike ride along any part of the trail.
- For another two (or three) hours before lunch, stop at **Hale Farm,** which is at the southern end of the CVNP. There is an admittance fee for this living-history farm, where you can wander the grounds and talk to the artisans who make the impressive wares for sale there. The buildings and history here are fascinating. Eat lunch and head south to Akron.
- Akron is a buzzing city with much to do. One possibility is to stop at the **Akron Art Museum.** Admission and parking here are free for visitors, and the museum always has engaging exhibits.
- After your visit at the museum, go to the **Cascade Locks Park** (also in Akron). Admittance is free here as well. This park has 15 canal locks in a 1-mile stretch (all necessary to climb over the Continental Divide), and you can follow them along on a trail; signs along the trail explain the locks and the history of the canal. There are also two historic buildings to visit in this park.
- Have dinner in Akron and round off your evening with a concert or a play at one of the many famous and popular performing arenas and halls. Check local event calendars for details.

THINGS TO SEE AND DO

Driving along the CanalWay Ohio Scenic Byway will certainly keep your senses engaged, but if you yearn to get out of the car and stretch your legs, or if you'd like to make a mini-vacation out of your trip, check out these attractions along the route.

AKRON ART MUSEUM. *70 E Market St, Akron (44308). Phone 330/376-9185. www.akronartmuseum.org.* Regional, national, and international art, 1850 to the present; also changing exhibits; sculpture garden. Open Tues-Sat; closed major holidays. **FREE**

AKRON CIVIC THEATRE. *182 S Main St, Akron (44308). Phone 330/535-3179 or -3178 (recording). www.akroncivic.com.* Lavishly designed by Viennese architect John Eberson to resemble a night in a Moorish garden, complete with blinking stars and floating clouds. The theater, built in 1929, is one of four atmospheric-type facilities of its size remaining in the country. Open daily.

AKRON ZOOLOGICAL PARK. *500 Edgewood Ave, Akron (44307). Phone 330/375-2525. www.akronzoo.com.* This 26-acre zoo features more than 300 birds, mammals, and reptiles from around the world. Exhibits include the Ohio Farmyard, where children can pet and feed the animals; Tiger Valley, with tigers, lions, and bears; a walk-through aviary; and an underwater viewing window for observing the river otters. Open daily May-Oct 10 am-5 pm; Nov-Apr 11 am-4 pm; closed holidays. **$$**

ALPINE VALLEY SKI AREA. *10620 Mayfield Rd, Chesterland (44026). Phone 440/285-2211 or 440/729-9775 (snow conditions). www.alpinevalleyohio.com.* The area has quad, double chairlifts; J-bar, two rope tows; patrol, school, rentals, snowmaking; cafeteria, lounge. Longest run 1/3 mile; vertical drop 240 feet. Open early Dec-early Mar. **$$$$**

ATWOOD LAKE PARK. *4956 Shop Rd NE, Mineral City (44656). Phone 330/343-6780.* A 1,540-acre lake with swimming, fishing, boating (25-hp limit), marinas; golf, tent and trailer sites (with showers). Restaurant, cottages, resort. Pets on leash only. Open daily. **$$**

BECK CENTER FOR THE ARTS. *17801 Detroit Ave, Cleveland (44107). Phone 216/521-2540. www.lkwdpl.org/beck/.* Open Sept-June; reduced schedule July-Aug.

BOSTON MILLS & BRANDYWINE. *7100 Riverview Rd, Peninsula (44264). Phone 330/657-2334 or 330/467-2242 (in Cleveland). www.bmbw.com.* Four triple, two double chairlifts; two handle tows; patrol, school, rentals, snowmaking; cafeteria, bar. Longest run 1,800 feet; vertical drop 240 feet. Open Dec-mid-Mar, daily. **$$$$**

CANAL FULTON AND MUSEUM. *103 Tuscarawas St, Canal Fulton (44616). Phone 330/854-3808; toll-free 800/HELENA-3.* St. Helena III, a replica of a mule-drawn canal boat of the mid-19th century, takes a 45-minute trip on the Ohio-Erie Canal. Leaves from Canal Fulton Park. Open June-Aug, daily; mid-May-late May and early-mid-Sept, weekends only. **$$**

THE CLEVELAND ARCADE. *401 Euclid Ave, Cleveland (44114). Phone 216/776-4461.* This five-story enclosed shopping mall, one of the world's first (1890), features more than 80 shops and restaurants. Open Mon-Sat.

CLEVELAND BOTANICAL GARDEN. *11030 East Blvd, Cleveland (44106). Phone 216/721-1600. www.cbgarden.org.* Herb, rose, perennial, wildflower, Japanese, and reading gardens. Grounds open daily 10 am-5 pm. **$$**

CLEVELAND METROPARKS. *4101 Fulton Pkwy, Cleveland (44144). Phone 216/351-6300. www.clemetparks.com.* Established in 1917, the system today consists of more than 20,000 acres of land in 14 reservations, their connecting parkways, and **Cleveland Metroparks Zoo.** Swimming, boating, and fishing; more than 100 miles of parkways provide scenic drives, picnic areas, and play fields; wildlife management areas and waterfowl sanctuaries; hiking and bridle trails, stables, golf courses; tobogganing, sledding, skating, and cross-country skiing areas; and eight outdoor education facilities offering nature exhibits and programs.

CLEVELAND METROPARKS ZOO. *3900 Wildlife Way, Cleveland (44109). Phone 216/661-6500. www.clemetzoo.com.* Seventh oldest zoo in the country, with more than 3,300 animals occupying 165 acres. Includes mammals, land and water birds; animals displayed in naturalized settings. More than 600 animals and 7,000 plants are featured in the 2-acre Rain Forest exhibit. Open daily 10 am-5 pm; Memorial Day-Labor Day, Sat-Sun 10 am-7 pm; closed Jan 1, Dec 25. **$$**

★ **THE CLEVELAND MUSEUM OF ART.** *11150 East Blvd, Cleveland (44106). Phone 216/421-7340; toll-free 888/CMA-0033. www.clemusart.com.* Extensive collections of approximately 30,000 works of art represent a wide range of history and culture; included are arts of the Islamic Near East, the pre-Columbian Americas, and European and Asian art; also African, Indian, American, ancient Roman, and Egyptian art. Concerts, lectures, special exhibitions, films; café. Museum entrance from East Blvd. Open Tues, Thurs, Sat-Sun 10 am-5 pm, Wed, Fri 10 am-9 pm; closed Mon, holidays. **FREE**

CLEVELAND MUSEUM OF NATURAL HISTORY. *1 Wade Oval, University Cir, Cleveland (44106). Phone 216/231-4600. www.cmnh.org.* Dinosaurs, mammals, birds, geological specimens, gems; exhibits on prehistoric Ohio, North American native cultures and ecology; Woods Garden, live animals; library. Open daily; closed holidays. **$$**

CLEVELAND PLAY HOUSE. *8500 Euclid Ave, Cleveland (44106). Phone 216/795-7000; toll-free 800/278-1CPH. www.clevelandplayhouse.com.* America's oldest regional, professional Equity

Ohio

❋ CanalWay Ohio Scenic Byway

theater presents traditional American classics and premiere productions of new works in five performance spaces; organized in 1915. Open Sept-June, Tues-Sun, also weekend matinees.

CULTURAL GARDENS. *Rockefeller Park from Superior Ave to St. Clair Ave along Martin Luther King, Jr. Blvd, Cleveland (44123). Phone 216/664-3534.* Chain of gardens combining landscape architecture and sculpture of 24 nationalities. **FREE**

CUYAHOGA RIVER GORGE RESERVATION. *Main and Market sts, Akron (44303). Phone 330/867-5511.* Part of the park system. On the north bank is the cave where Mary Campbell, the first white child in the Western Reserve, was held prisoner by Native Americans.

CUYAHOGA VALLEY LINE STEAM RAILROAD. *315 Clark Ave, Cleveland (44113). Phone 330/657-2000.* Scenic railroad trips between Independence (south of Cleveland) and Akron aboard a vintage train.

CUYAHOGA VALLEY NATIONAL PARK. *15610 Vaughn Rd, Akron (44141). Phone 330/650-4636. www.nps.gov/cuva/.* On 33,000 acres. Beautiful and varied area with extensive recreational facilities, many historic sites, and entertainment facilities; 20-mile-long, fully accessible Towpath Trail. Artistic events, performances, campfire programs, and nature walks. Three visitor centers (daily; closed holidays). Park open daily. Fee for some activities. **FREE**

CUYAHOGA VALLEY SCENIC RAILROAD. *1630 Mill St W, Akron (44264). Phone 330/657-2000; toll-free 800/468-4070. www.cvsr.com.* Scenic railroad trips through the Cuyahoga Valley National Recreation Area between Cleveland and Akron aboard vintage railroad cars pulled by first-generation ALCO diesels. Stations include Independence (south of Cleveland), Hale Farm, Akron Valley Business District, NPS Canal Visitor Center, Howard St, and Quaker Square in Akron. Reservations required. **$$$$**

DOVER LAKE WATERPARK. *1150 W Highland Rd, Sagamore Hills (44067). Phone 330/467-SWIM (Cleveland) or 330/655-SWIM (Akron). www.doverlake.com.* Swimming, beach, wave pool, water slides, tube slides; concessions, pavilions, chairlift ride, picnic grounds. Open mid-June-mid-Aug, Mon-Fri 11 am-7 pm (water attractions close at 6:30 pm). **$$$$**

DITTRICK MUSEUM OF MEDICAL HISTORY. *11000 Euclid Ave, Cleveland (44106). Phone 216/368-3648. www.cwru.edu/artsci/dittrick/home.htm.* Collection of objects relating to the history of medicine, dentistry, pharmacy, and nursing; doctor's offices of 1880 and 1930; exhibits on the development of medical concepts in the Western Reserve to the present. Also history of the X-ray and microscopes. Open Mon-Fri; closed holidays, day after Thanksgiving. **FREE**

EDGEWATER PARK. *8701 Lakeshore Blvd, NE, Cleveland (44108). Phone 216/881-8141. www.dnr.state.oh.us/parks/parks/clevelkf.htm.* A 119-acre park with swimming beach, fishing, boating (ramps, marina); biking; fitness course, picnic grounds (pavilions), playground, concessions. Scenic overlook. Open daily. **FREE**

EUCLID BEACH PARK. *16300 Lakeshore Blvd, Cleveland (44110). Phone 216/881-8141.* A 51-acre park with swimming beach, fishing access; picnic grounds, concession. **FREE**

FREDERICK C. CRAWFORD AUTO-AVIATION COLLECTION. *Magnolia Dr and E 108th, Cleveland (44110).* Antique cars and planes; motorcycles and bicycles; National Air Racing exhibit; Main Street, Ohio, 1890. Open daily; closed holidays.

FORT LAURENS STATE MEMORIAL. *County Rd 102, Bolivar (44612). 1/2 mile S of State Rte 212. Exit I-77. Phone 330/874-2059; toll-free 800/283-8914. www.ohiohistory.org/places/ftlauren/.* An 82-acre site of the only American fort in Ohio during the Revolutionary War. Named in honor of Henry Laurens, Continental Congress president. Built in 1778 as a defense against the

British and Native Americans. Picnicking; museum has artifacts and multimedia program on the American Revolution. Open Memorial Day-Labor Day, Wed-Sun 9:30 am-5 pm; after Labor Day-Oct, Sat 9:30 am-5 pm, Sun noon-5 pm. $

GOODTIME III BOAT CRUISE. *825 E 9th St Pier, Cleveland (44114). Phone 216/861-5110. www.goodtimeiii.com.* Two-hour sightseeing and dance cruises on the Cuyahoga River, lake, and harbor. Leaves the pier at E 9th St. Open mid-June-Sept, daily; limited schedule the rest of the year. $$$$

GOODYEAR WORLD OF RUBBER. *1201 E Market St, Akron (44316). Phone 330/796-7117.* Historic and product displays. A one-hour tour includes movies. The Goodyear Blimp. Open daily 8 am-4:30 pm; closed holidays. FREE

GREAT LAKES SCIENCE CENTER. *601 Erieside Ave, Cleveland (44114). Phone 216/694-2000. www.greatscience.com.* More than 350 hands-on exhibits explain scientific principles and topics specifically relating to the Great Lakes region. Also features an OmniMax domed theater. Open daily 9:30 am-5:30 pm; closed holidays. $$

HALE FARM AND VILLAGE. *2686 Oak Hill Rd, Bath (44210). Phone 330/666-3711; toll-free 800/589-9703.* Authentic Western Reserve house (circa 1825), other authentic buildings in a village setting depict northeastern Ohio's rural life in the mid-1800s; pioneer implements; craft demonstrations, special events; farming; costumed guides. Open late May-Oct, Tues-Sat 10 am-5 pm, Sun noon-5 pm. $$

HEALTH MUSEUM OF CLEVELAND. *8911 Euclid Ave, Cleveland (44106). Phone 216/231-5010. www.healthmuseum.org.* More than 200 participatory exhibits and displays on the human body, including Juno, the transparent talking woman. Also features the Wonder of New Life exhibit.

Events and educational programs, including corporate wellness, distance learning, and school programs. Open daily; closed holidays. $$

HOWER HOUSE. *60 Fir Hill, Akron (44304). Phone 330/972-6909.* A 28-room Victorian mansion (1871), Second Empire Italianate-style architecture, built by John Henry Hower; lavish furnishings from around the world. Open Feb-Dec, Wed-Fri and Sun afternoons; closed holidays. $$

JOHN BROWN HOME. *550 Copley Rd, Akron (44320). Phone 330/535-1120. www.ci.akron.oh.us/Tour/JBrown.htm.* Remodeled residence where the abolitionist lived from 1844 to 1846. Open Wed-Sun afternoons; closed holidays. $$

JOHN CARROLL UNIVERSITY. *20700 N Park Blvd, Cleveland (44118). Phone 216/397-1886. www.jcu.edu.* Started in 1886 and enrolling 4,500 students today, this university offers arts and sciences, business, and graduate schools in 21 Gothic-style buildings on its 60-acre campus. Large collection of G. K. Chesterton works.

KARAMU HOUSE AND THEATER. *2355 E 89th St, Cleveland (44106). Phone 216/795-7070. www.karamu.com.* Multicultural center for the arts. Classes/workshops in music, creative writing, dance, drama, and visual arts; dance, music, and theatrical performances (Sept-June); two theaters, art galleries. Fee for some activities.

Ohio

❇ *CanalWay Ohio Scenic Byway*

LAKE ERIE NATURE AND SCIENCE CENTER. *28728 Wolf Rd, Cleveland (44140). Phone 216/835-9912.* Features animals, marine tanks, nature displays, wildlife/teaching garden; planetarium show (fee). Science Center open daily; closed holidays. **FREE**

LAKE VIEW CEMETERY. *12316 Euclid Ave, Cleveland (44106). Phone 216/421-2665. www.lakeviewcemetery.com.* Graves of President James A. Garfield, Mark Hanna, John Hay, and John D. Rockefeller. Garfield Monument open Apr-mid-Nov, daily. Cemetery open daily year-round.

LEESVILLE LAKE. *Rte 212, Bowerston (44695). State Rte 39 to Rte 212. Phone 614/269-2131.* Petersburg Marina, on N shore, off OH 332; phone 330/627-4270. Fishing and boating on 1,000-acre lake (10-hp limit); tent and trailer sites (Apr-Oct). Open daily.

MASSILLON MUSEUM. *121 Lincoln Way E, Massillon (44646). Phone 330/833-4061.* Historical and art exhibits. Open Tues-Sat 9 am-5 pm, Sun 2-5 pm; closed holidays. **FREE**

NASA LEWIS VISITOR CENTER. *21000 Brookpark Rd, Cleveland (44135). Phone 216/433-2001.* The display and exhibit area features the space shuttle, space station, aeronautics and propulsion, planets and space exploration; also *Skylab 3*, an Apollo capsule, and communications satellites. Open daily; closed holidays. **FREE**

NATIONAL INVENTORS HALL OF FAME. *221 S Broadway St, Akron (44308). Phone 330/762-4463. www.invent.org.* Dedicated to the creative process; houses an interactive exhibit area, plus the hall of fame. Open Tues-Sat 10 am-4:30 pm, Sun noon-5 pm. **$$**

NATUREALM VISITORS CENTER. *1828 Smith Rd, Akron (44313). Phone 330/865-8065. www.summitmetroparks.org.* A 4,000-square-foot Gateway to Nature underground exhibit area surrounded by many sights and sounds of nature. Open daily; closed holidays. **FREE**

***NAUTICA QUEEN* BOAT CRUISE.** *1153 Main Avenue, Cleveland (44113). Phone 216/696-8888; toll-free 800/837-0604. www.nauticaqueen.com.* Lunch, brunch, and dinner cruises. For details, contact E 9th St Pier, North Coast Harbor, 44113. **$$$$**

OLDEST STONE HOUSE MUSEUM. *14710 Lake Ave, Cleveland (44107). Phone 216/221-7343. www.lkwdpl.org/histsoc/.* Built in 1838, this house has been authentically restored and furnished with early 19th-century artifacts; herb garden. Guided tour by a costumed host. Open Feb-Nov, Wed and Sun afternoons; closed holidays. **FREE**

PERKINS MANSION. *550 Copley Rd, Akron (44320). Phone 330/535-1120.* Greek Revival home (1837) built of Ohio sandstone by Simon Perkins, Jr., on 10 landscaped acres. Open Wed-Sun afternoons; closed holidays. **$$**

PLAYHOUSE SQUARE CENTER. *1501 Euclid Ave, Cleveland (44115). Phone 216/241-6000; toll-free 800/766-6048. www.playhousesquare.com.* Five restored theaters form the nation's second largest performing arts and entertainment center. Performances include legitimate theater, Broadway productions, popular and classical music, ballet, opera, children's theater, and concerts.

PORTAGE LAKES STATE PARK. *5031 Manchester Rd, Akron (44319). Phone 330/644-2220. www.dnr.state.oh.us/parks/parks/portage.htm.* Several reservoir lakes totaling 2,520 acres. Swimming, fishing, boating (launch); hunting, hiking, snowmobiling, picnicking (shelter), camping (campground located 5 miles from park headquarters), pet camping. **FREE**

PORTAGE PRINCESS CRUISE. *300 W Turkey Foot Lake Rd, Akron (44319). Phone 330/499-6891.* Cruise the glacier-made lakes on an enclosed riverboat. Open May-mid-Oct, daily. **$$$**

PUBLIC LIBRARY. *325 Superior Ave NE, Cleveland (44114). Phone 216/623-2800. www.cpl.org.* Business and Science Building adjoining, separated by a reading garden; special exhibits. Open Mon-Sat 9 am-6 pm; closed holidays.

QUAKER SQUARE. *135 S Broadway, Akron (44308). Phone 330/253-5970. www.quakersquare.com.* Shopping, hotel, restaurants, and entertainment center in the original mills and silos of the Quaker Oats Company. Historical displays include famous Quaker Oats advertising memorabilia. Open daily; closed holidays. **FREE**

✴ ROCK AND ROLL HALL OF FAME AND MUSEUM. *1 Key Plaza, Cleveland (44114). Phone 216/781-ROCK; toll-free 888/764-7625. www.rockhall.com.* A striking composition of geometric shapes, this building is now the permanent home of the Hall of Fame. More than 50,000 square feet of exhibition areas explore rock's ongoing evolution and its impact on culture. Interactive database of rock and roll songs; videos; working studio with DJs conducting live broadcasts; exhibits on rhythm and blues, soul, country, folk, and blues music. Open daily 10 am-5:30 pm; closed Thanksgiving, Dec 25. **$$$$**

ROCKEFELLER GREENHOUSE. *750 E 88th St, Cleveland (44108). Phone 216/664-3103.* Japanese, Latin American, and peace gardens; garden for the visually impaired; seasonal displays. Open daily. **FREE**

ROCKEFELLER PARK. *750 E 88th St, Cleveland (44108). Phone 216/881-8141.* Connects Wade Park and Gordon Park. A 296-acre park with lagoon area, playground, tennis courts, and picnic facilities.

SCHOENBRUNN VILLAGE STATE MEMORIAL. *1984 E High Ave, New Philadelphia (44663). Phone 330/339-3636; toll-free 800/752-2711. www.ohiohistory.org/places/schoenbr/.* Partial reconstruction of the first Ohio town built by Christian Native Americans under the leadership of Moravian missionaries; one of six villages constructed between 1772 and 1798. Picnicking; museum. Open Memorial Day-Labor Day, daily; after Labor Day-Oct, weekends only. **$$**

THE SHAFRAN PLANETARIUM & ASTRONOMY EXHIBIT HALL. *The Cleveland Museum of Natural History, 1 Wade Oval Dr, Cleveland (44106). Phone 216/231-4600. www.cmnh.org/exhibits/shafran.html.* Shows (daily). Observatory open Sept-May, Wed on cloudless nights; planetarium program on cloudy nights. Children's programs. **$$**

SPRING HILL. *1401 Spring Hill Ln NE, Massillon (44646). Phone 330/833-6749. www.massillonproud.com/springhill/.* Historic 19th-century home (1821), including basement kitchen and dining room, secret stairway, and original furnishings; on the grounds are a springhouse, smokehouse, woolhouse, and milkhouse; picnicking. Open June-Aug, Wed-Sun; Apr-May and Sept-Oct, by appointment. **$**

STAN HYWET HALL AND GARDENS. *714 N Portage Path, Akron (44303). Phone 330/836-5533. www.stanhywet.org.* Tudor Revival manor house built by F. A. Seiberling, co-founder of Goodyear Tire & Rubber; contains 65 rooms with antiques and art treasures dating from the 14th century. More than 70 acres of grounds and gardens. Open Feb-Mar, Tues-Sat 10 am-4 pm, Sun 1-4 pm; Apr-late Dec, daily 10 am-4:30 pm; closed holidays and the month of Jan. **$$**

STEAMSHIP *WILLIAM G. MATHER* MUSEUM. *1001 E 9th St Pier, Cleveland (44114). Phone 216/574-6262. wgmather.nhlink.net/wgmloc.html.* Former flagship of the Cleveland Cliffs Iron Company, this 618-foot steamship is now a floating discovery center. Built in 1925 to carry iron ore, coal, grain, and stone throughout the Great Lakes, she now houses exhibits and displays

Ohio

✼ CanalWay Ohio Scenic Byway

focusing on the heritage of these iron boats. Her forward cargo hold is an exhibit hall and also houses a theater and gift shop. Guided and self-guided tours of the vessels are available (depending upon the time of year); visitors will see the pilot house, crew and guest quarters, galley, guest and officers' dining room, and the four-story engine room. Open June-Aug, daily; May and Sept-Oct, Fri-Sun. **$$**

TEMPLE MUSEUM OF RELIGIOUS ART. *1855 Ansel Rd, Cleveland (44106). Phone 216/831-3233. www.ttti.org/museum.asp.* Jewish ceremonial objects; antiquities of the Holy Land region. Open daily, by appointment; closed Jewish holidays. **FREE**

TOWER CITY CENTER. *1100 Terminal Tower, 50 Public Sq, Cleveland (44113). Phone 216/621-6060. www.towercitycenter.com.* A former railroad station and terminal built in the 1920s. The Avenue, a three-level marble, glass, and brass complex of dining, entertainment, and retail establishments, has an 80-foot-high skylight, a 55-foot glass dome, 26 escalators and elevators, and a 40-foot-long fountain. An underground walkway connects Tower City to the Gateway Complex containing Jacobs Field and Gund Arena. Terminal Tower (circa 1930) was reborn again in the 1990s as the nucleus of Tower City Center. On the 42nd floor of this 52-story building is an observation deck open Sat and Sun. **$**

TROLLEY TOURS OF CLEVELAND. *2000 Sycamore St, Cleveland (44113). Phone 216/771-4484; toll-free 800/848-0173. www.lollytrolley.com.* One- and two-hour tours leave from Powerhouse at Nautica Complex. Advance reservations requested. **$$$$**

TUSCORA PARK. *161 Tuscora Ave, New Philadelphia (44663). Phone 330/343-4644. www.tuscora.park.net.* Swimming (fee); tennis, shuffleboard, picnicking, concession; amusement rides (fee per ride) include a 100-year-old carousel. Open late May-Labor Day, daily. **FREE**

THE UNIVERSITY OF AKRON. *302 Buchtel Mall, Akron (44325). Phone 330/972-7100. www.uakron.edu.* Third largest four-year university in Ohio (24,300 students); known for its Colleges of Polymer Science, Polymer Engineering, Global Business, and Fine and Applied Arts. Its science and engineering program is ranked in the top five nationally. The E. J. Thomas Performing Arts Hall is home to the Ohio Ballet and Akron Symphony. Bierce Library houses many collections, including the Archives of the History of American Psychology. Campus tours arranged through Admissions.

USS COD. *1089 E 9th St Pier, Cleveland (44114). Phone 216/566-8770. www.usscod.org.* World War II submarine credited with seven successful war patrols that sank more than 27,000 tons of Imperial Japanese shipping. Tours include all major compartments of this completely restored Gato-class submarine. Open May-Sept, daily.

WADE PARK. *Euclid Ave and 107th St, Cleveland (44108). Phone 216/621-4110.* More than 80 acres; lake; rose and herb gardens. **FREE**

WARTHER MUSEUM. *331 Karl Ave, Dover (44662). Phone 330/343-7513. www.warthers.com.* Collection of miniature locomotives by master carver Ernest Warther, carved of ebony, pearl, ivory, and walnut; largest model has 10,000 parts; collection of buttons in quilt patterns and arrow points. On the landscaped grounds are a telegraph station, operating hand car, and caboose. Tour of cutlery shop. Open daily Mar-Nov, 9 am-5 pm; Dec-Feb, daily 10 am-4 pm; closed holidays. **$$**

WESTERN RESERVE HISTORICAL SOCIETY MUSEUM AND LIBRARY. *10825 East Blvd, Cleveland (44106). Phone 216/721-5722. www.wrhs.org.* Changing exhibits; special programs; genealogy department; costume collection; American decorative arts. Open daily; closed holidays. **$$**

THE WILDERNESS CENTER. *9877 Alabama Ave SW, Wilmot (44689). Phone 330/359-5235; toll-free 877/359-5235. www.wildernesscenter.org.* Nature center on 573 acres includes six nature trails, 7 1/2-acre lake, 23-foot observation platform; picnicking. Interpretive building open Tues-Sun; closed holidays. Grounds open daily dawn-dusk. **FREE**

WILDWOOD PARK. *16975 Wildwood Dr, Cleveland (44119). Phone 216/881-8141. www.dnr.state.oh.us/parks/parks/clevelkf.htm.* An 80-acre park with fishing, boating (ramps); picnic grounds, playground, concession. **FREE**

THE WINERY AT WOLF CREEK. *2637 S Cleveland-Massillon Rd, Norton (44203). Phone 330/666-9285. www.wineryatwolfcreek.com.* Tasting room overlooks the vineyard and lake. Open Jan-Mar, Thurs noon-9 pm, Fri-Sat noon-11 pm, Sun 1-9 pm; Apr-Dec, Mon-Thurs noon-9 pm, Fri-Sat noon-11 pm, Sun 1-9 pm; closed holidays. **FREE**

ZOAR STATE MEMORIAL. *221 W 3rd St, Zoar (44697). Phone 330/874-3011. www.ohiohistory.org/places/zoar/.* A quaint village where the German religious separatists found refuge from persecution (1817); an experiment in communal living that lasted for 80 years. Number One House on Main St houses the historical museum and has Zoar Society pottery and furniture. Zoar Garden, in the center of the village, follows the description of New Jerusalem in the Bible. Restoration includes the garden house, blacksmith shop, bakery, tin shop, wagon shop, kitchen, magazine shop, and dairy. Open Memorial Day-Labor Day, Wed-Sat 9:30 am-5 pm, Sun and holidays noon-5 pm; Apr-May and after Labor Day-Oct, Sat 9:30 am-5 pm, Sun noon-5 pm. **$$**

PLACES TO STAY

If you choose to include an overnight stay in your trip along this Byway, Mobil Travel Guide recommends the following lodgings.

Akron

★ **COMFORT INN WEST.** *130 Montrose West Ave, Akron (44321). Phone 330/666-5050; toll-free 877/424-6423. www.choicehotels.com.* 132 rooms, 2 story. Complimentary continental breakfast. Check-out 11 am. TV; cable (premium), VCR available. In-room modem link. Laundry services. Health club privileges. Indoor pool. ¢

★★★ **CROWNE PLAZA HOTEL.** *135 S Broadway St, Akron (44308). Phone 330/253-5970; toll-free 800/2-CROWNE. www.crowneplaza.com.* Constructed from 19th-century silos and mills built for the Quaker Oats Company, this historic hotel is a landmark for the business sector of Akron. The guest rooms are round. 196 rooms, 8 story. Check-out noon. TV. Restaurant, bar; entertainment. In-house fitness room. Indoor pool. Airport transportation. Business center. **$**

★ **HOLIDAY INN EXPRESS.** *2940 Chenoweth Rd, Akron (44312). Phone 330/644-7126; toll-free 800/465-4329. www.holiday-inn.com.* 129 rooms, 2 story. Pet accepted; fee. Check-out noon. TV. Restaurant, bar; entertainment. Room service. Pool, poolside service. Airport transportation. ¢

★★ **RADISSON INN.** *200 Montrose West Ave, Akron (44321). Phone 330/666-9300; toll-free 800/333-3333. www.radisson.com.* 130 rooms, 4 story. Check-out noon. TV. In-room modem link. Restaurant, bar. Room service. In-house fitness room, sauna. Indoor pool. Airport transportation. **$**

Ohio

✳ *CanalWay Ohio Scenic Byway*

Cleveland

★★★ **BARICELLI INN.** *2203 Cornell Rd, Cleveland (44106). Phone 216/791-6500. www.baricelli.com.* This charming inn is known for its wonderful restaurant. 7 rooms, 3 story. Complimentary continental breakfast. Check-out 11 am, check-in 2 pm. TV. Dining room. Brownstone (1896) with individually decorated rooms. $$

★★★ **EMBASSY SUITES.** *1701 E 12th St, Cleveland (44114). Phone 216/523-8000; toll-free 800/EMBASSY. www.embassysuites.com.* Perched amidst Cleveland's thriving downtown district, this elegantly appointed hotel welcomes guests with a friendly and attentive staff. Located two blocks away is the Galleria and Playhouse Square, and just seven blocks away is the Rock and Roll Hall of Fame and Museum. 268 rooms, 13 story. Check-out noon. TV. In-room modem link. Laundry services. Restaurant, bar. In-house fitness room, sauna. Indoor pool. Outdoor tennis, lighted courts. Valet parking available. Concierge. $

★★★ **HYATT REGENCY AT THE ARCADE.** *420 Superior, Cleveland (44114). Phone 216/575-1234; toll-free 800/633-7313. www.hyatt.com.* 293 rooms, 9 story. Check-out noon. TV; cable (premium), VCR available. Restaurant. In-house fitness room. Business center. $$

★★★ **MARRIOTT AT KEY CENTER.** *127 Public Sq, Cleveland (44114). Phone 216/696-9200; toll-free 800/228-9290. www.marriott.com.* This hotel is located near the Cleveland Browns' stadium, Tower City Mall, tennis facilities, and several golf courses. 400 rooms, 25 story. Check-out noon. TV; cable (premium), VCR available. In-room modem link. Laundry services. Restaurant, bar. In-house fitness room, sauna. Indoor pool, whirlpool. Business center. Concierge. Luxury level. $$

★★★ **RENAISSANCE CLEVELAND HOTEL.** *24 Public Sq, Cleveland (44113). Phone 216/696-5600; toll-free 800/696-6898. www.renaissancehotels.com.* Located on Public Square downtown, this beautifully designed hotel is near great shopping and sightseeing areas. It has spacious guest suites, 62,000 square feet of meeting space, and the award-winning Mediterranean restaurant Sans Souci. 491 rooms, 14 story. Check-out noon. TV; cable (premium), VCR available. In-room modem link. Restaurant, bar; entertainment. In-house fitness room, sauna. Health club privileges. Indoor pool. Garage parking. Business center. Concierge. Luxury level. $$

★★★ **SHERATON CLEVELAND CITY CENTRE.** *777 NE St. Clair Ave, Cleveland (44114). Phone 216/771-7600; toll-free 800/21-1090. www.sheraton.com.* This hotel is located downtown overlooking Lake Erie, adjacent to the Galleria shopping and near Tower City, Municipal Stadium, and other attractions. 470 rooms, 22 story. Check-out noon. TV; cable (premium). In-room modem link. Restaurant, bar. In-house fitness room. Health club privileges. Airport transportation. Business center. Luxury level. $

★★★★ **THE RITZ-CARLTON CLEVELAND.** *1515 W Third St, Cleveland (44113). Phone 216/623-1300; toll-free 800/241-3333. www.ritzcarlton.com.* The Ritz-Carlton is Cleveland's premier destination. This hotel enjoys a coveted downtown location with views of Lake Erie and the Cuyahoga River, and all of the city's attractions and businesses are within walking distance. Visitors are hosted in grand style here, where thoughtful service grants every wish. The guest rooms, with commanding city and water views, are luxury defined to the last detail. The marble bathrooms are sumptuous, and guests are graciously provided with every imaginable service. All-day dining at the restaurant and bar is always a delight, its clean, modern design providing a soothing alternative to urban life. Seafood is a specialty here, and the sushi bar is an ever-popular choice. 208 rooms, 7 story. Pet accepted, some restrictions; fee. Check-out noon. TV; VCR available. In-room modem link. Restaurant, bar. Room service 24 hours. Afternoon tea. In-house fitness room, massage, sauna. Health club privileges. Indoor pool, whirlpool, poolside service. Business center. Concierge. Luxury level. $$

★★ **WYNDHAM PLAYHOUSE SQUARE HOTEL.** *1260 Euclid Ave, Cleveland (44115). Phone 216/ 615-7500; toll-free 800/996-3426. www. wyndham.com.* 205 rooms, 14 story. Check-out noon. TV. Restaurant, bar. In-house fitness room, sauna. Health club privileges. Indoor pool, whirlpool. Valet parking available. $$

New Philadelphia

★★★ **ATWOOD.** *2650 Lodge Rd, New Philadelphia (46620). Phone 330/735-2211.* This hilltop resort overlooking the lake provides a unique retreat, with each room offering a view of either the lake or the lovely countryside. Boat rentals are available through the hotel, and harbor cruises are available through the nearby marina. 104 rooms, 2 story. Check-out 1 pm, check-in 4 pm. TV. In-room modem link. Dining room, bar. Room service. Free supervised children's activities (mid-June-Sept). In-house fitness room, sauna. Game room. Indoor, outdoor pools; whirlpool. 9-hole par-3 golf, 18-hole par-70 golf, pro, driving range, putting green. Outdoor tennis, lighted courts, pro. Cross-country ski on site; sledding. Lawn games. Bicycle rentals. Motorboats, sailboats, canoes, rowboats. Airport transportation. Private airstrip, heliport. A Muskingum Watershed Conservancy District facility. $

★★ **HOLIDAY INN.** *131 Bluebell Dr SW, New Philadelphia (44663). Phone 330/339-7731; toll-free 800/465-4329. www.holiday-inn.com.* 107 rooms, 2 story. Check-out 11 am. TV; VCR available. Restaurant, bar. In-house fitness room, sauna. Indoor, outdoor pools; whirlpool. $

★ **TRAVELODGE.** *1256 W High Ave, New Philadelphia (44663). Phone 330/339-6671; toll-free 888/515-6375. www.travelodge.com.* 62 rooms, 2 story. Check-out noon. TV. Pool. ¢

PLACES TO EAT

A long day of driving is sure to make you hungry. At the end of your journey, take a table at one of the following restaurants.

Akron

★★★ **LANNING'S.** *826 N Cleveland-Massillon Rd, Akron (44333). Phone 330/666-1159. www.lannings-restaurant.com.* On the banks of Yellow Creek, this fine dining room, offering fresh fish and hand-cut steaks, has been in business for over 25 years. Everything is made in-house, including all dressings, sauces, soups, breads, and desserts. Seafood, steak menu. Closed Sun; major holidays. Dinner. Bar. Jacket required. Valet parking available. $$$

★★★ **TANGIER.** *532 W Market St, Akron (44303). Phone 330/376-7171.* This local gem hosts visiting dignitaries, as well as the best in contemporary music. Noted as one of Ohio's most unique locations, it features wonderful music by some of the top jazz and light rock artists, as well as an eclectic spin on Middle Eastern cuisine. Accommodates large parties and catered events. Mediterranean menu. Closed Sun; most major holidays. Lunch, dinner. Bar; entertainment. Children's menu. $$

Ohio

❋ *CanalWay Ohio Scenic Byway*

★ **THE GROTTO ON MERRIMAN.** *1841 Merriman Rd, Akron (44313). Phone 330/869-4981.* Continental menu. Dinner. Bar. Children's menu. Casual attire. Outdoor seating. $$
D

Cleveland

★★★ **BARICELLI INN.** *2203 Cornell Rd, Cleveland (44106). Phone 216/791-6500. www.baricelli.com.* Adjacent to University Circle, this inn and restaurant are perched on a bluff in a large, turn-of-the-century greystone. The Little Italy location has romantic, old-world charm, and the seasonal menu features thoughtful preparations of local ingredients. Continental menu. Closed Sun; most major holidays. Dinner. Outdoor seating. $$
D

★ **CLEVELAND GRILL.** *3359 W 117th St, Cleveland (44111). Phone 216/251-1025. www.clevelandgrill.com.* Mediterranean menu. Closed major holidays. Breakfast, lunch, dinner. Children's menu. Casual attire. $$
D

★★ **DON'S LIGHTHOUSE GRILLE.** *8905 Lake Ave, Cleveland (44102). Phone 216/961-6700. www.donslighthouse.com.* Continental menu. Closed Thanksgiving, Dec 25. Lunch, dinner. Bar. Valet parking available. $$
D

★ **GREAT LAKES BREWING CO.** *2516 Market Ave, Cleveland (44113). Phone 216/771-4404. www.greatlakesbrewing.com.* Seafood menu. Closed major holidays. Lunch, dinner. Bar. Microbrewery. Located in a historic 1860s brewery; turn-of-the-century pub atmosphere. Children's menu. Outdoor seating. $$
D

★★★ **JOHNNY'S BAR.** *3164 Fulton Rd, Cleveland (44109). Phone 216/281-0055. www.johnnyscleveland.com.* Italian menu. Closed Sun; most major holidays. Lunch, dinner. Bar. Former neighborhood grocery. Reservations required Fri-Sat. $$$
D

★ **KARL'S INN OF THE BARRISTER'S.** *1264 W 3rd St, Cleveland (44113). Phone 216/241-4141.* American, deli menu. Closed major holidays. Breakfast, lunch, dinner. Bar. Casual attire. $
D

★ **LE PETIT BISTRO.** *230 W Hurn Rd, Cleveland (44113). Phone 216/574-9660.* French menu. Closed major holidays. Lunch, dinner. Children's menu. Casual attire. $$
D

★★ **LION & LAMB.** *30519 Pine Tree Rd, Cleveland (44124). Phone 216/831-1213.* American menu. Closed Sun; major holidays. Lunch, dinner. Bar; entertainment. $$
D

★★★ **MORTON'S OF CHICAGO.** *1600 W 2nd, Cleveland (44113). Phone 216/621-6200. www.mortons.com.* The epitome of an all-American steakhouse. The food is always plentiful and of the highest quality. This chain promises fine dining with attentive service. Steak menu. Closed most major holidays. Lunch, dinner. Bar. $$$
D

★★★ **PARKER'S.** *2801 Bridge Ave, Cleveland (44113). Phone 216/771-7130. www.savvydiner.com.* Dining at this restaurant on the near west side is a dress-up occasion. Chef/owner Parker Bosley's seasonal menu is a testament to his avid support of local organic produce and farm-fed meats. American, French menu. Closed Sun; major holidays. Dinner. Bar. Totally nonsmoking. $$$
D

★★★ **RIVERVIEW ROOM.** *1515 W Third St, Cleveland (44113). Phone 216/623-1300. www.ritzcarlton.com.* Located in an elegant dining room on the sixth floor of the Ritz-Carlton. The riverfront and views of the city are a perfect backdrop for this extraordinary dining experience. Contemporary American menu. Breakfast, lunch, dinner, Sun brunch. Bar. Children's menu. Reservations required Sun brunch. Valet parking available. $$$
D

★★★ **SANS SOUCI.** *24 Public Sq, Cleveland (44113). Phone 216/696-5600. www.renaissancehotels.com.* Fine cuisine is served in this comfortable dining room of exposed beams, lush greens, and a stone hearth. The Renaissance Hotel space is sectioned into intimate rooms where diners feast on the chef's specialties. Mediterranean menu. Closed most major holidays. Lunch, dinner. Bar. $$$
D

★ **SWEETWATER'S CAFE SAUSALITO.** *1301 E 9th St, Cleveland (44114). Phone 216/696-2233. www.savvydiner.com.* Seafood menu. Closed Sun; Thanksgiving, Dec 25. Lunch, dinner. Pianist Fri-Sat (dinner). Valet parking available. $$
D

★ **THAT PLACE ON BELLFLOWER.** *11401 Bellflower Rd, Cleveland (44106). Phone 216/231-4469. www.thatplaceonbellflower.com.* Closed major holidays. Lunch, dinner. Bar. Children's menu. Converted turn-of-the-century carriage house. Valet parking available. Outdoor seating. $$$

★ **WATERMARK.** *1250 Old River Rd, Cleveland (44113). Phone 216/241-1600. www.watermark-flats.com.* Seafood menu. Buffet, lunch, dinner, Sun brunch. Bar. Children's menu. Former ship provision warehouse on the Cuyahoga River. Free valet parking. Outdoor seating. River cruises available. $$
D

Fairlawn

★ **HOUSE OF HUNAN.** *2717 W Market St, Fairlawn (44333). Phone 330/864-8215.* Chinese menu. Lunch, dinner. Bar. $$
D

Munroe Falls

★★ **TRIPLE CROWN.** *335 S Main, Munroe Falls (44262). Phone 330/633-5325. www.triplecrownrestaurant.com.* Steak menu. Closed major holidays. Lunch, dinner, Sun brunch. Bar. Children's menu. Horse racing memorabilia. $$
D

The Historic National Road

✳ OHIO AN ALL-AMERICAN ROAD
Part of a multistate Byway; see also IL, IN.

Quick Facts

LENGTH: 228 miles.

TIME TO ALLOW: 4 days.

BEST TIME TO DRIVE: Summer and fall, depending on which end (east or west) you will be spending most of your time. Summer is high season for festivals in Ohio. On the western section of the Byway, you find more agriculture, so summer and late summer have the most picturesque fields. The eastern section of the Byway offers hills and fall colors.

BYWAY TRAVEL INFORMATION: Ohio Historic Preservation Office: 614/298-2000.

SPECIAL CONSIDERATIONS: Excessive and heavy snow or rain falls occasionally in this part of the country. It may impair drivability but does not result in long-lived road closures.

BICYCLE/PEDESTRIAN FACILITIES: In rural areas, on-road bicycling is feasible because of relatively low traffic volumes and generally wide shoulders. Similarly, small towns accommodate both pedestrians and bicyclists because of low traffic volumes, slow traffic speeds, and sidewalks. The large urban areas along the Byway all provide pedestrian sidewalks, and some have dedicated bicycle lanes.

The story of the Ohio Historic National Road is the story of our nation's aspirations and desires. The National Road literally paved the way west through the newly formed states of Ohio, Indiana, and Illinois and provided a direct connection to the mercantile and political centers of the East Coast, helping to secure the influence and viability of these new settlements. As much as the road's boom times during the early and mid-19th century signified its importance to national commerce and expansion, its decline during the late 19th and early 20th centuries reveal the meteoric rise of the railroad as the primary means of transport and trade across the nation. Likewise, the renaissance of the National Road in the mid-20th century reflects the growing popularity of the automobile.

Today, the Byway is a scenic journey across Ohio. The steep, wooded hills and valleys of the eastern edge of the Byway give way to the gently rolling farmland of the western part of the Byway. Picturesque farms, hiking trails, craft industries, and historic sites and museums await you along this portion of the Historic National Road.

THE BYWAY STORY

The National Historic Road tells archaeological, cultural, historical, natural, recreational, and scenic stories that make it a unique and treasured Byway.

Archaeological

Prehistoric civilizations once dominated the land surrounding what is now known as the Historic National Road in Ohio, and today, remnants of these early people can be found

Ohio

The Historic National Road

just off of the Byway. One important aspect for these early cultures was making tools and weapons. They found flint for these tools at what is known today as Flint Ridge State Memorial. The Hopewell Culture frequented Flint Ridge because of the quality and beauty of the flint found there. This flint would have been a very important resource to that culture, the way that coal and iron ore are in Ohio today. Flint from Ohio has been found from the Atlantic seaboard to Louisiana. Flint was so important, in fact, that it has become the state gem of Ohio.

Located near Newark, the Moundbuilders State Memorial and Ohio Indian Art Museum tell about the Hopewell Indian civilization, which is most remembered for the large earthworks that were constructed there. Exhibits show the artistic achievements of these prehistoric cultures that lived in the area from around 10,000 BC to AD 1600. The Octagon Earthworks and Wright Earthworks are also located near Newark.

Cultural

The Historic National Road in Ohio hosts a rich tradition of culture. Museums, festivals, and other cultural facilities offer a chance to both explore more of the National Road's history and seek diversions from it. Outstanding performing arts venues are well represented along the Byway. In addition to these cultural features, the eastern section of the Byway is known for its selection of traditional and modern crafts.

Historical

The history of the Historic National Road in Ohio highlights the importance of this road in terms of development and settlement that it brought to the Ohio area. The history of the construction of the National Road is significant because it serves as an example of larger events that were transpiring in America simultaneously. Early pioneer settlement gave way to railroads, and finally the automobile became the most frequent traveler along the National Road.

Natural

While traveling the Historic National Road in Ohio, you are treated to a diverse Byway that traverses steep wooded hills and valleys in the east and gently rolling farmland in the west. This diversity of natural features offers refreshing and contrasting views along the Byway. The changing landscape determined how the National Road was constructed, as well as the types of livelihoods settlers engaged in throughout the history of the road.

Recreational

A tremendous network of large state parks, regional metropolitan parks, local parks, and privately run facilities provide a bountiful array of outdoor recreational facilities that give you a chance to enjoy the area's natural beauty. For the most part, the state parks are located on the eastern half of the Byway, and over 50,000 acres of state parks, forests, and wildlife areas are easily accessible from the Ohio Historic National Road. Along portions of the Byway, you can hike, camp, fish, hunt, and picnic while relishing the various species of local flora and fauna.

Scenic

The Historic National Road in Ohio is full of variation and diversity. Changing topography and landscapes provide scenic views for travelers of the Byway, from hilly ridges to long, unbroken views of the horizon. Small towns, unique and historic architecture, and stone bridges all provide scenes from the past along the Byway. Farms, unspoiled scenery, and large urban landscapes give you a sense of the great diversity of this Byway.

see page A6 for color map

THINGS TO SEE AND DO

Driving along the Historic National Road will certainly keep your senses engaged, but if you yearn to get out of the car and stretch your legs, or if you'd like to make a mini-vacation out of your trip, check out these attractions along the route.

BUCK CREEK STATE PARK. *1901 Buck Creek Ln, Springfield (45502). Phone 937/322-5284 or 937/325-2411. www.dnr.state.oh.us/parks/parks/ buckck.htm.* A 4,030-acre park with swimming, lifeguard, bathhouse (open Memorial Day-Labor Day), fishing, boating (launch, ramp), hiking, snowmobile trails, picnicking, concession, camping, and cabins. Standard fees. Open daily.

COSI COLUMBUS, OHIO'S CENTER OF SCIENCE AND INDUSTRY. *333 W Broad St, Columbus (43215). Phone 614/228-2674; toll-free 888/819-COSI. www.cosi.org.* Hands-on museum includes exhibits, programs, and demonstrations. Battelle Planetarium shows (open daily). Coal Mine, Hi-Tech Showcase, Free Enterprise Area, Foucault pendulum, Solar Front Exhibit Area, Computer Experience, Street of Yesteryear, Weather Station, KIDSPACE, and FAMILIESPACE. Open Mon-Sat 10 am-5 pm, Sun noon-6 pm; closed major holidays. **$$$**

DAVID CRABILL HOUSE. *818 N Fountain Ave, Springfield (45504). Phone 937/399-1245 or 937/324-0657.* Built by Clark County pioneer David Crabill in 1826; restored. Period rooms, log barn, smokehouse. Maintained by the Clark County Historical Society. Open Tues-Fri. **$**

DILLON STATE PARK. *5265 Dillon Hills Dr, Zanesville (43830). Phone 740/453-4377. www.dnr.state.oh.us/parks/parks/dillon.htm.* A 7,690-acre park with swimming, boating (ramps, rentals), picnicking (shelter), concession, camping, and cabins (by reservation). Open daily. **FREE**

FLINT RIDGE STATE MEMORIAL. *State Rte 668, Newark (43799). 4 miles N of I-70, 3 miles N of Brownsville, in Licking County. Exit eastbound I-70 at State Route 668 and westbound I-70 at exit 142. Phone 740/787-2476. www.ohiohistory.org/places/flint/.* Prehistoric Native American flint quarry; trails for the disabled and the visually impaired; picnic area. Museum open Memorial Day-Labor Day, Wed-Sat 9:30 am-5 pm; Sun, holidays noon-5 pm; after Labor Day-Oct, Sat 9:30 am-5 pm, Sun noon-5 pm. Park open Apr-Oct. **$**

FORT HILL STATE MEMORIAL. *Township Rd 256, Hillsboro (45133). Located off of State Rte 41, 5 miles N of Sinking Springs and 3 miles S of Cynthiana in Highland County. Phone 937/588-3221; toll-free 800/283-8905. www.ohiohistory.org/places/fthill/.* This is the site of a prehistoric Native American hilltop earth and stone enclosure. The identity of its builders has not been determined, but implements found in the vicinity point to the Hopewell people. A 2,000-foot trail leads to the ancient earthworks. Picnic area and shelter house. Open Memorial Day-Labor Day, Wed-Sat 9:30 am-5 pm; Sun, holidays noon-5 pm; after Labor Day-Oct, Sat 9:30 am-5 pm, Sun noon-5 pm. Park open Apr-Oct. **FREE**

GERMAN VILLAGE. *588 S 3rd St, Columbus (43215). Phone 614/221-8888. www.germanvillage .org.* Historic district restored as an old-world village with shops, old homes, and gardens; authentic foods. Bus tour available. Open daily. **FREE**

MOUNDBUILDERS STATE MEMORIAL. *99 Cooper Ave, Newark (43055). Phone 740/344-1920; toll-free 800/600-7174.* The Great Circle, 66 acres, has walls from 8 to 14 feet high with burial mounds in the center. Museum containing Hopewell artifacts is open Memorial Day-Labor Day, Wed-Sat 9:30 am- 5 pm; after Labor Day-Oct, Sat 9:30 am-5 pm, Sun noon-5 pm. Park open Apr-Oct. Museum **$**. Park **FREE**.

Ohio

The Historic National Road

NATIONAL ROAD-ZANE GREY MUSEUM. *8850 E Pike, Norwich (43767). Phone 740/872-3143. www.ohiohistory.org/places/natlroad/.* A 136-foot diorama traces the history of the National Road from Maryland to Illinois; display of vehicles that once traveled the road; Zane Grey memorabilia, reconstructed craft shops, and an antique art pottery exhibit. Open Mar-Apr and Oct-Nov, Wed-Sat 9:30 am- 5 pm, Sun noon-5 pm; May-Sept, Mon-Sat 9:30 am-5 pm, Sun and holidays noon-5 pm. **$$**

NEWARK EARTHWORKS. *S 21st St and OH 79, Newark (43055). Phone 740/872-3143; toll-free 800/600-7174. www.ohiohistory.org/places/newarkearthworks/.* The group of earthworks here was originally one of the most extensive of its kind in the country, covering an area of more than 4 square miles. The Hopewell used their geometric enclosures for social, religious, and ceremonial purposes. Remaining portions of the Newark group are Octagon Earthworks and Wright Earthworks, with many artifacts of pottery, beadwork, copper, bone, and shell exhibited at the nearby Moundbuilders State Museum.

OCTAGON EARTHWORKS. *33rd St and Parkview Rd, Newark (43055). Phone 740/344-1919; toll-free 800/600-7174. www.ohiohistory.org/places/newarkearthworks/octagon.cfm.* The octagon-shaped enclosure encircles 50 acres that includes small mounds and is joined by parallel walls to a circular embankment enclosing 20 acres. **FREE**

OHIO CERAMIC CENTER. *7327 Ceramic Rd, Roseville (43777). Phone 740/697-7021; toll-free 800/752-2604. www.ohiohistory.org/places/ohceram/.* Extensive displays of pottery are housed in five buildings. Exhibits include primitive stoneware and area pottery. Demonstrations of pottery-making during pottery festival (weekend in mid-July). Open mid-May-mid-Oct, Wed-Sat 9:30 am-5 pm, Sun noon-5 pm. **$**

OHIO HISTORICAL CENTER. *1982 Velma Ave, Columbus (43211). Phone 614/297-2300. www.ohiohistory.com.* Modern architectural design contrasts with the age-old themes of Ohio's prehistoric culture, natural history, and cultural history. Exhibits include an archaeology mall with computer interactive displays and life-size dioramas; a natural history mall with a mastodon skeleton and a demonstration laboratory; and a history mall with transportation, communication, and lifestyle exhibits. Ohio archives and historical library open Wed 9 am-5 pm, Thurs 1-9 pm, Sat 9 am-5 pm. Museum open Tues-Wed, Fri-Sat 9 am-5 pm, Thurs 9 am-9 pm, Sun and holidays noon-5 pm (hours vary in Dec); closed Jan 1, Thanksgiving, Dec 25. **$$**

OHIO VILLAGE. *17th Ave (exit 111), Columbus (43211). Phone 614/297-2300 (museum). www.ohiohistory.org/places/ohvillag/.* Reconstruction of a rural, 1860s Ohio community with one-room schoolhouse, town hall, general store, hotel, farmhouse, barn, doctor's house, and office. Costumed guides. Open to the public for Signature Events only. **$$**

PARK OF ROSES. *Acton and High sts, Columbus (43214).* This park contains over 10,000 rose bushes representing 350 varieties. Picnic facilities. Rose festival (early June). Musical programs Sun evenings in summer. Open daily. **FREE**

PENNSYLVANIA HOUSE. *1311 W Main St, Springfield (45504). Phone 937/322-7668.* Built as a tavern and stagecoach stop on the National Pike, this house includes period furnishings, pioneer artifacts, and button, quilt, and doll collections. Open the first Sun afternoon of each month; also by appointment; closed Easter and late Dec-Feb. **$$**

SANTA MARIA REPLICA. *Battelle Riverfront Park, Marconi Blvd and W Broad St, Columbus (43215). Phone 614/645-8760. www.santamaria.org.* A full-scale, museum-quality replica of Christopher Columbus's flagship, the *Santa Maria*. Costumed guides offer tours of the upper and lower decks. Visitors learn of life as a sailor on voyages in the late 1400s. Tours Apr-late May Wed-Fri 10 am-3 pm, Sat-Sun noon-5 pm; late May-Aug, Wed-Fri 10 am-5 pm, Sat-Sun noon- 6pm; Sept-late Oct, Wed-Fri 10 am-3 pm, Sat-Sun noon-5 pm; some holidays noon-5 pm. $

SERPENT MOUND STATE MEMORIAL. *State Rte 73, Pebbles (45660). Phone 937/587-2796; toll-free 800/752-2757. www.ohiohistory.org/places/serpent/.* The largest and most remarkable serpent effigy earthworks in North America. Built between 800 BC and AD 100 of stone and yellow clay, it curls like an enormous snake for 1,335 feet. An oval earth wall represents the serpent's open mouth. In the 61-acre area are an observation tower, a scenic gorge, a museum, and picnicking facilities. Site (open all year); museum (open Apr-Oct, daily 10 am-5 pm; closed Jan 1, Thanksgiving, Dec 25). **FREE**

THE WILDS. *14000 International Rd, Cumberland (43732). Phone 740/638-5030. www.thewilds.org.* A 9,154-acre conservation center dedicated to increasing the population of endangered species. Open May-Oct, Sat-Sun; June-Aug, Wed-Sun. $$$

ZANESVILLE ART CENTER. *620 Military Rd, Zanesville (43701). Phone 740/452-0741. www.zanesvilleartcenter.org.* American, European, and Asian art; children's art; early Midwestern glass and ceramics; photographs; special programs; and gallery tours. Open Tues-Wed, Fri 10 am-5 pm; Thurs 10 am-8 pm; Sat-Sun 1-5 pm; closed holidays. **FREE**

PLACES TO STAY

If you choose to include an overnight stay in your trip along this All-American Road, Mobil Travel Guide recommends the following lodgings.

Columbus

★★ **ADAM'S MARK.** *50 N 3rd St, Columbus (43215). Phone 614/228-5050; toll-free 800/444-ADAM. www.adamsmark.com.* 415 rooms, 21 story. TV; cable (premium). Check-out noon. Heated pool; whirlpool. Restaurant, bar. Room service. Convention facilities. Business services available. In-room modem link. Coin laundry. Exercise equipment; sauna. Valet parking $15. $

★ **BEST WESTERN SUITES.** *1133 Evans Way Ct, Columbus (43228). Phone 614/870-2378; toll-free 800/780-7234. www.bestwestern.com.* 66 suites, 2 story. Complimentary continental breakfast. Check-out noon. TV; cable (premium), VCR (movies). In-room modem link. Laundry services. Exercise equipment, sauna. Indoor pool; whirlpool. ¢

Ohio

The Historic National Road

Springfield

★ **FAIRFIELD INN BY MARRIOTT.** *1870 W 1st St, Springfield (45504). Phone 937/323-9554; toll-free 800/228-2800. www.fairfieldinn.com.* 124 rooms, 6 story. Check-out noon. TV; cable (premium), VCR available. In-room modem link. Restaurant, bar. Luxury level. ¢

★ **SPRINGFIELD INN.** *100 S Fountain Ave, Springfield (45502). Phone 937/322-3600.* 63 rooms, 3 story. Complimentary continental breakfast. Check-out noon. TV; cable (premium), VCR available. Some refrigerators. Microwave in suites. Health club privileges. Game room. Indoor pool; whirlpool. Business services. ¢

St. Clairsville

★ **KNIGHTS INN.** *51260 E National Rd, St. Clairsville (43950). Phone 740/695-5038; toll-free 800/843-5644. www.knightsinn.com.* 104 rooms. Pet accepted, some restrictions. Check-out noon. TV; cable (premium), VCR available (movies). Pool. Business services available. In-room modem link. Some in-room whirlpools; microwaves available. ¢

Vandalia

★ **SUPER 8.** *550 E National Rd, Vandalia (45377). Phone 937/898-7636; toll-free 800/800-8000. www.super8.com.* 94 rooms, 3 story. Check-out noon. TV; cable (premium). Pool. ¢

Worthington

★★★ **WORTHINGTON INN.** *649 High St, Worthington (43085). Phone 614/885-2600. www.worthingtoninn.com.* 26 rooms, 3 story. Complimentary full breakfast. Check-out noon, check-in 3 pm. TV. In-room modem link. Dining room, bar. Room service. Concierge. Renovated Victorian inn. $$

Zanesville

★ **AMERIHOST INN.** *230 Scenic Crest Dr, Zanesville (43701). Phone 740/454-9332; toll-free 800/434-5800. www.amerihostinn.com.* 60 rooms, 2 story. Complimentary continental breakfast. Check-out noon. TV; cable (premium). Restaurant. Business services available. In-room modem link. In-house fitness room; sauna. Indoor pool; whirlpool. In-room whirlpool, refrigerator, microwave in suites. ¢

★★ **HOLIDAY INN.** *4645 E Pike, Zanesville (43701). Phone 740/453-0771; toll-free 800/465-4329. www.holiday-inn.com.* 130 rooms, 2 story. Pet accepted. Check-out noon. TV; cable (premium). Restaurant, bars. Room service. Coin laundry. Business services available. In-room modem link. Sundries. In-house fitness room; sauna. Indoor pool; whirlpool, poolside service. Playground. Refrigerators, microwaves available. ¢

PLACES TO EAT

A long day of driving is sure to make you hungry. At the end of your journey, take a table at one of the following restaurants.

Cambridge

★ **BEARS DEN.** *13320 E Pike, Cambridge (43725). Phone 740/432-5285.* American menu. Closed Sun. Dinner. Elegant rotisserie home cooking in a rustic setting. $

★ **PLAZA RESTAURANT.** *1038 Wheeling Ave, Cambridge (43725). Phone 740/432-7997. www.plazarestaurant.com.* Indian menu. Closed Sun. Dinner. Serves homemade curries, pastries, and garam masala. $

Columbus

★★ **BEXLEY'S MONK.** *2232 E Main St, Columbus (43209). Phone 614/239-6665. www.bexleymonk.com.* Eclectic, seafood menu.

Closed major holidays. Lunch (Mon-Sat), dinner. Bar. Entertainment. Reservations accepted. $$$
[D]

★ **KATZINGER'S.** *475 S 3rd St, Columbus (43215). Phone 614/228-3354. www.katzingers.com.* Continental menu. Closed Easter, Thanksgiving, Dec 25. Breakfast, lunch, dinner. Children's menu. Outdoor seating. Totally nonsmoking. $$
[D]

★ **OLD MOHAWK.** *821 Mohawk St, Columbus (43206). Phone 614/444-7204. www.oldmohawk.com.* Closed some major holidays. Lunch, dinner. Bar. Building from the 1800s was once a grocery and tavern; exposed brick walls. $

★★★ **MORTON'S OF CHICAGO.** *2 Nationwide Plaza, Columbus (43215). Phone 614/464-4442. www.mortons.com.* For fresh lobster and steaks, put Morton's on your list. Professional service and a clubby setting offer a truly satisfying dining experience. Specializes in prime aged beef and fresh seafood. Closed major holidays. Dinner. Bar. Reservations accepted. Valet parking available. $$$
[D]

Springfield

★★ **CASEY'S.** *2205 Park Rd, Springfield (45504). Phone 937/322-0397.* American menu. Closed Sun, major holidays. Dinner. Bar. $$
[D]

★★ **KLOSTERMAN'S DERR ROAD INN.** *4343 Derr Rd, Springfield (45503). Phone 937/399-0822.* Continental menu. Closed most holidays. Lunch, dinner. Bar. Children's menu. $$
[D]

Zanesville

★★ **MARIA ADORNETTO.** *953 Market St, Zanesville (43702). Phone 740/453-0643.* Italian menu. Closed Sun, major holidays. Lunch, dinner. Bar. Contemporary dining room in a converted home. $$

★★ **OLD MARKET HOUSE INN.** *424 Market St, Zanesville (43701). Phone 740/454-2555. www.oldmarkethouseinn.com.* Italian, American menu. Closed Sun, major holidays. Dinner. Bar. $$
[D]

Ohio River Scenic Byway

❋ OHIO

Part of a multistate Byway; see also IL, IN.

Quick Facts

LENGTH: 452 miles.

TIME TO ALLOW: 2 days.

BEST TIME TO DRIVE: Spring and summer.

BYWAY TRAVEL INFORMATION: Ohio River Trails, Inc.: 513/553-1500.

SPECIAL CONSIDERATIONS: Storms seldom force road closings on rural state and federal highways like those utilized by the Ohio River Scenic Byway. However, delays are possible in heavily traveled areas around towns and cities during such storms. Also, although flooding along the Ohio River is a rare event, it is most likely to occur between November and April, and the Byway route could be affected due to its proximity to the river. In addition, short sections along the route may experience annual or bi-annual flooding, including the portion of US 52 east of Cincinnati known as Eastern Avenue. The Ohio Department of Transportation works with county managers to mitigate the effects of floods and other events that force road closings by temporarily redirecting traffic to alternate routes. Snowstorms are possible, particularly from mid-December through mid-March, and can cause road closures.

BICYCLE/PEDESTRIAN FACILITIES: The Ohio Department of Transportation keeps the Byway in excellent condition. Where the Byway passes through towns and cities, sidewalks accommodate pedestrian use. Some towns, such as Marietta, have made provisions for bicyclists as well. In addition, a portion of the American Discovery Trails System, which runs from California to Maryland, parallels the Ohio River Scenic Byway in western Hamilton County.

This Byway passes through 14 counties and encompasses the birthplaces of three US presidents. The Byway also includes the grand and historic US 52, known as the Atlantic and Pacific Highway and the US Grant Memorial Highway.

This scenic route follows the Ohio River, so you can enjoy majestic river views for the entire drive. The Ohio River is an historical American icon. Native Americans and early European settlers used it heavily, especially to access the West. It marked the boundary between the North and South during the slavery era and was later the gateway to freedom for many slaves. The river was also the means of progress for industrialists and merchants.

THE BYWAY STORY

The Ohio River Scenic Byway tells archaeological, cultural, historical, natural, recreational, and scenic stories that make it a unique and treasured Byway.

Archaeological

Several archaeological sites and museums are found along the Ohio River Scenic Byway. These include several Native American cultural sites in Tiltonsville, as well as the Mound Cemetery Chester.

The town of Portsmouth, which hosts the Southern Ohio Museum and Cultural Center, is where many travelers learn about the history of this area. There are also Native American cultural sites near here.

Ohio

✻ *Ohio River Scenic Byway*

Cultural

The Ohio River has been a center of human activity for centuries and, therefore, is a cultural hotbed. Academic and historical organizations recognize this fact and have made the river and the immediate area (the Byway's territory) the focus of their cultural studies. One such organization is the University of Kentucky's Ohio River Valley Series. This intercollegiate project specifically studies how the state's history (which is inextricable from culture) has been affected by the social use of the Ohio River.

Numerous grassroots organizations also investigate the river's role in the development of their towns and cities. One such group is the Ohio Historical Society's Ohio River Museum, which specifically examines how artifacts and technological improvements associated with the river affected area culture. Another of these organizations is Ohio River Trails, Inc., which focuses on the shoreline and scenic route in relation to area culture.

Historical

The Ohio River is most significant for its historical importance to the region. Described both as a river of beauty (the Iroquois named the river Oyo, which the French interpreted as *la belle riviere*) and a river of opportunity, the Ohio River symbolized a new future. Its history is a powerful reflection of the American experience; one of the valleys surrounding it was even called the Valley of Democracy.

The region's rich history includes the migration of people along the river into the four states that border it (Kentucky, Ohio, Indiana, and Illinois), as well as the three states that are located within the river's basin (Tennessee, Mississippi, and Alabama). The industry and technology of the area were important to the Ohio River as well. They changed the river's influence on the region through the development of steamboat travel and shipping, the creation of locks and dams to control navigation and flooding, the establishment of industries such as steel and coal plants and their associated landing docks, and the development of large chemical and electrical power generating facilities.

Natural

This Byway contains many natural features that surround both the Ohio River and the forests along the Byway. Many native animals and plants add to the scenic quality of the Byway. Natural sites along the Byway include Fernwood State Forest, located near the Mingo Junction on State Route 151; the Quaker Meeting House, located near Mount Pleasant on State Route 647; and Barkcamp State Park. This park has historical significance and is located by a lake just off the Byway turning from Bellaire on State Route 149 (65330 Barkcamp Park Road, Belmont, phone 614/484-4064). Campsites are available.

Other natural sites include Sunfish Creek State Forest, located past Powhatan Point on State Route 7, and Wayne National Forest, located right along the Byway, starting past the city of Hannibal covering the area to Marietta. Two rustic covered bridges are located in this forest.

Travelers also enjoy visiting Shade River State Forest and Forked Run State Park, both of which are located right next to each other just past the city of Belpre. There is camping available at Forked Run State Park. Also, the Wayne National Forest begins again near the town of Gallipolis and ends near Hanging Rock along the Byway. Camping is available at the Vesuvius Recreation Area in the forest.

Finally, the Shawnee State Forest is located along the Byway past Portsmouth. Camping is available in the nearby Shawnee State Park to the west of the forest.

Recreational

Several state forests and parks along this Byway offer opportunities for outdoor recreation and camping. Camping is available at the Barkcamp State Park, in the Forked Run State Park near Belpre, and in the Shawnee State Park near Portsmouth. The Wayne National Forest begins near the town of Gallipolis and ends near Hanging Rock along the Byway. You can camp at the Vesuvius Recreation Area in the forest as well.

Scenic

This scenic Byway leads travelers through the colorful tapestry surrounding the Ohio River and also through many national and state parks. During the fall, the golden leaves of the forests reflect in the shining waters of the Ohio River, giving travelers a sense of serenity as they drive the Byway. Families coming in the summer feel the energy of nature's fresh, green growth as their children play along the banks of the Ohio.

HIGHLIGHTS

Starting from Cincinnati, you can travel alongside the Ohio River without flood walls obstructing your view. Here, you see a more pastoral life of the Ohio, with no industry or other large buildings to interrupt the flow of fields, nature, and water.

- **North Bend:** Your first stop is at the nation's great monument, the **Harrison Tomb,** located off Cliff Road, west of US 50, North Bend. The 60-foot marble obelisk in this 14-acre park pays tribute to William Henry Harrison, the ninth US president.
- **Cincinnati:** Continue to Cincinnati and visit the historic **Harriet Beecher Stowe House.** It was here that Stowe learned of the injustices of slavery and wrote her famous novel, *Uncle Tom's Cabin.*
- **Point Pleasant:** On the way to New Richmond, the scenery changes from the cityscape of Cincinnati to lush green forests, blue skies, and a rolling river. You are going on to Point Pleasant, the birthplace of the 18th US president, Ulysses S. Grant (US 52 and State Route 232): a one-story, three-room cottage.
- **Georgetown:** Follow Grant's life back on US 52, heading north, taking State Route 231 to historic Georgetown, where you can visit **Grant's boyhood home** on 219 East Grant Avenue. From Georgetown, head south again on State Route 68 to State Route 62, back to the Byway (US 52).
- **Ripley:** Stop in the town of Ripley and visit the **Ripley Museum** (219 N Second Street; call 513/392-4660 for group tours) and **Rankin House** (Rankin St, off State Route 52; call 513/392-1627 for group tours). Abolitionists John and Jean Rankin hid some 2,000 escaped slaves in this way station on the Underground Railroad. Harriet Beecher Stowe stopped here to speak with Rankin about the problems of slavery before writing her novel; she used some of the stories she heard in her book. The **Parker home,** also in Ripley, was another home involved in the Underground Railroad; Parker was himself a former slave, inventor, and businessman. His home, located on 330 N Front Street (phone 937/392-4004), is in the process of being restored.
- **Portsmouth:** Continue on US 52 to the town of Portsmouth, where you'll see artistic flood-wall murals that beautify the Byway and protect the city from the rising waters of the Ohio River.
- **Gallipolis:** Continue to the **French Art Colony** of Gallipolis. Here, you can learn about the rich history of the local area and state. While traveling through, note the French-style homes along the riverbanks.
- **Pomeroy:** Traveling farther east along the Byway, you come upon the city of Pomeroy, which has been featured on *Ripley's Believe It or Not* for its unusual **courthouse,** which is built into the side of a cliff and is accessible on all three levels from the outside.
- **Marietta:** Continue on the Byway to Marietta, the place where Ohio began, the first city

Ohio

❋ *Ohio River Scenic Byway*

founded in the Northwest Territory. The early days of Marietta are remembered at the **Campus Martius Museum,** which offers displays of riverboats and other antiquities. You can also stop at the **Ohio River Museum** next to the Martius Museum. Be sure to tour the historic town itself—beautiful old buildings and antiques are waiting to be discovered.

- **Steubenville:** Heading north up the river, you run into the town of Steubenville, where you witness the **Old Fort Steuben** reconstruction (100 S Third Street, phone 614/264-6304). This fort was under the command of Captain John Francis Hamtramck for the protection of the surveyors of the Northwest Territory. Demonstrations, land office tours, and food are available here. In downtown Steubenville, you'll see murals depicting the 1850s and 1920s city life of the town. The murals are painted on the sides of many of the great buildings in the area. While here, explore the historic bed-and-breakfasts and unusual paddleboat restaurant dedicated to river living. Also, in late August, check out the Steubenville Marina off State Route 7, for the Steubenville Regatta and Racing Association's Rumble on the River. Call ahead to check for times at the Steubenville Convention & Visitors Bureau, phone toll-free 800/510-4442.

- **Wellsville:** Continuing on State Route 7, you come to Wellsville and the **Wellsville River Museum,** a three-story building constructed in 1870. Period furniture and paddlewheel displays are featured in the various rooms.

- **East Liverpool:** Continue on the Byway to the **Museum of Ceramics** in the town of East Liverpool. The town has been called Crockery City and is known for its artistic place settings. The city's ceramics museum operates out of the former post office. From here, you can continue following the Byway into Pennsylvania if you wish.

THINGS TO SEE AND DO

Driving along the Ohio River Scenic Byway will certainly keep your senses engaged, but if you yearn to get out of the car and stretch your legs, or if you'd like to make a mini-vacation out of your trip, check out these attractions along the route.

1810 HOUSE. *1926 Waller St, Portsmouth (45662). Phone 740/354-3760.* Original homestead built by hand. Nine rooms with period furniture. Guided tours. Open May-Dec, Sat and Sun afternoons; weekdays by appointment. **FREE**

AIRPORT PLAYFIELD. *Beechmont and Wilmer aves, Cincinnati (45226). Phone 513/321-6500.* Baseball fields, 18- and 9-hole golf courses, driving range, miniature golf, tennis courts, paved biking and hiking trails, and bike rentals. Land of Make Believe playground with wheelchair-accessible play equipment; jet plane, stagecoach; Spirit of '76 picnic area. Summer concerts. Open May-Sept, daily.

BB RIVERBOATS. *1 Madison Ave, Cincinnati (41011). Phone 859/261-8500; toll-free 800/261-8586. www.bbriverboats.com.* Variety of cruises include sightseeing, lunch, and dinner cruises (reservations required). Open daily. **$$$$**

BICENTENNIAL COMMONS AT SAWYER POINT. *801 E Pete Rose Way, Cincinnati (45202). Phone 513/352-4000.* Overlooks with different views of the Ohio River; 4-mile Riverwalk has a geologic river timeline. Performance pavilion and amphitheater. Tennis pavilion with eight courts; skating pavilion; three sand volleyball courts; fitness area with exercise stations. Picnicking, playground. Dining area with umbrella tables. Some fees.

BEAVER CREEK STATE PARK. *12021 Echi Dell Rd, East Liverpool (43920). Phone 330/385-3091. www.dnr.state.oh/parks/parks/beaverck.htm.* This 3,038-acre forested area contains many streams, the ruins of the Sandy and Beaver Canal, and one well-preserved lock. Gaston's

Mill (circa 1837) has been restored. Fishing, canoeing; hunting, hiking, bridle trails, picnicking, primitive camping $$-$$$. Open daily. **FREE**

BREWERY ARCADE. *224 2nd St, Portsmouth (45662).* A former brewery (1842), restored.

BOB EVANS FARM. *791 Farm View Rd, Bidwell (45614). Phone 740/245-5305; toll-free 800/994-FARM. www.bobevans.com.* A 1,100-acre farm. Canoeing (fee); hiking, horseback riding (fee), special weekend events (fee); Craftbarn and Farm Museum; craft demonstrations; domestic animals, farm crops. Open Memorial Day weekend-Sept 1, daily 9 am-5 pm, with activities on Sat-Sun. **FREE**

CAMPUS MARTIUS, MUSEUM OF THE NORTHWEST TERRITORY. *601 2nd St, Marietta (45750). Phone 740/373-3750; toll-free 800/86-0145.* Rufus Putnam home, which was part of the original Campus Martius Fort (1788); furnished with pioneer articles. On the grounds is the Ohio Company Land Office (1788); restored and furnished. Open Mar-Apr and Oct-Nov 9:30 am-5 pm, Sat-Sun noon-5 pm; May-Sept, Mon-Sat 9:30 am-5pm; Sun noon-5 pm; closed Thanksgiving. $$

CAREW TOWER. *441 Vine St, Cincinnati (45202). Phone 513/241-3888 or 513/579-9735.* Cincinnati's tallest building (48 stories). Observation tower. Open Mon-Thurs 9 am-5 pm; Fri-Sat 9:30 am-9 pm; Sun 11 am-5 pm; closed holidays. $

CINCINNATI ART MUSEUM. *953 Eden Park Dr, Cincinnati (45202). Phone 513/721-ARTS. www.cincinnatiartmuseum.org.* The museum houses paintings, sculpture, prints, photographs, costumes, decorative and tribal arts, and musical instruments, representing most major civilizations for the past 5,000 years. Also examples of Cincinnati decorative arts, such as art furniture and Rookwood pottery. Continuous schedule of temporary exhibits. Restaurant, gift shop. Open Tues, Thurs-Sun 11 am-5 pm, Wed to 9 pm; closed Mon, holidays. Tours for the visually impaired (call for appointment). $$

CINCINNATI FIRE MUSEUM. *315 W Court St, Cincinnati (45202). Phone 513/621-5553. www.cincyfiremuseum.com.* Restored firehouse (1907) exhibits firefighting artifacts preserved since 1808; hands-on displays; emphasis on fire prevention. Open Tues-Fri 10 am-4 pm, Sat-Sun from noon; closed holidays. $

CINCINNATI HISTORY MUSEUM. *1301 Western Ave, Cincinnati (45203). Phone 513/287-7000. www.cincymuseum.org.* Permanent exhibit on the Public Landing of Cincinnati; also temporary exhibits. Library (open Mon-Sat; free). Open Mon-Sat 10 am-5 pm, Sat 11 am-6 pm; closed Thanksgiving, Dec 25. $$$

CINCINNATI PLAYHOUSE IN THE PARK. *962 Mt Adams Cir, Cincinnati (45202). Phone 513/421-3888; toll-free 800/582-3208 (in OH). www.cincyplay.com.* Professional regional theater presenting classic and contemporary plays and musicals on two stages: the Robert S. Marx Theater and the Thompson Shelterhouse. Dinner available before each performance. $$$$

CINCINNATI ZOO AND BOTANICAL GARDEN. *3400 Vine St, Cincinnati (45220). Phone 513/281-4700. www.cincyzoo.org.* Features more than 700 species in a variety of naturalistic habitats, including its world-famous gorillas and white Bengal tigers. The Cat House features 16 species of cats; Insect World is a one-of-a-kind exhibit. The Jungle Trails exhibit is an indoor/outdoor rain forest. Rare okapi, walrus, Komodo dragons, and giant eland are also on display. Participatory children's zoo. Animal shows (summer). Elephant and camel rides. Picnic areas, restaurant. Gates open daily at 9 am; close at 4 pm in fall and winter, at 5 pm in summer Mon-Fri, and at 6 pm in summer Sat-Sun. $$$

CINERGY CHILDREN'S MUSEUM. *1301 Western Ave, Cincinnati (45203). Phone 513/287-7000. www.cincymuseum.org.* More than 200 hands-on displays for preschoolers to preteens. Special performances, along with interactive and educational programs. Open Mon-Sat 10 am-5 pm, Sun 11 am-6 pm; closed Thanksgiving, Dec 25. $$

Ohio

✤ *Ohio River Scenic Byway*

CITY HALL. *801 Plum St, Cincinnati (45202). Phone 513/352-3000.* Houses many departments of city government. The interior includes a grand marble stairway with historical stained-glass windows at the landings and murals on the ceiling (1888). Open Mon-Fri.

CIVIC GARDEN CENTER OF GREATER CINCINNATI. *2715 Reading Rd, Cincinnati (45206). Phone 513/221-0981. www.civicgardencenter.org.* Specimen trees; perennials, dwarf evergreens, herbs, and raised vegetable gardens; greenhouse; gift shop; library. Open Mon-Fri 9 am-4 pm, Sat 9 am-3 pm; closed Sun, holidays.

CONTEMPORARY ARTS CENTER. *115 E 5th St, Cincinnati (45202). Phone 513/721-0390 or 513/345-8400.* Changing exhibits and performances of recent art. Open Mon-Sat 10 am-6 pm, Sun noon-5 pm; closed holidays. $

CREEGAN COMPANY. *510 Washington St, Steubenville (43952). Phone 740/283-3708.* Country's largest designer and manufacturer of animations, costume characters, and décor offers guided tours (approximately one hour) through its three-story factory and showroom. Open daily. $

DELTA QUEEN* AND *MISSISSIPPI QUEEN. *At the foot of Broadway, Cincinnati (45202). Phone toll-free 800/543-1949. www.deltaqueen.com.* Each paddlewheeler makes three 8-night cruises on the Ohio, Mississippi, Cumberland, and Tennessee rivers. $$$$

EAST FORK STATE PARK. *3294 Elklick Rd, Bethel (45106). Phone 513/734-4323. www.dnr.state.oh.us/parks/parks/eastfork.htm.* This 10,580-acre park includes rugged hills, open meadows, and reservoir. Swimming beach, fishing, boating; hiking (overnight hiking areas with permit from park office) and bridle trails, picnicking, camping.

EDEN PARK. *950 Eden Park Dr, Cincinnati (45202).* More than 185 acres initially called the Garden of Eden. Ice skating on Mirror Lake. The Murray Seasongood Pavilion features spring and summer band concerts and other events. Picnicking. Four overlooks with scenic views of the Ohio River, the city, and Kentucky hillsides.

FENTON ART GLASS COMPANY. *1 mile off I-77 (exit 185), Marietta (45750). Phone 304/375-7772.* Handmade pressed and blown glassware. Free 30-minute guided tours (Mon-Fri; closed holidays, also the first two weeks in July). Must wear shoes; no children under age 2. Gift shop and outlet on premises; also a museum with a film of the tour (Sun noon-5 pm, Mon-Fri 8 am-8 pm, Sat 8 am-5 pm; Jan-Mar, Mon-Fri 8 am-5 pm; closed holidays). **FREE**

FLOODWALL MURALS PROJECT. *Front St and Ohio River, Portsmouth (45662). Phone 740/353-1116. www.portsmouthmuralproducts.com.* Thirty-four murals depicting area history completed by Robert Dafford. Open daily. **FREE**

FORT ANCIENT STATE MEMORIAL. *6123 State Rte 350, Oregonia (45054). Phone 513/932-4421; toll-free 800/283-8904.* Fort Ancient is one of the largest and most impressive prehistoric earthworks of its kind in the United States. The Fort Ancient earthworks were built by the Hopewell people between 100 BC and AD 500. This site occupies an elevated plateau overlooking the Little Miami River Valley. Its massive earthen walls, more than 23 feet high in places, enclose an area of 100 acres; within this area are earth mounds once used as a calendar of event markers and other archaeological features. Relics from the site and the nearby prehistoric Native American village are displayed in Fort Ancient Museum. Hiking trails, picnic facilities. Open daily, 10 am-5 pm; closed Mon, Tues Mar-Apr, Oct-late Nov. $$

FOUNTAIN SQUARE PLAZA. *5th St, Cincinnati (45202).* Center of downtown activity whose focal point is the Tyler Davidson Fountain, cast in Munich, Germany, and erected in Cincinnati in 1871. The sculpture, whose highest point is the open-armed Genius of Water, symbolizes the many values of water. A bandstand pavilion enables lunch-hour audiences to enjoy outdoor performances. Horse-drawn carriage tours of downtown also begin at the square.

FRENCH ART COLONY. *530 1st Ave, Gallipolis (45631). Phone 740/446-3834.* Monthly exhibits. Open Tues-Sun; closed holidays. **FREE**

HAMILTON COUNTY COURTHOUSE. *1000 Main St, Cincinnati (45202). Phone 513/946-5879.* A good example of adapted Greek Ionic architecture; contains one of America's most complete law libraries. Open Mon-Fri; closed holidays.

HARRIET BEECHER STOWE HOUSE. *2950 Gilbert Ave, Cincinnati (45206). Phone 513/632-5120.* The author of *Uncle Tom's Cabin* lived here from 1832 to 1836. Completely restored with some original furnishings. Open Tues-Thurs 10 am-4 pm.

HEBREW UNION COLLEGE—JEWISH INSTITUTE OF RELIGION. *3101 Clifton Ave, Cincinnati (45220). Phone 513/221-1875.www.huc.edu.* First institution (1875) of Jewish higher learning in the United States. Graduate school offers a variety of programs. Klau Library includes Dalsheimer Rare Book Building with rare remnants of Chinese Jewry and collections of Spinoza and Americana. American Jewish Archives Building is dedicated to study and preservation of American Jewish historical records. Archaeological exhibits and Jewish ceremonial objects are in the Skirball Museum Cincinnati Branch. Guided tours (by appointment).

HERITAGE VILLAGE MUSEUM. *11450 Lebanon Rd, Sharonville (45241). Phone 513/563-9484. www.heritagevillagecincinnati.org.* A 30-acre historic village recaptures life in southwestern Ohio prior to 1880. Eleven buildings (1804-1880) reconstructed and authentically restored and refurnished. Special exhibits and events; period craft demonstrations; guided tours. Open May-Oct, Wed-Sat noon-4 pm, Sun 1-5 pm; Apr, Nov, Dec Sat noon-4 pm, Sun 1-5 pm; closed Jan-Mar. **$$**

HISTORIC LOVELAND CASTLE. *12025 Shore Dr, Loveland (45140). Phone 513/683-4686. www.lovelandcastle.com.* A one-fifth scale, medieval stone castle built by one man over a period of 50 years. Open daily 11 am-5 pm; closed Dec 25. **$**

JEFFERSON COUNTY COURTHOUSE. *301 Market St, Steubenville (43952). Phone 740/283-4111.* Contains the first recorded county deed, signed by George Washington, and portraits of Steubenville personalities. A statue of Edwin Stanton, Lincoln's secretary of war, is on the lawn. Open Mon-Fri; closed holidays. **FREE**

JOHN HAUCK HOUSE MUSEUM. *812 Dayton St, Cincinnati (45214). Phone 513/721-3570. www.heritagevillagecincinnati.org.* Ornate 19th-century stone-front townhouse in a historic district. Restored home contains period furnishings, memorabilia, antique children's toys, and special displays. Open Fri, the last two Sun of month, by appointment; also special Christmas hours; closed holidays. **$**

Ohio

Ohio River Scenic Byway

KROHN CONSERVATORY. *1501 Eden Park Dr, Cincinnati (45202). Phone 513/421-5707.* Floral conservatory of more than 5,000 species of exotic plants, including a five-story indoor rain forest complete with a 20-foot waterfall and a major collection of unusual epiphytic plants. Individual horticultural houses contain palm, desert, orchid, and tropical collections. Themed flower and garden shows six times annually. Guided tours available. Gift shop. Open daily; extended hours for holiday shows. **FREE**

LAKE VESUVIUS. *6518 OH 93, Ironton (45659). Phone 740/534-6500.* The stack of Vesuvius (1833), one of the earliest iron blast furnaces, still remains. Swimming, fishing, boating (dock; May-Sept); hiking, picnicking, camping. Open daily. **$$**

MEIER'S WINE CELLARS. *6955 Plainfield Pike, Cincinnati (45236). Phone 513/891-2900; toll-free 800/346-2941. www.meierswinecellars.com.* Country wine store, tasting room. Tours (45 minutes each). Open June-Oct, Mon-Sat 9 am-5 pm; closed Jan 1, Thanksgiving, Dec 25. **FREE**

MOUND CEMETERY. *5th and Scammel sts, Marietta (45750).* A 30-foot-high conical mound stands in the cemetery where 24 Revolutionary War officers are buried.

MOUNT ADAMS NEIGHBORHOOD. *Area directly SW of Eden Park, on a hill overlooking Cincinnati and the Ohio River, Cincinnati (45202).* Mount Adams is the Montmartre of Cincinnati: its narrow streets, intimate restaurants, boutiques, and art stores give the area a European flavor.

MOUNT AIRY FOREST AND ARBORETUM. *5083 Colerain Ave, Cincinnati (45223). Phone 513/352-4080.* Cincinnati's largest park was the first municipal reforestation project in the United States. More than 1,450 acres include 800 acres of conifers and hardwoods, 300 acres of native hardwoods, and 241 acres of grasslands. The 120-acre arboretum (guided tours by appointment, phone 513/541-8176) includes specialty gardens, floral displays, and extensive plant collections. The area is used by students and amateur and professional gardeners as a testing area for observation of growth, habits, and tolerance of plants. Nature trails, picnic areas, lodges; 2 acres reserved for a dog park. Open daily. **FREE**

EAST LIVERPOOL MUSEUM OF CERAMICS. *400 E 5th St, East Liverpool (43920). Phone 330/386-6001. www.themuseumofceramics.org.* History museum contains a collection of regional pottery and porcelain, bone china, and life-size dioramas; multimedia presentation. Open Mar-Apr, Sat 9:30 am-5 pm, Sun from noon; May-Nov, Wed-Sat 9:30 am-5 pm, Sun from noon; closed Thanksgiving. **$$**

MUSEUM OF NATURAL HISTORY & SCIENCE. *1301 Western Ave, Cincinnati (45203). Phone 513/287-7000. www.cincymuseum.org.* Natural history of the Ohio Valley. Wilderness Trail with Ohio flora and fauna and full-scale walk-through replica of a cavern with 32-foot waterfall; Children's Discovery Center. Open Mon-Sat 10 am-5 pm, Sun 11 am-6 pm; closed Thanksgiving, Dec 25. **$$**

MUSKINGUM PARK. *Putnam and Front sts, Marietta (45750). Phone 740/373-5178.* A riverfront common where Arthur St. Clair was inaugurated the first governor of the Northwest Territory in 1788; monument to westward migration sculpted by Gutzon Borglum.

OHIO RIVER MUSEUM STATE MEMORIAL. *601 2nd St, Marietta (45750). Phone 740/373-3717.* Exhibits on the history and development of inland waterways. Open May-Sept, daily; Mar-Apr and Oct-Nov, Wed-Sun; closed Thanksgiving. The steamboat *W. P. Snyder, Jr.* (1918) is moored on the Muskingum River; guided tours (April-Oct). **$$**

OMNIMAX THEATER. *1301 Western Ave, Cincinnati (45203). Phone 513/287-7000. www.cincymuseum.org.* A 260-degree domed screen five stories high and 72 feet wide. Films change every six months. Open daily; closed Thanksgiving, Dec 25. **$$**

OUR HOUSE STATE MEMORIAL. *434 1st St, Gallipolis (45631). Phone 740/446-0586 (museum). www.ohiohistory.org/places/ourhouse.* Built as a tavern in 1819; restored. Lafayette stayed here. Open June-Aug, Tues-Sat 10 am-5 pm, Sun from 1 pm; May, Sept-Oct, Sat 10 am-5 pm, Sun from 1 pm. **$$**

★ **PARAMOUNT'S KINGS ISLAND.** *6300 Kings Island Dr, Cincinnati (45034). Phone 513/754-5800; toll-free 800/288-0808. www.pki.com.* Premier seasonal family theme park. 150-acre facility with more than 100 rides and attractions, including The Outer Limits thrill ride and Flight of Fear, an indoor roller coaster. Open late May-early Sept, daily; mid-Apr-late May, weekends; also selected weekends in early Sept-Oct; hours vary. **$$$$**

HALL CHINA COMPANY. *10 Anna St, East Liverpool (43920). Phone 330/385-2900. www.hallchina.com.* Pottery tours by appointment (761 Dresden Ave, phone 330/385-4293). Open Mon-Sat 9 am-5 pm.

MARIETTA TROLLEY TOURS. *127 Ohio St, Marietta (45750). Phone 740/374-2233.* One-hour narrated tours of Marietta aboard a turn-of-the-century style trolley. Open July-Aug, Tues-Sun; mid-late June, Wed-Sun; May-mid-June, Thurs-Sun; and Sept-Oct, weekends. **$$**

PUBLIC LANDING. *At the foot of Broadway, Cincinnati (45202).* Where the first settlers touched the shore and the first log cabin was built. Center of river trade; look for paddlewheelers, Cinergy Field, and six Ohio River bridges to northern Kentucky.

ROSSI PASTA. *114 Greene St, Marietta (45750). Phone 740/373-5155; toll-free 800/227-6774. www.rossipasta.com.* Pasta makers for many fine stores offer visitors the opportunity (limited) to watch the process in the factory. Open Mon-Fri; closed holidays. Also a retail outlet. **FREE**

SACRA VIA STREET. *3rd and Sacra Via, Marietta (45750). Phone 740/373-5178.* Built by Mound Builders as a "sacred way" to the Muskingum River.

SHAWNEE STATE FOREST. *13291 US 52, West Portsmouth (45663). Phone 740/858-6685. www.dnr.state.oh.us/forestry/forests/stateforests/shawnee.htm.* A 63,000-acre forest with six lakes. Hunting in season. Bridle trail. Open daily. **FREE**

SHAWNEE STATE PARK. *940 2nd St, Portsmouth (45662). Phone 740/858-4561 or -6652. www.dnr.state.oh.us/parks/parks/shawnee.htm.* Swimming, fishing, and boating (ramps, dock) on a 68-acre lake; 18-hole golf, picnicking (shelter); lodge, camping (fee), teepees, cabins. Nature center. Open daily. **FREE**

SOUTHERN OHIO MUSEUM AND CULTURAL CENTER. *825 Gallia St, Portsmouth (45662). Phone 740/354-5629.* Changing exhibits in visual arts; performing arts; workshops; guided tours. Open Tues-Sun; closed holidays, also Jan and Aug. Free admission Fri. **$**

TRAILSIDE NATURE CENTER. *3251 Brookline Dr, Cincinnati (45220). Phone 513/751-3679.* Discovery center has displays on local birds, mammals, insects, and geology. Weekend nature walks and program (all year). Open Tues-Sat; also Sun afternoons. **FREE**

UNION CEMETERY. *1720 W Sunset Blvd, Steubenville (43952). Phone 740/283-3384.* Contains many Civil War graves, including the Fighting McCook plot, where members of a family that sent 13 men to fight in the Union Army are buried. Open daily.

★ **UNION TERMINAL.** *1301 Western Ave, Cincinnati (45203). www.cincymuseum.org.* Famous Art Deco landmark (1933) noted for its mosaic murals, Verona marble walls, terrazzo floors, and large-domed rotunda. Three museums on site: Cincinnati History Museum, Museum of Natural History & Science, and Cinergy Children's Museum. Also an OmniMax theater, café, and museum shops.

***VALLEY GEM* STERNWHEELER.** *123 Strecker Hill, Marietta (45750). Phone 740/373-7862. www.valleygemsternwheeler.com.* Excursions on

Ohio

❋ *Ohio River Scenic Byway*

the Ohio and Muskingum rivers aboard the sternwheeler *Valley Gem*. Fall foliage trips in Oct. Open June-Aug, Tues-Sun; May and Sept-Oct, weekends only. $$-$$$

WAYNE NATIONAL FOREST. *19700 US Hwy 33, Athens (45764). Phone 740/753-0101. www.fs.fed.us/r9/wayne/.* Three sections make up this 202,967-acre area of southeast Ohio. Private lands are interspersed within the federal land. One section is the east side of Ohio, northeast of Marietta; the second is northeast of Athens; and the third is in the southern tip of the state, southwest of Gallipolis. The forest lies in the foothills of the Appalachian Mountains. It's characterized by rugged hills covered with diverse stands of hardwoods, pine, and cedar; lakes, rivers, and streams; springs, rock shelters, covered bridges, trails, and campgrounds are located in the forest. A Ranger District Office of the forest is also located here. Open daily. **FREE**

WILLIAM HOWARD TAFT NATIONAL HISTORIC SITE. *2038 Auburn Ave, Cincinnati (45219). Phone 513/684-3262. www.nps.gov/wiho/.* Birthplace and boyhood home of the 27th president and chief justice of the United States. Four rooms with period furnishings; other rooms contain exhibits on Taft's life and careers. Open daily; closed Jan 1, Thanksgiving, Dec 25. **FREE**

PLACES TO STAY

If you choose to include an overnight stay in your trip along this Byway, Mobil Travel Guide recommends the following lodgings.

Cincinnati

★★ **BEST WESTERN MARIEMONT INN.** *6880 Wooster Pike, Cincinnati (45227). Phone 513/271-2100; toll-free 800/780-7234. www.bestwestern.com.* 60 rooms, 3 story. Check-out noon. TV; cable (premium). In-room modem link. Laundry services. Restaurant, bar. Room service. ¢

★★ **CROSS COUNTRY INN CLERMONT.** *4004 Williams Dr, Cincinnati (45255). Phone 513/528-7702; toll-free 800/621-1429. www.crosscountryinns.com.* 128 rooms, 2 story. Check-out noon. TV; cable (premium). Pool. ¢

★★★ **CROWNE PLAZA HOTEL.** *15 W 6th St, Cincinnati (45202). Phone 513/381-4000; toll-free 800/2-CROWNE. www.crowneplaza.com.* Located downtown and connected to the skywalk system, this hotel is within walking distance of Riverfront Stadium, restaurants, shopping, and cultural attractions. 321 rooms, 20 story. Check-out 11 am. TV; cable (premium). Restaurant, bar. In-house fitness room, sauna. Valet parking available. Business center. Concierge. $

★★ **GARFIELD SUITES HOTEL.** *2 Garfield Pl, Cincinnati (45202). Phone 513/421-3355; toll-free 800/367-2155. www.garfieldsuiteshotel.com.* 152 rooms, 16 story. Pet accepted; fee. Check-out noon. TV; cable (premium), VCR available. In-room modem link. Laundry services. Restaurant, bar. Room service. In-house fitness room. Valet parking available. $$

★★ **HYATT REGENCY.** *151 W 5th St, Cincinnati (45202). Phone 513/579-1234; toll-free 800/633-7313. www.cincinnati.hyatt.com.* Located in the downtown area, this hotel is across from the Cincinnati Convention Center and is connected to shopping, dining, and entertainment by an enclosed skywalk. 485 rooms, 22 story. Check-out noon. TV; cable (premium). Restaurant, bar. In-house fitness room, sauna. Indoor pool, whirlpool, poolside service. Business center. Concierge. Luxury level. $$

★★ **MILLENNIUM HOTEL.** *141 W 6th St, Cincinnati (45202). Phone 513/352-2130. www.millenniumhotels.com.* 450 rooms, 32 story. Check-out noon. TV; cable (premium).

In-room modem link. Restaurant, bar. In-house fitness room. Pool. Business center. Concierge. Luxury level. $

★ **QUALITY HOTEL & SUITES.** *4747 Montgomery Rd, Cincinnati (45212). Phone 513/351-6000; toll-free 292-2079. www.qualityhotelandsuites.com.* 148 rooms, 8 story. Complimentary breakfast. Check-out noon. TV; cable (premium). In-room modem link. Restaurant, bar. Room service. Pool, poolside service. Business center. ¢

★★★★ **THE CINCINNATIAN HOTEL.** *601 Vine St, Cincinnati (45202). Phone 513/381-3000; toll-free 800/942-9000. www.cincinnatianhotel.com.* Since 1882, The Cincinnatian Hotel has been a fixture on the local scene. The first luxury hotel in the city, it continues to provide the finest accommodations and highest levels of service to its guests. The accommodations are lovingly maintained and incorporate modern technology, like high-speed Internet access, and multiline telephones. Balconies and fireplaces add an inviting touch to the gracious, tailored interiors. The eight-story atrium of the Cricket Lounge creates an airy ambience, perfect for whiling away the day. A harpist entertains those partaking in the wonderful afternoon tea, and the mellow notes of a piano entertain diners in the evening. The hotel thoughtfully provides box lunches for guests leaving the hotel for the day, although most are sure to book a table for dinner that evening at the Palace. The fine dining and impeccable service make it one of the top tables in town. 146 rooms, 8 story. Check-out noon. TV; cable (premium), VCR available. Restaurant, bar; entertainment. Room service 24 hours. In-house fitness room, sauna. Covered parking. Concierge. Totally nonsmoking. $$

★★ **THE VERNON MANOR HOTEL.** *400 Oak St, Cincinnati (45219). Phone 513/281-3300; toll-free 800/543-3999. www.vernon-manor.com.* 177 rooms, 7 story. Check-out noon. TV; cable (premium). In-room modem link. Laundry services. Restaurant, bar. In-house fitness room. Valet parking available. Business center. $

★★★ **THE WESTIN.** *21 E 5th St, Cincinnati (45202). Phone 513/621-7700; toll-free 800/228-3000. www.westin.com.* With a riverside location in the downtown area, this hotel is near Riverside Park, shopping, restaurants, a theater, and more. It offers a rooftop fitness center and pool along with many other amenities. 450 rooms, 17 story. Check-out noon. TV; cable (premium), VCR available. In-room modem link. Restaurant, bar. Room service 24 hours. In-house fitness room, steam room. Indoor pool, whirlpool, poolside service. Valet parking available. Business center. Luxury level. $$

East Liverpool

★★ **THE STURGIS HOUSE.** *122 W 5th St, East Liverpool (43920). Phone 330/382-0194. www.sturgishouse.com.* 6 rooms, 2 story. Located in a restored Victorian mansion with private baths and continental breakfast. $

Gallipolis

★★ **HOLIDAY INN.** *577 OH 7 N, Gallipolis (45631). Phone 740/446-0090; toll-free 800/465-4329. www.holiday-inn.com.* 100 rooms, 2 story. Pet accepted, some restrictions. Check-out noon. TV. Laundry services. Restaurant, bar. Room service. Pool, children's pool. ¢

Marietta

★ **BEST WESTERN.** *279 Muskingum Dr, Marietta (45750). Phone 740/374-7211; toll-free 800/780-7234. www.bestwestern.com.* 47 rooms, 1-2 story. Complimentary continental breakfast. Check-out noon. TV. Health club privileges. On the Muskingum River; free dockage. ¢

211

Ohio

※ *Ohio River Scenic Byway*

★ **COMFORT INN.** *700 Pike St, Marietta (45750). Phone 740/374-8190; toll-free 877/424-6423. www.choicehotels.com.* 120 rooms, 4 story. Complimentary continental breakfast. Check-out 11 am. TV; cable (premium), VCR available. Restaurant, bar. In-house fitness room. Indoor pool, poolside service. Free airport transportation. ¢

★★ **HOLIDAY INN.** *701 Pike St, Marietta (45750). Phone 740/374-9660; toll-free 800/465-4329. www.holiday-inn.com.* 109 rooms, 2 story. Check-out noon. TV; cable (premium), VCR available. In-room modem link. Restaurant, bar; entertainment. Room service. Pool, children's pool. ¢

★★ **LAFAYETTE.** *101 Front St, Marietta (45750). Phone 740/373-5522; toll-free 800/331-9336. www.historiclafayette.com.* 78 rooms, 5 story. Pet accepted; fee. Check-out noon. TV. Restaurant, bar. Health club privileges. Free airport transportation. On the Ohio River. ¢

Portsmouth

★★ **RAMADA INN.** *711 Second St, Portsmouth (45662). Phone 740/354-7711; toll-free 800/272-6232. www.ramada.com.* 119 rooms, 5 story. Pet accepted. Complimentary continental breakfast. Check-out noon. TV; cable (premium). Restaurant. Room service. In-house fitness room. Health club privileges. Indoor pool, children's pool, whirlpool, poolside service. Dockage. ¢

Steubenville

★★ **HOLIDAY INN.** *1401 University Blvd, Steubenville (43952). Phone 740/282-0901; toll-free 800/465-4329. www.holiday-inn.com.* 120 rooms, 2 story. No elevator. Check-out noon. TV; cable (premium), VCR available. In-room modem link. Laundry services. Restaurant, bar. Room service. Heated pool. ¢

PLACES TO EAT

A long day of driving is sure to make you hungry. At the end of your journey, take a table at one of the following restaurants.

Cincinnati

★ **AGLAMESIS BROS.** *3046 Madison Rd, Cincinnati (45209). Phone 513/531-5196. www.aglamesis.com.* Closed Jan 1, Easter, Dec 25. Lunch, dinner. Old-time ice cream parlor established in 1908. $

★★ **CHATEAU POMIJE.** *2019 Madison Rd, Cincinnati (45208). Phone 513/871-8788. www.chateaupomije.com.* Continental menu. Closed Sun; major holidays. Lunch, dinner. Outdoor seating. $$

★ **FORE & AFT.** *7449 Forbes Rd, Cincinnati (45233). Phone 513/941-8400.* Steak menu. Closed Jan 1, Dec 24-25. Lunch, dinner. Bar. Children's menu. Outdoor seating. Floating barge on the Ohio River; nautical memorabilia. $$

★★★ **GRAND FINALE.** *3 E Sharon Ave, Cincinnati (45246). Phone 513/771-5925. www.grandfinale.info.* This restaurant is located in a historic Victorian landmark. Fresh seafood, hand-trimmed steaks, rack of lamb, frog legs, crepes, homemade breads, and pastries are offered. The courtyard is open year-round. Continental menu. Closed Mon; Dec 25. Lunch, dinner. Bar. Children's menu. Remodeled turn-of-the-century saloon. Outdoor seating. $$

★★★ **HERITAGE.** *7664 Wooster Pike (OH 50), Cincinnati (45227). Phone 513/561-9300; toll-free 800/814-9963. www.theheritage.com.* An interesting menu is offered in this restored 1827 farmhouse with corner booths and cozy close-set tables. The snappy Cajun and Creole menu is short but changes daily. Desserts are homemade. Regional American menu. Closed some major holidays. Dinner, Sun brunch. Bar. Children's menu. Valet parking available. Outdoor seating. Own herb garden. $$
D

★★ **LA NORMANDIE TAVERN & CHOPHOUSE.** *118 E 6th St, Cincinnati (45202). Phone 513/721-2761. www.maisonette.com.* Closed Sun; major holidays. Lunch, dinner. Bar. Valet parking available. $$$
D

★ **LENHARDT'S AND CHRISTY'S.** *151 W McMillan St, Cincinnati (45219). Phone 513/281-3600. www.lenhardtsandchristys.com.* German, Hungarian menu. Closed Sun, Mon; July 4; also the first two weeks of Aug, two weeks at Christmas. Lunch, dinner. Bar. Former Moerlin brewery mansion. $$

★★★★★ **MAISONETTE.** *114 E 6th St, Cincinnati (45202). Phone 513/721-2260. www.maisonette.com.* Bathed in classic French charm since 1949, with original oil paintings adorning the walls, loveseat-styled banquettes for cozy dining for two, and elegant table settings, this is the Casablanca of restaurants: a classic that never goes out of style. The menu makes a case for authentic, elegant French food all over again. The kitchen prepares light, flavorful, and luxurious meals from shimmering seasonal produce, dressed up with garnishes chosen with as much care as the main ingredients. Medallions of veal wrapped in country bacon with Spanish pepper and chorizo risotto and roasted domestic rack of lamb with tomato crust and chickpea crepes are house specialties. French cuisine. Closed Sun; holidays. Lunch, dinner. Bar. Jacket required. Valet parking available (dinner). $$$$
D

★★ **NATIONAL EXEMPLAR.** *6880 Wooster Pike, Cincinnati (45227). Phone 513/271-2103.* Steak menu. Closed Dec 25. Breakfast, lunch, dinner. Bar. Children's menu. Casual attire. $$
D

★★★ **THE PALACE.** *601 Vine St, Cincinnati (45202). Phone 513/381-6006. www.palace cincinnati.com.* The regional American menu changes seasonally to ensure that only the freshest ingredients available grace the table. The wine list boasts 350 choices, one sure to be perfect for any meal. American menu. Breakfast, lunch, dinner. Bar. Valet parking available. $$$
D

★★★ **THE PHOENIX.** *812 Race St, Cincinnati (45202). Phone 513/721-8901. www.thephx.com.* Dinner is served in the President's Room, which is adorned with two elegant chandeliers. The historical east wall showcases a magnificent hand-carved library breakfront. Steak menu. Closed Sun-Tues; major holidays. Dinner. Built in 1893; white marble staircase, 12 German stained-glass windows from the 1880s. Valet parking available. Totally nonsmoking. $$
D

★★★ **PRECINCT.** *311 Delta Ave, Cincinnati (45226). Phone 513/321-5454; toll-free 877/321-5454. www.jeffruby.com.* This former police precinct offers the best in steakhouses—from aged angus beef to the perfect "eye," broiled to perfection and seasoned with a secret spice mix. Steak menu. Closed some major holidays. Dinner. Bar. In an 1890s police station. Valet parking available. $$$

Ohio

✺ *Ohio River Scenic Byway*

★★ **THE RESTAURANT AT THE PALM COURT.** *35 W 5th St, Cincinnati (45202). Phone 513/421-9100.* At the Palm Court of the Hilton Netherland Plaza Hotel, this museum-like dining room offers friendly, accommodating service. The splendid space has a vaulted, muraled ceiling; wood paneling; and brass accents. Contemporary American menu. Lunch, dinner. Bar. Valet parking available. $$$
[D] [SC]

★ **TEAK.** *1049 St. Gregory St, Cincinnati (45202). Phone 513/665-9800. www.teakrestaurant.com.* Thai menu. Closed major holidays. Lunch, dinner. Bar. Outdoor seating. $$
[D]

★ **VINEYARD CAFÉ.** *2653 Erie Ave, Cincinnati (45208). Phone 513/871-6167.* Eclectic menu. Closed Thanksgiving, Dec 25. Lunch, dinner. Bar. Children's menu. Casual attire. Outdoor seating. $$
[D]

Marietta

★★★ **THE GUN ROOM.** *101 Front St, Marietta (45750). Phone 740/373-5522. www.historiclafayette.com.* The traditional menu at this Lafayette Hotel restaurant is as much a draw as the room's ornate 19th-century riverboat décor and antique gun collection. The adjacent Riverview Lounge offers great views of the Ohio River. Continental, seafood menu. Dinner, Sun brunch. Bar. Children's menu. $$$
[D]

The Native American Scenic Byway
✳ SOUTH DAKOTA

Quick Facts

LENGTH: 101 miles.

TIME TO ALLOW: 2.5 hours.

BEST TIME TO DRIVE: Year-round; high season is summer.

BYWAY TRAVEL INFORMATION: Lower Brule Sioux Tribe Tourism Office: 605/473-0561.

SPECIAL CONSIDERATIONS: Because most tourist traffic occurs during the summer months, several of the interpretive sites (those with trained staff and living-history interpreters) are closed during the off-season. The speed limit along the Byway is 55 mph, and gasoline is available in Oacoma, Chamberlain, Fort Thompson, and Fort Pierre.

BICYCLE/PEDESTRIAN FACILITIES: Bicyclists and pedestrians are well accommodated along this route. In Chamberlain and Pierre, the two largest population centers on the Byway, walking/biking paths are located along the river. Low traffic levels and wide highways are conducive to safe bicycle travel along the entire Byway.

Along this Byway, the lakes and streams of the Great Plains tell the story of the Sioux's connection with the land. Journey through the Crow Creek and Lower Brule Sioux Indian reservations and view their unique cultures firsthand. Buffalo roam on the high plains that sharply contrast the nearby bottomlands and the hills and bluffs along the river. The Big Bend of the Missouri River features prominently in the geography of the area. Dams along the river created Lake Francis Case and Lake Sharpe, which are excellent recreational areas as well as spots of beauty.

THE BYWAY STORY

The Native American Scenic Byway tells archaeological, cultural, historical, natural, recreational, and scenic stories that make it a unique and treasured Byway.

Archaeological

A significant number of archaeological studies are being conducted along the Native American Scenic Byway corridor. One of the major interpretive sites along the Byway, the Crow Creek Massacre site near the southern border of the Crow Creek Sioux Reservation, is a well-known archaeological site. In 1978, archaeologists from the University of South Dakota and the Smithsonian Institute uncovered the remains of nearly 500 people, victims of the largest prehistoric massacre known in North America. The event occurred around 1325. Archaeologists have several ideas about what happened at the Crow Creek site, who killed the people, and why. One hypothesis is that there were too many people for the land to support and that village farmers got involved in battles over land.

S. Dakota

❋ The Native American Scenic Byway

Cultural
Travel through the tribal grounds of one of the most significant Native American tribes in the country. You encounter two Sioux reservations as you drive the Byway: the Crow Creek Sioux Reservation and the Lower Brule Sioux Reservation.

When Lewis and Clark made their journey through this part of the country, they came with the intention of forming a peaceful relationship with the people there. Remnants of that journey remain today, as travelers explore a new landscape and a new culture.

The Native American Scenic Byway is a journey through the heart of the Sioux Nation. Excellent cultural interpretive sites give you an in-depth, educational, and entertaining experience with Native American culture.

Historical
Many interesting historical events are associated with the Native American Scenic Byway, and history and culture are intimately connected. The history of the Sioux and the other indigenous people who preceded them is a major part of Native American culture. A major change in the Sioux culture was initiated by their contact with Lewis and Clark, who went up the river in 1804 and returned in 1806. Their voyage of discovery is one of the major interpretive themes of this Byway. In the area of the corridor, Lewis and Clark first observed mule deer, coyotes, antelopes, jack rabbits, magpies, and several other animals, plants, and birds. Lewis and Clark were followed by fur traders, who in turn were followed by steamboats, which were followed by settlers. Fur trading posts were converted into military posts, and the great Sioux Reservation was reduced to its present size. Communities were built and the river was dammed. Through it all, the land remains, giving evidence of the past.

Natural
One of the prominent natural features of the Native American Scenic Byway is the buffalo that roam the area. Native American culture is inseparably connected to the land and its bounty. The traditional Sioux culture was dependent upon the buffalo because it provided food, shelter, and implements. The elimination of the buffalo almost eliminated the Sioux culture, but buffalo are now being returned to the reservations.

Another prominent natural feature of the Byway is the Missouri River, the main route of commerce through central North America for hundreds of years. The Big Bend of the Missouri is one of the best-known geographic features of the continent. Native plants are also an integral part of Native American ceremony and lifestyle.

Recreational
The damming of the river along this Byway created Lake Francis Case and Lake Sharpe. These impounded waters provide excellent recreational opportunities for visitors to the Byway. All types of water sports are available, and the fishing is among the best in the country. The recreational facilities along the lakes are managed by the US Army Corps of Engineers, and nice camping facilities are also available. The communities at either end of the Byway offer well-developed biking, hiking, and nature trails. In addition, both Sioux reservations have well-managed hunting and fishing programs that are available to the public.

Scenic
The scenery along the Byway is unique and spectacular. The hills and bluffs along the river are rugged and prominent. The bottomlands where creeks and streams enter the river are characterized by woodlands and wetlands. The high plains have a unique appeal of openness and freedom. Visitors often have the opportunity to view wildlife along the Byway: it is a thrill to see buffalo on the hills and on the prairie.

HIGHLIGHTS

This itinerary takes you the entire length of the Byway, from north to south. If you're traveling in the other direction, read this list from the bottom and work your way up.

- You can begin traveling the Native American Scenic Byway at the north terminus at the South Dakota capital of **Pierre**. While in Pierre, tour the **Cultural Heritage Center,** which showcases South Dakota history. An extensive collection tells the story of the Great Sioux Nation. Also in Pierre is the **state capitol building,** which has been deemed one of the most fully restored capitol buildings in the United States. The building looks much as it did when completed in 1910.

- Only 6 miles northwest of Pierre is the **Oahe Dam,** the world's largest earth-filled dam. Originally built to control flooding from the Missouri River, the dam is today a popular site for water recreation activities of all sorts.

- The **Missouri River** runs parallel to the Byway. You can enjoy numerous water recreational activities at many points along the Byway, including swimming, camping, boating, and fishing.

- Located 4 miles east of Pierre is the **Farm Island Recreation Area.** The park is nestled along the shores of the Missouri River and provides a great place for a variety of outdoor recreation activities. Camping, boating, and fishing are popular here. Swimmers enjoy the beach of the Missouri River.

- Traveling south of Pierre, you pass through the **Fort Pierre National Grassland.** The land is publicly owned and administered by the US Forest Service. The grassland is noted for crop fields of sorghum, wheat, and sunflowers.

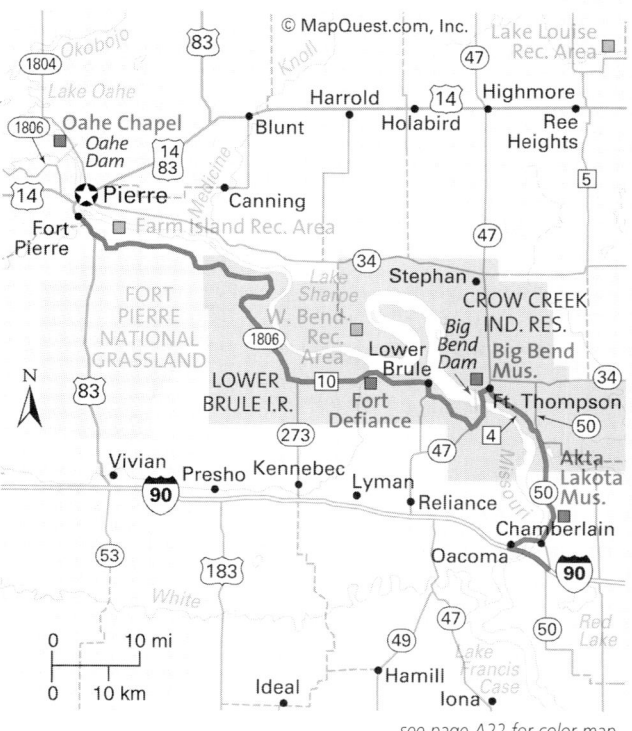

see page A22 for color map

- Only 30 miles east of Pierre begins the **Lower Brule Indian Reservation.** Visitors to the reservation enjoy the tribal casino, as well as visits to Lake Sharpe and the Big Bend Dam.

- **Lake Sharpe** stretches some 80 miles from the Oahe Dam in the north to the Big Bend Dam in Fort Thompson. It is easily accessed at the West Bend State Recreation area 50 miles east of Pierre on the Byway.

- **Big Bend Dam** is located approximately 50 miles east of Pierre. Built to control flooding of the Missouri River and to provide hydroelectric power, the dam has created vast reservoirs that today provide ample opportunity for fishing and water recreation. Swimming, sailing, scuba diving, and fishing are popular activities in the area. Take the tour of the powerhouse to learn about the engineering feat of creating the dam. One-hour tours of Big Bend Powerhouse are available at no charge. Picnic areas, docks, and marinas are available. A visitor center is located at the dam.

- The **Crow Creek Indian Reservation** begins near the Big Bend Dam and continues until the southern terminus of the Byway. Tribal casinos,

S. Dakota

❋ The Native American Scenic Byway

water sports, fishing, entertainment, and hunting are available on the reservation.

- The **Akta Lakota Museum** is located in Chamberlain, the southern terminus of the Byway, and is devoted to preserving and promoting Sioux culture. The museum is located at the St. Joseph's Indian School for Native American Youth.

THINGS TO SEE AND DO

Driving along the Native American Scenic Byway will certainly keep your senses engaged, but if you yearn to get out of the car and stretch your legs, or if you'd like to make a mini-vacation out of your trip, check out these attractions along the route.

AKTA LAKOTA MUSEUM. *N Main St, Chamberlain (57326). Phone 605/734-3452; toll-free 800/798-3452. www.stjo.org/museum/.* Features a large collection of Sioux artifacts and handcrafts, as well as several dioramas that depict daily life on the prairie. Also a large collection of Native American paintings and sculpture. The gift shop carries an extensive selection of books of Native American history and culture, as well as locally made jewelry and art. Open Mon-Fri 8 am-5 pm; Memorial Day-Labor Day, daily. **FREE**

AMERICAN CREEK RECREATIONAL AREA. *Chamberlain (57325). Phone 605/734-5151.* On Lake Francis Case; swimming (May-Oct, fee), sand beaches, water-skiing; fishing; boat docks. Picnicking, camping (May-Oct; fee). Park open daily; ranger, May-Oct.

BIG BEND DAM-LAKE SHARPE. *Chamberlain (57325). Phone 605/245-2255.* One of a series of six dams on the Missouri River built by the US Army Corps of Engineers as units in the Pick-Sloan Plan for power production, flood control, and recreation. Guided tours of the powerhouse (open June-Aug, daily; rest of year, by appointment). Visitor center at dam site (open mid-May-mid-Sept, daily). Many recreation areas along the reservoir have swimming, fishing; boating (docks, ramps, fee); winter sports; picnicking; camping (May-mid-Sept, fee). **FREE**

CULTURAL HERITAGE CENTER. *900 Governors Dr, Pierre (57501). Phone 605/773-3458. www.state.sd.us/state/capitol/cultural/cultural.html.* Pioneer life, Native American, mining, and historic exhibits; the Verendrye Plate, a lead plate buried in 1743 at Fort Pierre by French explorers, the first known Europeans in South Dakota. Open daily Mon-Fri 9 am-4:30 pm, Sat-Sun and holidays 1-4:30 pm; closed Jan 1, Thanksgiving, Dec 25. **$**

FARM ISLAND STATE RECREATION AREA. *1301 Farm Island Rd, Pierre (57501). Phone 605/224-5605. www.state.sd.us/gfp/sdparks/fmisland/fmisland.htm.* On 1,184 acres. Swimming, bathhouse; fishing; boating, canoeing. Hiking. Picnicking, playground. Interpretive program. Camping (electrical hookups, dump station). Open May-Sept, 6 am-11 pm; Oct-Apr 6 am-9 pm.

FIGHTING STALLIONS MEMORIAL. *500 E Capitol Ave, Pierre (57501). www.state.sd.us/state/capitol/stallions/stallions.html.* Replica of a carving by Korczak Ziolkowski honors Governor George S. Mickelson and seven others who died in an airplane crash in 1993.

FLAMING FOUNTAIN. *Northwest shore of Capitol Lake, Pierre (57501).* The artesian well that feeds this fountain has a natural sulfur content so high that the waters can be lit. The fountain serves as a memorial to war veterans.

OAHE DAM AND RESERVOIR. *28563 Power House, Pierre (57501). Phone 605/224-5862.* This large earth-fill dam (9,300 feet long, 245 feet high) is part of the Missouri River Basin project. Lobby with exhibits (from late May-early Sept, daily; rest of year Mon-Fri; free). Recreation areas along the reservoir offer water-skiing, fishing, and boating. Nature trails. Picnicking. Primitive and improved camping (mid-May-mid-Sept; fee). Open daily.

SOUTH DAKOTA DISCOVERY CENTER AND AQUARIUM. *805 W Sioux Ave, Pierre (57501). Phone 605/224-8295. www.sd-discovery.com.* Hands-on science and technology exhibits; aquarium features native fish. Open Memorial Day-Labor Day, Mon-Sat 10 am-5 pm, Sun 1-5 pm; Labor Day-Memorial Day, Sun-Fri 1-5 pm, Sat 10 am-5 pm. **$**

STATE CAPITOL. *500 E Capitol Ave, Pierre (57501). Phone 605/773-3765.* Built of Bedford limestone, local boulder granite, and marble. Guided tours. Open Mon-Fri. **FREE**

STATE NATIONAL GUARD MUSEUM. *301 E Dakota, Pierre (57501). Phone 605/224-9991. www.state.sd.us/military/Military/museum.htm.* Historical guard memorabilia. Open Mon, Wed, Fri 8 am-5 pm; tours by appointment.

PLACES TO STAY

If you choose to include an overnight stay in your trip along this Byway, Mobil Travel Guide recommends the following lodgings.

Chamberlain

★ **BEST WESTERN LEE'S MOTOR INN.** *220 W King Ave, Chamberlain (57325). Phone 605/734-5575; toll-free 800/780-7234. www.bestwestern.com.* 60 rooms, 2 story. Complimentary continental breakfast. Check-out 11 am. TV. In-house fitness room. Game room. Indoor pool, whirlpool. ¢

★ **RIVERVIEW INN.** *128 N Front St, Chamberlain (57325). Phone 605/734-6057.* 29 rooms, 2 story. Closed Nov-Apr. Check-out 10:30 am. TV. Laundry services. Sauna. Indoor pool, whirlpool. ¢

★ **SUPER 8.** *I-90; Lakeview Heights Rd, Chamberlain (57325). Phone 605/734-6548; toll-free 800/800-8000. www.super8.com.* 56 rooms, 2 story. Check-out 11 am. TV. ¢

Oacoma

★ **OASIS KELLY INN.** *1100 E Hwy 16, Oacoma (57365). Phone 605/734-6061; toll-free 800/635-3559.* 69 rooms, 2 story. Pet accepted, some restrictions. Complimentary continental breakfast. Check-out 11 am. TV; cable (premium). Laundry services. Bar. Sauna. Pool, whirlpool. Miniature golf. Airport transportation. Pond. On the river. ¢

Pierre

★★ **BEST WESTERN RAMKOTA HOTEL.** *920 W Sioux Ave, Pierre (57501). Phone 605/224-6877; toll-free 800/780-7234. www.bestwestern.com.* 151 rooms, 2 story. Pet accepted. Check-out noon. TV. Laundry services. Restaurant, bar. Room service. In-house fitness room, sauna. Game room. Indoor pool, children's pool, whirlpool. Free airport transportation. At the Missouri River. ¢

★ **CAPITOL INN & SUITES.** *815 E Wells Ave, Pierre (57501). Phone 605/224-6387; toll-free 800/658-3055.* 83 rooms, 2 story. Pet accepted, some restrictions. Check-out 11 am. TV. Pool. ¢

S. Dakota

✤ The Native American Scenic Byway

★ **DAYS INN AIRPORT.** *520 W Sioux Blvd, Pierre (57501). Phone 605/224-0411; toll-free 800/544-8313. www.daysinn.com.* 81 rooms, 2 story. Complimentary continental breakfast. Check-out 11 am. TV. Cross-country ski 1 mile. ¢

★★ **KINGS INN HOTEL AND CONVENTION CENTER.** *220 S Pierre St, Pierre (57501). Phone 605/224-5951; toll-free 800/232-1112.* 104 rooms, 2 story. Pet accepted, some restrictions; fee. Check-out 11 am. TV. Restaurant, bar. Room service. Sauna. Whirlpool. Cross-country ski 1 mile. ¢

★ **SUPER 8.** *320 W Sioux Ave, Pierre (57501). Phone 605/224-1617; toll-free 800/800-8000. www.super8.com.* 78 rooms, 3 story. Pet accepted, some restrictions; fee. Check-out 11 am. TV; cable (premium), VCR available. ¢

PLACES TO EAT

A long day of driving is sure to make you hungry. At the end of your journey, take a table at one of the following restaurants.

★ **AL'S OASIS.** *I-90, exit 260, Chamberlain (57365). Phone 605/734-6051. www.alsoasis.com.* American menu. Closed Jan 1, Thanksgiving, Dec 25. Breakfast, lunch, dinner. Bar. Children's menu. $$

★★★ **LA MINESTRA.** *106 E Dakota Ave, Pierre (57501). Phone 605/224-8090. www.laminestra.com.* Set in the historic downtown district in an 1896 building, La Minestra features original wainscoting and tin walls and ceilings. The restaurant serves upscale Italian cuisine in a wonderful ambience. Closed Sun. Lunch (Mon-Fri), dinner. Bar. Reservations not accepted. $$$

Peter Norbeck Scenic Byway
❈ SOUTH DAKOTA

Quick Facts

LENGTH: 68 miles.

TIME TO ALLOW: 2 to 4 hours.

BEST TIME TO DRIVE: Spring through autumn. High season is May through September.

BYWAY TRAVEL INFORMATION: South Dakota Department of Transportation, Custer: 605/673-4948; Custer State Park: 605/255-4515.

SPECIAL CONSIDERATIONS: Portions of the Byway have steep grades, sharp curves, and tunnels with height and width restrictions, with speed limits ranging from 20 to 45 mph. Custer State Park and Mount Rushmore are fee sites. Watch out for wildlife and other Byway visitors.

RESTRICTIONS: Portions of US 16A south of Mount Rushmore and SD 87 east of Sylvan Lake are closed during the winter.

BICYCLE/PEDESTRIAN FACILITIES: You'll find numerous hiking trails in the Black Hills National Forest and in Custer State Park, both of which are bordered by the Byway. For cyclists, the Mickelson Trail is a good choice for touring the area. It is a Rails to Trails path that runs parallel to the Byway next to Highway 16/385. Centennial Trail is also an option for hikers and bikers. The trail crosses the Byway in three places and offers beautiful scenic views of the Peter Norbeck Wilderness.

This 68-mile scenic route honors a South Dakota conservationist, governor, and US senator. Peter Norbeck first saw the Black Hills in 1905 after crossing the prairie on rugged, unimproved roads. His first visit began a lifelong love affair with the hills.

Norbeck played a key role in establishing Custer State Park and was an ardent supporter of the Mount Rushmore project. Norbeck Overlook, Norbeck Wildlife Preserve, and now this Byway all bear his name and memorialize his conservation achievements. Many of the traditions he began continue today. A mountain-size statue of Chief Crazy Horse is being completed to become another gem along this Byway.

The pristine beauty of the other pine-clad mountains with their dramatic granite pinnacles and outcrops impressed Norbeck and impress travelers even more today as natural lands and wonders continue to disappear. Norbeck chose to travel the routes of the Iron Mountain Road and the Needles Highway on foot and horseback in order to ensure that the natural features would be preserved. As you drive the Peter Norbeck Scenic Byway, let your imagination roam with the buffalo herds as you catch a vision of this amazing place.

THE BYWAY STORY

The Peter Norbeck Scenic Byway tells cultural, historical, natural, recreational, and scenic stories that make it a unique and treasured Byway.

Cultural

The earliest modern inhabitants of the Black Hills were the Crow Indians. Not much is known about these residents, except that they were forced to move west by the Cheyenne Indians. Ironically, the Cheyenne held the Black Hills for only a few decades. Westward-expanding Sioux Indians reached the hills around 1775. Old stories tell of great battles between the Sioux and Cheyenne tribes over the land. The last battle, which saw the defeat of the Cheyenne, is said to have taken place near Battle Mountain in Hot Springs, South Dakota. The Sioux reigned over the High Plains without any opposition for nearly a century.

In 1803, the Louisiana Purchase from France opened the West to white opportunists. The Dakotas, however, were regarded as Indian territory and were left virtually untouched by settlers and travelers. In an effort to ensure the safety of the travelers from Indian attack, the United States purchased Fort Laramie and negotiated a settlement with the Plains Indians. The United States agreed to pay for the safe passage of Americans through Indian country and granted that the Black Hills were sovereign territory of the Lakota (another name for the Sioux), promising to keep white settlers out of the hills.

There had been rumors of gold in the Black Hills, and George A. Custer and 1,000 of his men were in Dakota Territory on a military reconnaissance when the rumors became fact. Horatio Ross, a miner accompanying the troops, discovered gold in the bed of French Creek in the Black Hills. On July 30, 1874, newspapers around the country picked up the story: gold in the Black Hills! No single event shaped the destiny of the Black Hills of South Dakota like the 1874-1892 gold rush did. Opportunists pursuing golden dreams from all over the country soon gathered in Cheyenne, waiting for the hills to be opened for settlement. At first, the government arrested trespassers sneaking into the hills of the Sioux. But anxious settlers wouldn't wait. Within two years, nearly 10,000 people had invaded the hills, setting up small towns like Hill City, Silver City, Central, Deadwood, and Lead. Supplies were expensive and difficult to get, and miners took home around $20 to $125 a day. Some were very successful, leaving with large fortunes; others found nothing.

In 1876, the Sioux War broke out as the tension between the settlers and the Sioux escalated. After several defeats by the US Calvary, the Sioux, with neighboring tribes in Wyoming and Montana, united along the Big Horn River to battle Custer's troops. The overconfident Custer advanced into what is now known as the Battle of Little Bighorn. Custer and his troops were massacred, and the Sioux emerged victorious. As more US troops arrived on the plains, however, the tide turned. By October, most Indians were disarmed and forced onto reservations. The government bought the Black Hills from the Sioux for the price of $6 million, along with food and clothing supplies for the next 80 years. Even though this purchase occurred, tension between the Sioux and the white population remained high as small bands of Indians continued to attack the outlying settlers. The spring of 1877 became one of the bloodiest in the Black Hills' history, and settlers begged the United States for help. The government agreed to build Fort Meade just west of Bear Butte Mountain to permanently house soldiers who would protect the settlers.

Outnumbered and outgunned, the downhearted Sioux turned to their religion for deliverance. By performing the sacred Ghost Dance, the Lakota envisioned divine intervention that would drive out the white settlers and restore the great buffalo herds to the plains. When this message reached the nearby Black Hills towns, settlers became wary that another war was about to break out. Troops were sent to the reservation to quell the Ghost Dancers. Hundreds of Sioux men, women, and children died at Wounded Knee Creek when a minor melee triggered nervous troops to open fire in a tragic accident of war. The Wounded Knee incident of 1890 marked the end of bloodshed between Indians and whites.

Historical

Two centuries after French explorers recounted their impressions of the Black Hills, these isolated mountains attracted another visitor whose name would become intimately linked to them. In 1905, Peter Norbeck, son of Norwegian immigrants and a native of the Dakota Territory, traveled to the Black Hills from his home on the eastern Dakota prairies. Norbeck was a burly, unpretentious well driller and politician, a man of common origin but uncommon destiny. He soon became the unlikely steward of the Black Hills and a nationally prominent conservationist and legislator. In 1908, Norbeck began a distinguished political career as state senator. It carried him to the state capitol as lieutenant governor and governor, and then on to Washington, DC as a US senator.

Beginning with his first visit to the region in 1905 and continuing for several decades, one of Norbeck's greatest concerns was creating a great state park befitting the extraordinary beauty and diversity of the Black Hills. He envisioned a preserve that would serve as a sanctuary for wildlife and encompass features like the Needles, Harney Peak, and the Sylvan Lake area. His tireless efforts led to the establishment of Custer State Park in 1919 (the second largest state park in the United States) and the Norbeck Wildlife Preserve in 1920. Norbeck was keenly aware that the newly established state park had to be made accessible to visitors. He spent many days afoot and on horseback identifying potential road routes. He was guided by a persistent wish to preserve natural beauty, to build roads from which the public could obtain the grandest views, and to share this special place he loved with as many people as possible.

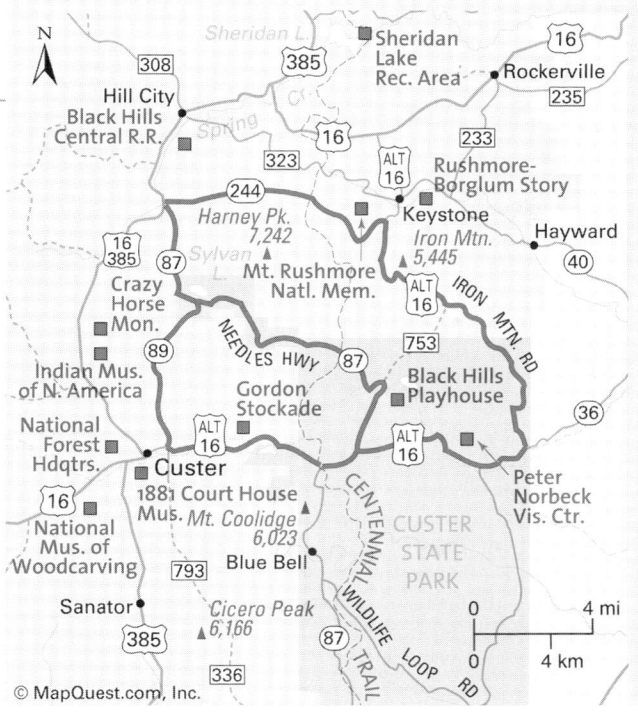

see page A23 for color map

In 1919, work began on the Needles Highway, following a route through rugged terrain that was often impassable on horseback. The road wound its serpentine way through ancient, upthrust fingers of stone, massed like stone sentinels protecting the heart of the Black Hills. The route was, to conventional engineering wisdom, impossible to build. However, Peter Norbeck was an unconventional man. When Norbeck asked an engineer named Johnson whether he could build the road, Johnson replied, "If you can furnish me enough dynamite, I can." In response, Norbeck supplied 150,000 pounds. By 1921, the spectacular Needles Highway was completed.

In addition to his work on the Needles Highway, Norbeck aided in raising funds for the Shrine of Democracy and Mount Rushmore, and he also helped engineer Iron Mountain Road. His philosophy about this road was, "You're not supposed to drive here at 60 miles an hour. To do the scenery half justice, people should drive 20 or under, [and] to do it full justice they should get out and walk." Dedicated on June 12, 1991, the Peter Norbeck Scenic Byway is a permanent memorial to a man who said, "I would rather be remembered as an artist than as a United States senator."

S. Dakota

❋ Peter Norbeck Scenic Byway

Natural

The Peter Norbeck Scenic Byway lies predominately within the crystalline (or granite) core of the central Black Hills. Many researchers refer to this area as the Harney Range. For the most part, sedimentary formations have eroded away from this range, often exposing massive granite mountainsides, outcrops, and spires, especially at higher elevations. At lower elevations, meadows interrupt stands of ponderosa pine and aspen, while narrow streams lined with grasses or hardwoods, such as bur oak and willow, tumble through park-like settings. This is a place where the buffalo still roam and skies are not cloudy all day.

Lakes and trees transform the Harney Range from sheer rock to a living, breathing landscape. Travelers will find ponderosa pine covering more than 90 percent of the area. The mature ponderosas, called yellow-barks, majestically tower over the road to create a canyon of trees. Granite outcrops, spruce trees, aspens, and meadows decorate the remaining 10 percent of the area. Lakes by the name of Horsethief, Lakota, Center, Legion, Bismarck, Stockade, and Sylvan offer beautiful features just next to the Byway, and visitors love to see them. Streams called Spring, Sunday, Palmer, Willow, Iron, Spokane, Grace Coolidge, Galena, and French Creeks crisscross the Byway, catching the eye and creating scenery along the drive.

Some of the higher and more prominent peaks near or adjacent to the Byway include Harney Peak (7,242 feet) and the nearby Needles area of granite spire, Calamity Peak (5,625 feet), Mount Coolidge (6,023 feet), Iron Mountain (5,445 feet), and Mount Rushmore (6,040 feet). A drive among these giants leaves drivers in awe of nature's sculptures. The Lakota called the Black Hills Paha Sapa, or "Hills That Are Black." Held in reverence by Native American tribes of the Northern Plains and traversed in their time by trappers, mountain men, miners, and tourists, the Black Hills have attracted generations of people in pursuit of interests as varied as the place itself.

Recreational

A variety of activities await Byway travelers on the Peter Norbeck Scenic Byway. Landmarks are coupled with the beauty of the wilderness areas, and cultural monuments and natural wonders are equally accessible along the Byway, which is encompassed by a state park and a national forest. With buffalo herds, giant sculptures, and mountain peaks just out the window, sightseeing is a favorite activity. But sightseeing is only where recreation begins on the Byway.

One of the best ways to see the Peter Norbeck Scenic Byway is to see it as Peter Norbeck did—on foot. Trails go throughout Custer State Park and the Black Hills National Forest. For cyclists and hikers, Centennial Trail runs into the backcountry areas of the Byway. The trail is known to be heavily forested and a great place to enjoy views of nearby mountain ranges. No matter how you travel the trails along the Byway, you'll be captivated by the sight of boulders and spires rising up from the ground to turn into mountains.

On the Peter Norbeck Scenic Byway, local culture is mixed with national memorials. Many visitors travel the Byway for the sole purpose of seeing a national landmark. Mount Rushmore can be seen from the Byway, but travelers will want to stop and hike the Presidential Trail to see the most impressive views of the monument. Byway visitors will also be delighted to discover the Black Hills Playhouse. This outdoor theater presents several productions during the summer to entice weary travelers at the end of a long day of sightseeing.

Roads with names like the Needles Highway and Iron Mountain Road are just waiting to be explored. The Needles Highway is a natural attraction for scenic drivers, and rock climbers find the granite spires of the area irresistible. Visitors camping in Custer State Park or staying in one of the Byway communities often take side trips to see some of the most astounding cave systems in the United States. Both Jewel

Cave National Monument and Rushmore Cave are just a short drive from the Byway. These caves offer adventures for both professional and amateur spelunkers.

Scenic

On the Peter Norbeck Scenic Byway, the road itself is a scenic feature that travelers enjoy year after year. Many come just for a glimpse of the majestic Mount Rushmore. Traveling through granite spires and past forests, lakes, and buffalo herds, natural beauty doesn't seem to be in short supply, either. Before it became a vacation area, the wilderness of southwestern South Dakota was recognized for unique scenery of rocky mountainsides and thick pine forests. As the Byway twists and curves past gulches, grasslands, mountains, and lakes, you get a look at the wilderness that captured the hearts of Peter Norbeck and many other explorers.

As a monument to the Midwest, the Peter Norbeck Scenic Byway immortalizes ideals for a fantastic land enjoyed and respected by travelers from far and wide. It was recognized immediately as the perfect place for one of the largest monuments in the country. Mount Rushmore is considered one of America's scenic wonders. Drivers love to catch this landmark as they drive through one of the Byway tunnels. Travelers stop to try to capture these majestic sculptures on film when the sun is almost gone and the contours of each face are most pronounced.

Pass by mountain lakes or stop and enjoy this placid setting for a while. The Byway passes a handful of lakes that are perfect for shoreline picnics or a little fishing. Piles of giant granite boulders make Sylvan Lake unique, and its inviting shoreline make it a perfect place to stop and rest on the hike to Harney Peak. On calm days, the lake reflects the boulders above it,

creating a great bridge of stone with water and sky on either side. At the edge of Horsethief Lake, another Byway stop, you find towering ponderosa forests, where an afternoon of hiking and exploring is the best way to enjoy the scenery.

As you drive among the monuments, lakes, and natural splendor of the Peter Norbeck Scenic Byway, you'll see why the road is a favorite for driving and for recreation. It's a landscape so beautiful that travelers prefer to be in the midst of it all.

HIGHLIGHTS

As you drive the two loops that make up the Peter Norbeck Scenic Byway, be aware that portions of the Byway have steep grades, sharp curves, tunnels, and bridges. Watch out for wildlife, other recreationists, and traffic.

- You may want to begin your tour at the **General Custer Expedition Campsite,** at the southwest corner of the Byway. The site was base camp for the 1874 Custer Expedition that explored and mapped the Black Hills and discovered gold at this location.

- Traveling eastbound on 16A, you encounter the **Gordon Stockade,** a replica of the 1874

S. Dakota

❋ Peter Norbeck Scenic Byway

structure built by the first settlers in the hills. Also in the area is **Stockade Lake,** which provides camping and picnic facilities. You enter **Custer State Park** here.

- Continuing farther east along the Byway, turn left onto Highway 87 at Legion Lake heading north to the next point of interest. Shortly after turning left, you encounter the **Forest Ecology Wayside,** which conveys the natural history of the Black Hills forests and fire.

- Shortly after passing the Forest Ecology Wayside, you pass the turn-off to the **Black Hills Playhouse.** At this point, the Byway makes a sharp turn to the northwest and leaves Custer State Park. This part of the Byway is known as the **Needles Highway** because of the geological formations in the rock that resemble needles. In this area, you should notice the **Cathedral Spires,** a spectacular rock formation in the 1.7-billion-year-old Harney Peak granite. Also in the area is the **Needles Eye,** a unique erosion feature.

- Sylvan Lake is located at the point where Highways 87 and 89 intersect. At this junction, take a right turn heading north on the Byway. **Sylvan Lake** is the oldest reservoir in the Black Hills, constructed in 1898. Take time to swim, fish, hike, camp, or rock climb in the area. From here, the Byway continues north for several miles before making a sharp turn to the east onto Highway 244.

- Only a few miles onto Highway 244, you enter the **Norbeck Wildlife Preserve.** The preserve is home to deer, elk, mountain goats, bighorn sheep, and many bird species. **Black Elk Wilderness** is named for the Lakota spiritual leader.

- Only a few miles to the east is **Horsethief Lake,** a favorite spot for camping and for hikes into the backcountry. At this point, the Byway begins traveling southeast toward Washington Profile.

- **Washington Profile** is a spectacular side view of the Washington head on Mount Rushmore. This area is frequently visited by mountain goats.

- **Mount Rushmore** was carved by Gutzon Borglum between 1927 and 1941 and stands as a memorial to significant figures of our present form of government.

- Traveling southeast along the Byway, you encounter a portion of the highway known as **Pigtail Bridges.** These unique bridges, designed by Norbeck and built by the Civilian Conservation Corps in the 1930s, span steep climbs in short distances. These bridges get their name because they consist of a series of spiral curves, much like the shape of a pig's tail.

- **Norbeck Overlook** is located at the southern end of the area of Pigtail Bridges. At this spot, you can stop at the overlook to view the panoramic scenes of Harney Peak, the outcrops of ancient granite, and the 9,800-acre Black Elk Wilderness.

- The Byway continues south for several miles, entering Custer State Park once again and then traveling through the **Galena Fire Wayside.** The area focuses on the 1988 wildfire that burned more than 16,000 acres and illustrates fire recovery. From here, the Byway continues south before making a sharp turn west.

- At the point where the Byway makes the sharp westward turn, you find yourself at **Game Lodge,** which was built as the gamekeeper's residence and later served as the 1927 Coolidge summer White House. Only a short distance from Game Lodge is the **Peter Norbeck Visitor Center,** offering information on Norbeck's life and Black Hills history.

- West of the lodge is the **Park Office.** You can continue west once again to the intersection of Highways 16A and 87, where the tour began.

THINGS TO SEE AND DO

Driving along the Peter Norbeck Scenic Byway will certainly keep your senses engaged, but if you yearn to get out of the car and stretch your legs, or if you'd like to make a mini-vacation out of your trip, check out these attractions along the route.

1875 LOG CABIN. *Way Park, Custer (57730).* The oldest cabin in the Black Hills, preserved as a pioneer museum. $

★ **1880 TRAIN.** *222 Railroad Ave, Hill City (57745). Phone 605/574-2222. www.1880train.com.* Steam train runs on a Gold Rush-era track; vintage railroad equipment. Two-hour round trip between Hill City and Keystone through a national forest and mountain meadowlands. Vintage car restaurant, gift shop. Open mid-May-mid-Oct, daily. $$$$

BIG THUNDER GOLD MINE. *604 Blair St, Keystone (57751). Phone 605/666-4847; toll-free 800/843-1300.* Authentic 1880s gold mine. Visitors may dig gold ore or pan it by the stream. Guided tours; historic film. Open May-Oct, daily. $$$

★ **BLACK HILLS.** *Custer (57730). www.blackhills.com.* Magnificent forests, mountain scenery, ghost towns, Mount Rushmore National Memorial, Harney Peak (the highest mountain east of the Rockies), Crazy Horse Memorial, swimming, horseback riding, rodeos, hiking, skiing, and the Black Hills Passion Play make up only a partial list of attractions. Memories of Calamity Jane, Wild Bill Hickock, and Preacher Smith (all buried in Deadwood) haunt the old Western towns. There are parks, lakes, and picturesque mountain streams. Bison, deer, elk, coyotes, mountain goats, bighorn sheep, and smaller animals make this area home. Black Hills National Forest includes 1,247,000 acres—nearly half of the Black Hills. The forest offers 28 campgrounds, 20 picnic grounds, and one winter sports area. Daily fees are charged at most campgrounds. Headquarters are in Custer.

For information and a map ($) of the National Forest, write to the Forest Supervisor, Train 2, Box 200, Custer 57730. Two major snowmobile trail systems, one in the Bearlodge Mountains and the other in the northern Black Hills, offer 330 miles of some of the best snowmobiling in the nation. There are also 250 miles of hiking, bridle, and mountain biking trails. Pactola Visitor Center, on US 385 at Pactola Reservoir, west of Rapid City, has information and interpretive exhibits on Black Hills history, geology, and ecology (open Memorial Day-Labor Day).

BORGLUM HISTORICAL CENTER. *342 Winter St, Keystone (57751). Phone 605/666-4449; toll-free 800/888-4369. www.rushmoreborglum.com.* Exhibits on Gutzon Borglum and the carving of Mount Rushmore. Newsreels of carving in progress. Original models and tools, collection of his paintings, unpublished photos of the memorial, sculptures, historical documents. Full-size replica of Lincoln's eye. Open May-Oct, daily. $$$

COSMOS OF THE BLACK HILLS. *Hwy 16 W, Rapid City (57701). Phone 605/343-9802.* Curious gravitational and optical effects. Guided tours every 12 minutes. Open Apr-Oct, daily. $$$

★ **CRAZY HORSE MEMORIAL.** *5 miles N of Custer off US 16, 385, Custer (57730). Phone 605/673-4681 or 605/673-2828 (museum). www.crazyhorse.org.* This large sculpture, still being carved from the granite of Thunderhead Mountain, was the life work of Korczak Ziolkowski (1908-1982), who briefly assisted Gutzon Borglum on Mount Rushmore. With funds gained solely from admission fees and contributions, Ziolkowski worked alone on the memorial, refusing federal and state funding. The work is being continued by the sculptor's wife, Ruth, and several of their children. The sculpture will depict Crazy Horse—the stalwart Sioux chief who helped defeat Custer and the US Seventh Cavalry—astride a magnificent horse. It is meant to honor not only Crazy Horse and the unconquerable human spirit, but also all Native American tribes.

S. Dakota

❋ Peter Norbeck Scenic Byway

Crazy Horse's emerging head and face are nearly nine stories tall. When completed in the round, the mountain carving will be 563 feet high and 641 feet long—the largest sculpture in the world. To date, 8.4 million tons of granite have been blasted off the mountain. Audiovisual programs and displays show how the mountain is being carved.

The Indian Museum of North America is on the grounds and houses some 20,000 artifacts in three wings. A Native American educational and cultural center opened in 1997. The visitor complex also includes the sculptor's log studio/home and workshop filled with sculpture, fine arts, and antiques. A restaurant is open daily (in season). The memorial is open summer 7 am-dark; winter 8 am-4:30 pm; closed Dec 25. **$$**

CUSTER STATE PARK. *5 miles E on US 16 A, Custer (57730). Phone 605/255-4515. www.custerstatepark.info/.* This is one of the largest state parks in the United States—73,000 acres. A mountain recreation area and game refuge, the park has one of the largest publicly owned herds of bison in the country (more than 1,400), as well as Rocky Mountain bighorn sheep, mountain goats, burros, deer, elk, and other wildlife. Four man-made lakes and three streams provide excellent fishing and swimming. Near the park is the site of the original gold strike of 1874 and a replica of the Gordon stockade, built by the first Gold Rush party in 1874. Peter Norbeck Visitor Center (open May-Oct, daily) has information about the park and naturalist programs, which are offered daily (May-Sept). Paddleboats; horseback riding, hiking, bicycle rentals, jeep rides, camping, hayrides, chuck-wagon cookouts. A park entrance license is required. **$$**

The **Black Hills Playhouse,** in the heart of the park, is the scene of productions for 11 weeks (mid-June-late Aug, schedule varies); phone 605/255-4141. **$$$**

FLINTSTONES BEDROCK CITY. *Hwy US 16 and 385, Custer (57730). Phone 605/673-4079.* Adventures with the modern stone-age family: Fred, Wilma, Barney, Betty, Pebbles, Bamm Bamm, and Dino. Village tour; train ride; concessions. Campground. Open mid-May-mid-Sept, daily. **$$$**

★ JEWEL CAVE NATIONAL MONUMENT. *14 miles W on US 16, Custer (57730). Phone 605/673-2288. www.nps.gov/jeca/.* On a high rolling plateau in the Black Hills is Jewel Cave, with an entrance on the east side of Hell Canyon. More than 100 miles of passageways make this cave system the second longest in the United States. Formations of jewel-like calcite crystals produce unusual effects. The surrounding terrain is covered by ponderosa pine. Many varieties of wildflowers bloom from early spring through summer on the 1,274-acre monument. There is a guided 1 1/4-hour scenic tour of the monument (May-Sept) and a 1 1/2-hour historic tour (Memorial Day-Labor Day; age 6 or older only). There is also a four- to five-hour spelunking tour (June-Aug); advance reservations required; minimum age 16. All tours are recommended only for those in good physical condition. For spelunking, wear hiking clothes and sturdy, lace-up boots. Visitor center. Open daily. **$$$**

★ MOUNT RUSHMORE NATIONAL MEMORIAL. *25 miles SW of Rapid City off US 16 A and SD 244, Custer (57730). Phone 605/574-2523. www.nps.gov/moru/.* The faces of four great American presidents—Washington, Jefferson, Lincoln, and Theodore Roosevelt—stand out on a 5,675-foot mountain in the Black Hills of South Dakota, as grand and enduring as the contributions of the men they represent. Senator Peter Norbeck was instrumental in the realization of the monument. The original plan called for the presidents to be sculpted to the waist. It was a controversial project when sculptor Gutzon Borglum began his work on the carving in 1927. With crews often numbering 30 workers, he continued through 14 years of crisis and heartbreak and had almost finished by March 1941, when he died. Lincoln Borglum, his son, brought the project to a close in October of that year. Today, the memorial is host to almost 3 million visitors a year.

An orientation center, administrative and information headquarters, gift shop, and a snack bar (open daily) are on the grounds; also Buffalo Room Cafeteria. Evening program followed by sculpture lighting and other interpretive programs (mid-May-mid-Sept, daily). Sculptor's studio museum (summer). $$$

MOUNTAIN MUSIC SHOW. *1160 Camino Cruz Blanca, Custer (87501). Phone 605/673-2405.* Country music show with comedy; family entertainment. Open late May-Labor Day, daily. $$

NATIONAL MUSEUM OF WOODCARVING. *2 miles W on US 16 W, Custer (57730). Phone 605/673-4404. www.blackhills.com/woodcarving/.* Features woodcarvings by an original Disney animator and other professional woodcarvers; Wooden Nickel Theater, museum gallery, carving area, and snack bar. Open May, Sept-Oct, daily 9 am-5 pm; June-Aug, daily 8 am-8 pm.

PARADE OF PRESIDENTS WAX MUSEUM. *609 Hwy 16 A, Keystone (57543). Phone 605/666-4455.* Nearly 100 life-size wax figures depict historic scenes from the nation's past. Open May-Sept, daily. $$

RUSHMORE AERIAL TRAMWAY. *203 Cemetery Rd, Keystone (57751). Phone 605/666-4478. www.rushmoretramway.com.* This 15-minute ride allows a view of the Black Hills and Mount Rushmore across the valley. Open late May-early Sept, daily. $

RUSHMORE CAVE. *13622 Hwy 40, Keystone (57751). Phone 605/255-4467 or 605/255-4384. www.beautifulrushmorecave.com.* Guided tours May-Oct, daily. $$

RUSHMORE HELICOPTER SIGHTSEEING TOURS. *301 Cemetery Rd, Keystone (57751). Phone 605/666-4461.* Helicopter rides over Mount Rushmore and nearby points of interest. Open mid-May-late Sept, daily. $$$$

WADE'S GOLD MILL. *12401 Deerfield Rd, Hill City (57745). Phone 605/574-2680.* This authentic mill shows four methods of recovering gold. For a fee, you can pan for gold yourself. Open Memorial Day-Labor Day, daily. $$

PLACES TO STAY

If you choose to include an overnight stay in your trip along this Byway, Mobil Travel Guide recommends the following lodgings.

Custer

★★ **BAVARIAN INN.** *Hwy 16 and 385 N, Custer (57730). Phone 605/673-2802; toll-free 800/657-4312. www.custer-sd.com/bavarian.* 65 rooms, 2 story. Pet accepted, some restrictions; fee. Check-out noon, check-in 2 pm. TV; cable (premium). Sauna. Game room. Two pools, whirlpool. Outdoor tennis, lighted courts. Lawn games. ¢

★★ **CUSTER MANSION BED & BREAKFAST.** *35 Centennial Dr, Custer (57730). Phone 605/673-3333; toll-free 877/519-4948. www.custermansionbb.com.* 6 rooms, 2 story. No A/C. No room phones. Complimentary full breakfast. Check-out 11 am, check-in 3-7 pm. TV in sitting room. Totally nonsmoking. ¢

★★ **STATE GAME LODGE.** *Hwy 16 A; Custer State Park, Custer (57730). Phone 605/255-4541; toll-free 800/658-3530. www.custerresorts.com.* 68 rooms, 2 and 3 story. No A/C in cabins, lodge, motel units. Some room phones. Closed Oct-Apr. Pet accepted, some restrictions; fee. Check-out 10 am, check-in 2 pm. TV. Restaurant, bar. Hiking trails. $

❋ *Peter Norbeck Scenic Byway*

★ **SUPER 8.** *415 W Mt Rushmore Rd (US 16), Custer (57730). Phone 605/673-2200; toll-free 800/800-8000. www.super8.com.* 54 rooms, 2 story. Complimentary continental breakfast. Check-out 11 am, check-in 3 pm. TV. Game room. Indoor pool. ¢

★★ **SYLVAN LAKE RESORT.** *HC 83, Box 74, Custer (57730). Phone 605/574-2561; toll-free 800/658-3530. www.custerresorts.com.* 33 rooms, 3 story. No A/C. Closed Oct-Apr. Pet accepted, some restrictions; fee. Check-out 10 am, check-in 2 pm. TV. Restaurant. Paddleboats. Hiking. $

Hill City

★★ **BEST WESTERN GOLDEN SPIKE INN & SUITES.** *106 Main St, Hill City (57745). Phone 605/574-2577; toll-free 800/780-7234. www.bestwestern.com.* 80 rooms, 2 story. Closed Dec-Mar. Pet accepted, some restrictions; fee. Check-out 11 am, check-in 3 pm. TV. In-room modem link. Restaurant. In-house fitness room. Game room. Pool, whirlpool. Bicycle rentals. $

★ **COMFORT INN.** *678 Main St, Hill City (57745). Phone 605/574-2100; toll-free 877/424-6423. www.choicehotels.com.* 55 rooms, 2 story. Complimentary continental breakfast. Check-out noon, check-in 2 pm. TV; cable (premium). Indoor pool, whirlpool. $

Keystone

★ **BEST WESTERN FOUR PRESIDENTS LODGE.** *24075 US 16 A, Keystone (57751). Phone 605/666-4472; toll-free 800/780-7234. www.bestwestern.com.* 49 rooms, 3 story. No elevator. Closed Nov-Mar. Complimentary continental breakfast. Check-out 11 am, check-in 3 pm. TV; cable (premium). In-room modem link. In-house fitness room. $

★ **RUSHMORE EXPRESS.** *610 US 16 A, Keystone (57751). Phone 605/666-4483; toll-free 800/635-3559 or 800/323-6476. www.rushmoreexpress.com.* 44 rooms, 2 story. Closed Nov-Apr. Pet accepted, some restrictions; fee. Complimentary continental breakfast. Check-out 11 am, check-in 3 pm. TV. Laundry services. In-house fitness room, sauna. Whirlpool. ¢

PLACES TO EAT

A long day of driving is sure to make you hungry. At the end of your journey, take a table at one of the following restaurants.

★ **SKYWAY.** *511 Mt Rushmore Rd (US 16), Custer (57730). Phone 605/673-4477.* American menu. Closed Thanksgiving, Dec 25. Dinner. Bar. Children's menu. Casual attire. $$

★★ **ALPINE INN.** *225 Main St, Hill City (57745). Phone 605/574-2749.* American menu. Closed Sun; Thanksgiving; also Dec 23-mid-Jan. Lunch, dinner. Casual attire. Totally nonsmoking. $

The Great River Road

❄ WISCONSIN

Part of a multistate Byway; see also IL, IA, MN.

Quick Facts

LENGTH: 249 miles.

TIME TO ALLOW: 10 hours.

BEST TIME TO DRIVE: The Great River Road is a delight to drive in any season. High season is summer.

BYWAY TRAVEL INFORMATION: La Crosse Area Convention and Visitor Bureau: toll-free 800/658-9480; Byway local Web site: www.wigreatriverroad.org.

SPECIAL CONSIDERATIONS: Fuel stations and food services are available in the 33 river towns along the Byway. These towns are found on average about every 10 miles.

RESTRICTIONS: No seasonal road closures are anticipated.

BICYCLE/PEDESTRIAN FACILITIES: Nearly 217 miles of the Wisconsin Great River Road make up the on-road portion of the bikeway. This portion has been rated as acceptable for accommodating cyclists.

Wisconsin's Great River Road flanks the majestic and magnificent Mississippi River as it leisurely winds its way along 250 miles of Wisconsin's western border. Along the way, the road is nestled between the river on one side and towering bluffs on the other, becoming one of the most scenic drives in mid-America. Most of the time, the road parallels the river, but when the road does meander a short way from the river, it treats its guests to vistas of rolling farmland, as well as beautiful forested valleys and coulees.

The 33 quaint river towns along this Byway proudly reveal their culture and heritage through their festivals and the 19th- and 20th-century architecture of their homes, business blocks, storefronts, mansions, and more. The corridor is rich in history, and Great River Road travelers can learn about the early Native American occupants, the French fur traders and explorers, the lead mining boom, the steamboat era, and the lumber barons. Travelers can stop at the 30 or more state historical markers and archaeological sites and at the many local museums.

Recreational opportunities await you around each bend of the Great River Road at more than 50 local parks and beaches, in addition to 12 state and three national recreational resources. Observation decks at four lock and dams provide the opportunity to watch the barges and riverboats pass through. You may even see steamboats like the *Delta Queen* cruise by.

Wisconsin

The Great River Road

Travel the Wisconsin Great River Road. It is a marvelous mix of natural beauty and history and is an area to be enjoyed by all ages—leisurely. In the new-green brilliance of spring or the white mantle of winter, in the summer sunlight or the amber hues of autumn, Wisconsin's Great River Road is a delight in any season.

THE BYWAY STORY

The Great River Road tells archaeological, cultural, historical, natural, recreational, and scenic stories that make it a unique and treasured Byway.

Archaeological

Based on many archaeological studies, it is estimated that Wisconsin was inhabited by people nearly 12,000 years ago, who hunted prehistoric animals like mammoths and mastodons. The primitive cultures of the Wisconsin Great River Road developed and evolved to leave behind many artifacts and monuments to the past—not the least of which were the great mounds of grand designs and animal shapes. Today, Wisconsin archaeologists research these ancient cultures, as well as the not-so-ancient cultures of the first explorers and traders who came to settle along the Wisconsin Great River Road.

The Wisconsin Great River Road runs through an area called the Driftless Area because it was not covered by glaciers during the last ice age. Because of this, many exposed rocks gave ancient inhabitants of the area an opportunity to create rock art that is still being discovered today. In caves and outcroppings, petroglyphs and pictographs have been found as displays of a way of life long forgotten. Designs of animals and people are left behind in the rocks for archaeologists to study.

Archaeological displays can be found in many local museums. You can view mound groups and village sites at Wyalusing State Park, Diamond Bluff, La Crosse, Prairie du Chien, along Lake Pepin, and at Trempealeau in Perrot State Park. In the city of La Crosse, archaeological enthusiasts can view displays at the Riverside Museum that catalog the earliest times in Wisconsin to the present. The Archaeology Center of the University of Wisconsin-La Crosse offers a closer look at artifacts from recent investigations, as well as detailed information about archaeology in Wisconsin. As the headquarters for the Mississippi Valley Archaeology Center, the University of Wisconsin-La Crosse Archaeology Center offers you an opportunity to view displays explaining techniques and prehistory.

Archaeological resources are plentiful along the entire length of the Wisconsin Great River Road corridor. There are 33 archaeological sites that are currently listed in the National Register of Historic Places. Excavations in the corridor have revealed pottery, ceramics, arrowheads, and tools, while burial mounds are prevalent throughout the corridor, ranging from individual sites to large groups. In Onalaska, an entire prehistoric village was uncovered, revealing structures and artifacts that indicated the lifestyle of the earliest inhabitants of the area. At Prairie du Chien, archaeologists have uncovered an American military garrison that was established in 1829. Visiting Trempealeau and Perrot State Park, you'll find remaining ancient mounds in the shapes of animals. With so many sites like these along the Wisconsin Great River Road, travelers develop a sense of respect for those who have come before to the mighty Mississippi.

Cultural

The varied past and present cultures of the corridor are recorded and revealed in the 33 river towns, the many state historical markers, and the archaeological sites found along the Byway. The residents of the corridor take pride in preserving their heritage, as evidenced by the many festivals. Some of these festivals include La Crosse's Riverfest and Octoberfest, Villa Louis's Carriage Classic, Prairie du Chien's Fur Trade Rendezvous, Alma's Mark Twain Days, and Pepin's Laura Ingalls Wilder Days. Travelers lucky enough to encounter one of these festivals get a taste of the true flavor of Wisconsin's Great River Road.

Well-maintained early architecture of homes and storefronts is evident throughout the corridor as well. Architecture from the 19th and 20th centuries is scattered throughout the towns and cities of the Byway; many of them reflect the varied architectural trends of the early days of settlement. Greek Revival, Italianate, and Queen Anne are just some of the styles you'll notice as you drive the Byway. As you watch for unique architecture, you will also want to notice the mail-order houses that were constructed in a matter of days after arriving by train. Today, unique buildings and art forms continue to surface on the Great River Road. At Prairie Moon Sculpture Garden, for example, a unique form of art typical to the Midwest is displayed.

Historical

Indians were the first people to live in this region, as evidenced by artifacts from archaeological sites and the presence of burial mounds—many of which survive today. European explorers and missionaries arrived in 1673. This area was first claimed by the French, and then later by the British. The United States gained control of the Northwest Territory in 1794, but many British traders maintained their lucrative posts until after the War of 1812. You can stop and ponder this area's history at the 30-plus historical markers found along the Byway, such as Fort Antoine, the Battle of Bad Ax, and the War of 1812. You also learn more at the Fur Trade Museum and Villa Louis in Prairie du Chien.

If you're a history hunter driving the Byway, you may notice evidence of a culture very different from America's present culture. Thousands of mounds can be found throughout the area that display the culture of the Hopewell Indians who once lived here. The culture of this people evolved over the years, and they began to establish large, permanent villages. Known as the Oneota people, they were able to farm the river valley using hoes made from bison shoulder blades. The way of life these people developed can be seen in the many museums along the road that display tools and artwork of past cultures. By the time the first Europeans arrived, this culture had disappeared, replaced by a group of Sioux.

French missionary Jacques Marquette and explorer Louis Jolliet were the first Europeans to come through the area. They were searching for a waterway that would connect the Atlantic Ocean and the Gulf of Mexico. Later, French forts were established, and commerce and trade between the European and native cultures ensued. The area changed hands from the Indians to the French to the British and finally to the Americans, but not without struggle. From the beginning, Native Americans fought to retain their ancestral lands, but to no avail. Settlement began in Wisconsin soon after the Black Hawk War between the Sauk Indians and American troops. In 1848, Wisconsin became a state.

Wisconsin thrived as a state for lumbering and sawmills. Because the Mississippi flows alongside it, Wisconsin was in a good position for the steamboat industry to develop. Steamboat races and wrecks were as legendary then as they are today. Remnants of the new Mississippi culture can be seen along the Great River Road:

see page A5 for color map

Wisconsin

The Great River Road

abandoned quarries and old building ruins are just some of the things you may spot that remind you of an earlier day. Although the cities and communities have grown and changed, Mississippi River heritage remains.

Natural

Many natural wonders are found along this Byway. For example, the Mississippi River/Wisconsin Great River Road corridor incorporates four national features: the Upper Mississippi River National Wildlife and Fish Refuge, the Trempealeau National Wildlife Refuge, the Genoa National Fish Hatchery, and the St. Croix River National Scenic River. There are also 12 state-recognized natural areas featuring state parks and wildlife areas, as well as 17 state-designated scientific areas located along the corridor.

Recreational

Every season offers spectacular recreational opportunities on the Upper Mississippi River along the Wisconsin Great River Road. The upper Mississippi River provides excellent boating and sailing, and there are over 50 local parks, beaches, recreational areas, and water access sites along the route. On Lake Pepin, a huge lake in the Mississippi River, boaters have access to numerous boat landings, marinas, and docking. Fishing is a favorite activity because of the variety of fish species, ranging from catfish to walleye. The sandbars along the river provide places for public camping, picnicking, or just getting off the road.

Bird-watchers enjoy seeing bald eagles as the magnificent birds catch their dinners. Berry-picking and mushroom-hunting are also popular activities, and many travelers choose to experience a farm vacation by milking a cow. Shopping and antique hunting in quaint river towns may be of interest to shoppers, while golfers will find enticing, scenic golf courses.

The Wisconsin Great River Road provides safe accommodations for bikers—with alternate choices of separate bike trails and local roads or streets. Depending on the cyclist's preference and skills, the rider has a choice of touring the Great River Road on-road or off-road. Canoe and bike rentals are available so that every visitor may enjoy these forms of recreation.

Winter in Wisconsin provides ice fishing and wind sailing on the river. Many travelers prefer cross-country or downhill skiing, snowshoeing, and snowmobiling through deep valleys and scenic bluffs. Summertime is excellent for dinner cruises on the river, or you can rent a houseboat and explore on your own. The river valley features hiking and biking trails, picnic areas, and camping opportunities in the numerous parks and campgrounds located along the Byway.

Scenic

Wisconsin's Great River Road meanders through the Mississippi River corridor. This corridor forms the southern half of the state's western border. The corridor's dramatic landscape was created by the melting ice age glaciers, which carved the magnificent river valley.

Many segments of the Wisconsin Great River Road parallel the Mississippi River, some of which gracefully snuggle between the bluffs and the river. You're afforded numerous vistas of the mighty Mississippi and its valley and vast backwaters. You are also accommodated by 20-plus waysides, scenic overlooks, and pull-out areas. Travel the Great River Road during all four seasons to experience its different year-round splendors.

HIGHLIGHTS

The following is an itinerary for the Victory to Bridgeport section—which is only one portion—of Wisconsin's Great River Road.

- You begin your tour in **Victory.** This small settlement along the Great River Road has a picturesque setting: snuggled next to the river on one side, with bluffs acting as a backdrop on the other. Five settlers laid out this village in 1852 and named it Victory to commemorate the final battle of the Black Hawk War fought south of the village 20 years earlier. Victory prospered during the wheat boom of the 1850s, but today, it is only a remnant of its past.
- The village of **De Soto** is 4 miles south of Victory. This river town has the distinction of being named after the famous Spanish explorer Fernando de Soto, the first European to see the Mississippi River. It was platted in 1854 on the site of a small outpost of the American Fur Company. Today, this community is a shadow of its past, when it peaked with sawmills, grain dealers, blacksmiths, dressmakers, breweries, and hotels. Learn from the locals how the wing dams constructed in the Mississippi diverted the river closer to their community.
- Eight miles south of De Soto lies **Ferryville.** This little river town clings to the bluffs along the river and is the longest one-street village in the world. It was first called Humble Bush but was renamed Ferryville when platted in 1858. The name reflects the founder's intentions to establish ferry service across the Mississippi to Iowa. In 1878, after being devastated by a tornado, it was written that "today a passerby can see no evidence of a village…" Ferryville still clings to the bluffs and portrays a true river town experience to its visitors.
- Your next stop, **Lynxville,** is 8 miles past Ferryville. Because of the stable depth of the river at Lynxville, it was a reliable and popular landing during the steamboat era of the mid- to late 1800s. Although the steamboats are gone, this quaint little river town remains as the host community to Lock and Dam 9.
- With about 6,000 residents, **Prairie du Chien** is the largest town on your tour. Stop at the Wisconsin Tourist Information Center to find out about the many area attractions of this second oldest settlement in Wisconsin. It became a trade center as early as the 1670s with the arrival of Marquette and Jolliet. Hercules Dousman built **Villa Louis,** now owned and operated by the State Historical Society, an opulent 1870s estate with one of the nation's finest collections of Victorian decorative arts. The Villa Louis Historical marker at this site provides an overview of the origin and history of this luxurious mansion. Medical history from the 1800s and an exhibit of medical quackery is displayed at the **Fort Crawford Medical Museum.** Some warehouses built in the early 19th century by the American Fur Company still survive on historic St. Feriole Island, as do remnants of the old American Fort built to protect this outpost. Tour the town in a horse and carriage or view the Mississippi aboard an excursion boat.
- Driving 7 miles southeast of Prairie du Chien, you arrive in **Bridgeport.** The name of this village is most fitting. In the late 1800s, a ferry carried grain and other farm products across the Mississippi River to a railroad in Minnesota. Today, Bridgeport is near the highway bridge crossing the Wisconsin National Scenic River and the gateway to Wyalusing State Park and Sentinel Ridge, where the Woodland Indians left behind hundreds of earthen mounds.

Wisconsin

✳ *The Great River Road*

THINGS TO SEE AND DO

Driving along the Great River Road will certainly keep your senses engaged, but if you yearn to get out of the car and stretch your legs, or if you'd like to make a mini-vacation out of your trip, check out these attractions along the route.

CITY BREWERY TOUR. *925 S 3rd St, La Crosse (54601). Phone 608/785-4200; toll-free 800/433-BEER. www.citybrewery.com.* One-hour guided tours. Open Mon-Sat; closed holidays. **FREE**

FORT CRAWFORD MUSEUM. *717 S Beaumont Rd, Prairie du Chien (53821). Phone 608/326-6960. www.fortcrawfordmuseum.com.* Relics of 19th-century medicine, Native American herbal remedies, drugstore, dentist and physicians' offices. Educational health exhibits; Dessloch Theater displays "transparent twins." Dedicated to Dr. William Beaumont, who did some of his famous digestive system studies at Fort Crawford. Open May-Oct, daily 10 am-4 pm; until 5 pm Jun-Aug. **$**

GOOSE ISLAND COUNTY PARK. *W6488 County Rd GI, Stoddard (54658). Phone 608/788-7018.* Beach, fishing, boat ramps; hiking trails, picnicking, camping (electric hookups). Open mid-Apr-mid-Oct, daily. **$$**

GRANDDAD BLUFF. *400 La Crosse St, La Crosse (54601). Phone 608/789-7533.* At 1,172 feet, this is the tallest of the crags that overlook the city; it provides a panoramic view of the winding Mississippi, the tree-shaded city, and the Minnesota and Iowa bluffs. Picnic area. There is a surfaced path to the shelter house and the top of the bluff for the disabled. Open May-late Oct, daily. **FREE**

HIXON HOUSE. *429 N 7th St, La Crosse (54601). Phone 608/782-1980 or 608/782-1990. www.lchsonline.org/hixon.html/.* A 15-room home (circa 1860); Victorian and Asian furnishings. Visitor information center and gift shop in a building that once served as a wash house. Open Memorial Day-Labor Day, daily 1-5 pm. **$**

✳ **KICKAPOO INDIAN CAVERNS AND NATIVE AMERICAN MUSEUM.** *54850 Rhein Hollow Rd, Wauzeka (53826). Phone 608/875-7723. www.kickapooindiancaverns.com.* Largest caverns in Wisconsin, used by Native Americans for centuries as a shelter. Sights include a subterranean lake, Cathedral Room, Turquoise Room, Stalactite Chamber, and Chamber of the Lost Waters. Guided tours mid-May-Oct, daily 9 am-4 pm. **$$$**

***LA CROSSE QUEEN* CRUISES.** *Boat Dock, Riverside Park, W end of State St, La Crosse (54603). Phone 608/784-2893 (boat dock) or -8523 (group reservations). www.greatriver.com/laxqueen/.* Sightseeing cruises on the Mississippi River aboard a 150-passenger, double-decker paddlewheeler. Also dinner cruises (Fri night, Sat, and Sun). Charters (approximately Apr-Oct). Open early May-mid-Oct, daily. **$$$-$$$$**

MOUNT LA CROSSE SKI AREA. *2 miles S on WI 35, La Crosse (54601). Phone 608/788-0044; toll-free 800/426-3665. www.mtlacrosse.com.* Area has three chairlifts, rope tow; patrol, rentals, school; snowmaking; night skiing; cafeteria, bar. Longest run 1 mile; vertical drop 516 feet. Half-day rates on weekends, holidays. Cross-country trails (Dec-mid-Mar, daily). Open Thanksgiving-mid-Mar, daily; closed Dec 25. **$$$$**

NELSON DEWEY STATE PARK. *12190 County Rd VV, Cassville (53806). Phone 608/725-5374; toll-free 888/WI-PARKS (reservations). www.cassville.org/nelsondewey.html.* This 756-acre park offers nature and hiking trails and camping (hookups). Open daily.

PUMP HOUSE REGIONAL ARTS CENTER. *119 King St, La Crosse (54601). Phone 608/785-1434. www.thepumphouse.org.* Regional art exhibits; performing arts (weekends). Open Tues-Fri noon-5 pm, Sat 10 am-3 pm; closed Mon, holidays. **FREE**

STONEFIELD. *12195 County Rd VV, Cassville (53806). Phone 608/725-5210. www.cassville.org/stonefield.html.* Named for a rock-studded, 2,000-acre farm that Dewey (the first elected governor of Wisconsin) established on the bluffs of the Mississippi River. State Agricultural Museum contains a display of farm machinery. The site also features a re-creation of an 1890 Stonefield Village, including blacksmith, general store, print shop, school, church, and 26 other buildings. Horse-drawn wagon rides (limited hours; fee). Open Memorial Day-Labor Day, daily 10 am-4 pm; Labor Day-mid-Oct, Sat-Sun. $$

SWARTHOUT MUSEUM. *112 S 9th St, La Crosse (54601). Phone 608/782-1980.* Changing historical exhibits ranging from prehistoric times to the 20th century. Open Tues-Sat 10 am-5 pm, Sat-Sun 1-5 pm; closed holidays. **FREE**

VILLA LOUIS. *St. Feriole Island, 521 N Villa Louis Rd, Prairie du Chien (53821). Phone 608/326-2721. www.wisconsinhistory.org/sites/villa/.* Built on the site of Fort Crawford and restored to its 1870 splendor. Contains original furnishings and a collection of Victorian decorative arts. Surrounded by extensive grounds, bounded by the Mississippi River. Tours include the Fur Trade Museum. Open May-Oct, daily 9 am-5 pm. $$

WYALUSING STATE PARK. *13081 State Park Ln, Bagley (53801). Phone 608/996-2261. www.dnr.state.wi.us/org/land/parks/specific/wyalusing/.* A 2,654-acre park at the confluence of the Mississippi and Wisconsin rivers. Sentinel Ridge (500 feet) provides a commanding view of the area; valleys, caves, waterfalls, springs; Native American effigy mounds. Swimming beach nearby, fishing, boating (landing), canoeing; 18 miles of nature, hiking, and cross-country ski trails; picnicking, playground, concession, camping (electric hookups, dump station). Nature center; naturalist programs (summer). Open daily. $$

PLACES TO STAY

If you choose to include an overnight stay in your trip along this Byway, Mobil Travel Guide recommends the following lodgings.

La Crosse

★★ **BEST WESTERN MIDWAY HOTEL.** *1835 Rose St, La Crosse (54603). Phone 608/781-7000; toll-free 800/780-7234. www.bestwestern.com.* 121 rooms, 2 story. Check-out noon. TV; cable (premium), VCR available. In-room modem link. Restaurant, bar; entertainment Tues-Sat. Room service. In-house fitness room, sauna. Indoor pool, whirlpool. Downhill, cross-country ski 10 miles. On the river; dockage. ¢

★ **HAMPTON INN.** *2110 Rose St, La Crosse (54603). Phone 608/781-5100; toll-free 800/426-7866. www.hamptoninn.com.* 101 rooms, 2 story. Complimentary continental breakfast. Check-out noon. TV. ¢

★ **NIGHT SAVER INN.** *1906 Rose St, La Crosse (54603). Phone 608/781-0200; toll-free 800/658-9497. www.visitor-guide.com/nights.* 73 rooms, 2 story. Complimentary continental breakfast.

Wisconsin

✻ The Great River Road

Check-out 11 am. TV; cable (premium), VCR available. In-room modem link. In-house fitness room. Downhill ski 8 miles, cross-country ski 1 mile. ¢

★★ **RADISSON HOTEL.** *200 Harborview Plz, La Crosse (54601). Phone 608/784-6680; toll-free 800/272-6232. www.radisson.com.* 169 rooms, 8 story. Pet accepted. Check-out noon. TV. Restaurant, bar; entertainment. In-house fitness room. Indoor pool. Downhill, cross-country ski 8 miles. Free airport transportation. Overlooks the Mississippi River. $

★ **SUPER 8.** *1625 Rose St, La Crosse (54603). Phone 608/781-8880; toll-free 800/800-8000. www.super8.com.* 82 rooms, 2 story. Complimentary continental breakfast. Check-out 11 am. TV. Laundry services. Indoor pool, whirlpool. Downhill, cross-country ski 10 miles. ¢

Prairie du Chien

★ **BEST WESTERN QUIET HOUSE & SUITES.** *Hwys 18 and 35 S, Prairie du Chien (53821). Phone 608/326-4777; toll-free 800/780-7234. www.bestwestern.com.* 42 rooms, 2 story. Pet accepted, some restrictions; fee. Check-out 11 am. TV; cable (premium). In-room modem link. In-house fitness room. Indoor pool, whirlpool. $

★ **BRISBOIS MOTOR INN.** *533 N Marquette Rd, Prairie du Chien (53821). Phone 608/326-8404; toll-free 800/356-5850. www.brisboismotorinn.com.* 46 rooms. Pet accepted, some restrictions; fee. Check-out 11 am. TV; cable (premium). Pool. Cross-country ski 2 miles. Free airport transportation. ¢

★ **PRAIRIE MOTEL.** *1616 S Marquette Rd, Prairie du Chien (53821). Phone 608/326-6461; toll-free 800/526-3776.* 32 rooms. Pet accepted, some restrictions; fee. Check-out 11 am. TV; cable (premium). Pool. Cross-country ski 2 miles. Lawn games. Miniature golf. ¢

PLACES TO EAT

A long day of driving is sure to make you hungry. At the end of your journey, take a table at one of the following restaurants.

Cashton

★ **BACK DOOR CAFÉ.** *1223 Front St, Cashton (54619). Phone toll-free 888/322-5494.* American menu. Closed Sun. Lunch, dinner. Inside a beautiful restored Victorian home with an elegant dining room and antique furnishings; family-oriented home-style cuisine. $

La Crosse

★★ **FREIGHTHOUSE.** *107 Vine St, La Crosse (54601). Phone 608/784-6211.* Closed Easter, Thanksgiving, Dec 24-25. Dinner. Bar. Former freight house of the Chicago, Milwaukee, and St. Paul Railroad (1880). Outdoor seating. $$$

★ **PICASSO'S CAFÉ.** *600 N 3rd St, La Crosse (54601). Phone 608/784-4485.* Mediterranean menu. Closed Sun. Lunch, dinner. Located in beautiful downtown La Crosse, serving California-inspired Mediterranean cuisine in a hip, sophisticated atmosphere. $

★★ **PIGGY'S.** *328 S Front St, La Crosse (54601). Phone 608/784-4877.* Steak menu. Closed Memorial Day, Labor Day, Dec 24-25. Lunch, dinner. Bar. Children's menu. $$$

Onalaska

★★ CIATTI'S ITALIAN RESTAURANT. 9348 US Hwy 16, Onalaska (54601). Phone 608/781-8686. Italian menu. Lunch, dinner. One of the area's most highly regarded restaurants; innovative northern Italian cuisine in a modern setting. **$**

★★ SEVEN BRIDGES RESTAURANT. 910 2nd Ave N, Onalaska (54650). Phone 608/783-6103. American menu. Closed Mon. Dinner. Featuring expansive river views and innovative local cuisine, this restaurant has been a favorite since it opened. **$**

★★ TRADITIONS RESTAURANT. 201 Main St, Onalaska (54650). Phone 608/783-0200. American menu. Closed Sun, Mon. Lunch, dinner. Located in the historic former State Bank Building and serving exciting regional cuisine. **$$**

NOTES

NOTES

NOTES

NOTES

NOTES

NOTES

NOTES